OCCUPATIONAL LOW BACK PAIN

Edited by
Malcolm H. Pope, Ph.D.,
John W. Frymoyer, M.D.,
and Gunnar Andersson, M.D.

PRAEGER

PRAEGER SPECIAL STUDIES • PRAEGER SCIENTIFIC

New York • Philadelphia • Eastbourne, UK
Toronto • Hong Kong • Tokyo • Sydney

Library of Congress Cataloging in Publication Data
Main entry under title:

Occupational low back pain.

Bibliography, p.
Includes index.
1. Backache. 2. Occupational diseases. I. Pope, M. H. (Malcolm Henry), 1941– . II. Frymoyer, John W. III. Andersson, Gunnar, 1942– . [DNLM: 1. Backache. 2. Occupational diseases. WE 755 015]
RD768.023 1984 362.1'9756 84-6815
ISBN 0-03-063288-9 (alk. paper)

Published in 1984 by Praeger Publishers
CBS Educational and Professional Publishing
a Division of CBS Inc.
521 Fifth Avenue, New York, NY 10175 USA

456789 052 987654321

Printed in the United States of America
on acid-free paper

This book is dedicated to Alf Nachemson, M.D., Ph.D.

Preface

Low back disorders have long eluded a purely medical diagnosis. They are always job related, at least in the sense that the nature of the specific job influences the possibility and timing of a return to work. Thus an increase in disability pensions due to low back pain, now to be found in many Western societies, can probably not be explained in medical terms only. Conditions within a given society such as availability of work and national insurance laws, are of obvious importance.

There are several good reasons for attending to people's need for health care closer to their place of work than has previously been the case. Demands made by the job—and the effects these have on the course of an illness and the chances of a return to work—are one such reason. There is often much to be gained by treating illnesses in the environment in which they have developed. The results of clinical investigations and treatment can then be used to make direct improvements in the work environment. As health costs escalate, Western societies increasingly pin their hopes on a broader approach to medical care and an emphasis on prevention of illnesses and disabilities.

Another reason why people's health needs should be met close to their place of work is that preventive measures call for considerable personal involvement. Individual habits must often be changed, and sometimes the entire lifestyle. This requires a learning process which must develop within the group with whom an individual shares customs and behavior patterns.

Health care at the place of work may fulfill an even more important function than that of dealing with job-related illness. In today's society, many people find the friendships they make at work to be among the most central and lasting friendships they have.

It is a great advantage to dispense health care within the environment where work is done and among those groups of people who do the work. Continuity is thus achieved, together with a greater understanding of the overall situation of the person seeking help. In addition, it is possible more directly to influence attitudes and habits.

It is my hope that this book will play a part in the creation of better working conditions and in the development of new methods for combatting occupational low back disorders.

Pehr G. Gyllenhammar
Med Dr h c
Chairman and Chief Executive Officer, AB Volvo

Contents

Introduction and Definitions

John W. Frymoyer, Malcolm H. Pope, and Gunnar B.J. Andersson

Low back pain (LBP) has been called the nemesis of medicine and the albatross of industry. Although evidence of low back diseases can be identified in primitive man, most of the advances and understanding of the LBP problem have occurred in the past 50 years, and particularly in the past decade. As the social and economic costs of LBP have risen, interest has also increased, not only into the causation of LBP, but into methods of prevention. Despite this increased awareness, there has been widely dispersed literature for those interested in the specific problem of industrial LBP. Information has been disseminated in journals of orthopaedics, biomechanics, engineering, industrial medicine and hygiene, physical and occupational therapy, and the law.

In 1981, we held a symposium on industrial low back pain, which has now been repeated three times. Lessons were learned from these symposia. A diverse group is vitally interested in LBP. Physicians and surgeons, industrial nurses, physical and occupational therapists, engineers, representatives of the insurance industry, labor and management have all been enthusiastic attendees. As anticipated, such a diverse group often had differences in perspective and also in the language by which they described their experiences with industrial LBP. We have learned a great deal from all of these disciplines, most importantly, that the ultimate solution to industrial LBP must involve all interested parties and is not solely a problem to be solved by doctors, labor or management. We have also learned the importance of developing a communication system so that all participants have a commonality of understanding.

This book, then, represents a first attempt to bring together in one place the scientific knowledge and its practical applications to the solution of industrial LBP. This information has been organized with the full recognition that it is to be used by a diverse readership. This book is organized into seven parts. Part I presents the basic knowledge necessary to understand fully the problem of occupational low back pain (LBP). Chapters in this section describe the structure and function of the lumbar spine (the functional biomechanics), the occupational biomechanics of the lumbar spine (how overloading can occur), and clinical classification. The occupational biomechanics chapter (Chapter 2) is, by necessity, quite technical; however, it forms the scientific basis for workplace design and evaluation. The concepts contained within this chapter will be useful to all readers. Part II concentrates on the definition of the problem in terms of its overall epidemiology or impact and cost. Part III is an extension of this theme that breaks the problem down into considerations of the workplace and the worker. The intention here is to provide insight into how to treat both the worker and the workplace. Part IV deals with patient care and includes a chapter on evaluation and one on treatment, education and rehabilitation. These chapters are designed to be understandable to those without a medical background. Part V focuses on prevention with chapters on the overall concepts, considerations in worker selection and a chapter on education and training. Further chapters in this section give practical advice about how to evaluate the workplace and how to design (or redesign) it. Part VI deals with the important area of the legal aspects of LBP and contains useful information on impairment ratings, Worker's Compensation, and hiring practices. The final section, Part VII, is designed to set the stage for future work and summarize our present knowledge.

Depending on the reader's experience and professional background, some sections of this book may represent new concepts or terminology. Although we have attempted to build a basis of understanding and definition within the text, a glossary of terms is provided as an additional aid to the reader. In every instance, we have tried to keep the illustrative material simple, and to avoid the use of more specialized illustrations such as spinal radiographs. Because there is a diversity of opinion contained within much of the scientific literature, we have incorporated these controversies, rather than attempted to preach a single-minded point of view. However, each chapter summary represents our carefully considered opinion as to the most useful theories and practices. For those who wish to explore further, fairly extensive bibliographies have been included.

Finally, it is clear that low back diseases are not a phenomenon of North America, but affect every country, particularly those that are highly industrialized. We have attempted to give an international perspective in all the material presented, with one exception. The law is so specific to each country that a review of compensation law, for example, in European countries alone would require a single volume. We therefore hope that our international readership will understand this obvious omission.

The participants in the symposia on LBP, and now the contributors to this book, have each been challenged by the problem of LBP and bring to this effort a diversity of backgrounds and scientific experience. We hope you will enjoy their efforts, and we welcome your comments or criticisms.

OCCUPATIONAL LOW BACK PAIN

PART I
The Basis

1 Structure and Function of the Lumbar Spine

Malcolm H. Pope, Thomas R. Lehmann, and John W. Frymoyer

OVERVIEW OF SPINE ANATOMY AND FUNCTION

The spine is a flexible multi-curved column. It is divided into five regions (Fig. 1.1): cervical (neck), thoracic (rib cage region, sometimes referred to as the dorsal region), lumbar (the low back, between the rib cage and the pelvis), sacral (five fused vertebrae that serve as an attachment to the pelvic girdle), and coccygeal (the tail, coccyx, which is not well developed in humans). Normally the cervical region has seven vertebrae, the thoracic twelve and the lumbar five. The curves of the normal spine are illustrated in Figure 1.1. As seen in the side profile, the cervical and lumbar regions are in lordosis (lordotic curves); the thoracic, sacral, and coccygeal regions are in kyphosis (kyphotic curves).

In humans, the thoracic spine is partially splinted by the rib cage. The sacrum is absolutely rigid and the coccyx has no functional role in humans. By contrast the cervical spine and lumbar spine not only support heavy loads but are quite flexible. It should not be surprising that these regions produce the most symptoms because of their more demanding functional roles.

The basic structure of the spine is similar in all of its subdivisions (Fig. 1.2). However, individual structures are more developed or less developed depending on the functional needs of that region. The descriptions to follow will be confined to the lumbar spine where the components are generally larger because they support more body weight.

BASIC ANATOMY AND FUNCTION OF THE SPINE

The spine has four major interrelated functions: 1) support, 2) mobility, 3) housing and protection, and 4) control. As a support, the spine functions

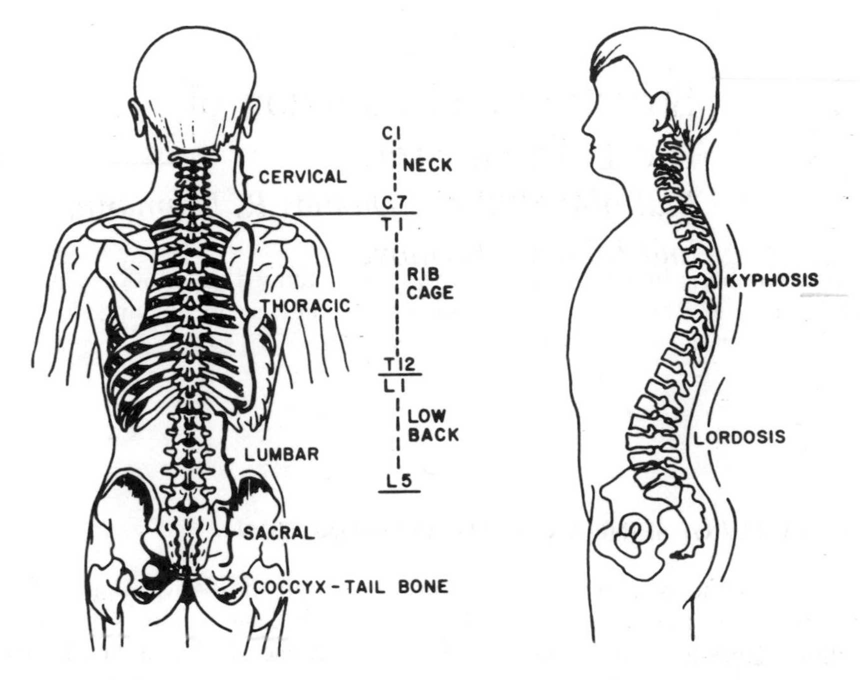

Figure 1.1
The regions and curves of the normal spine.

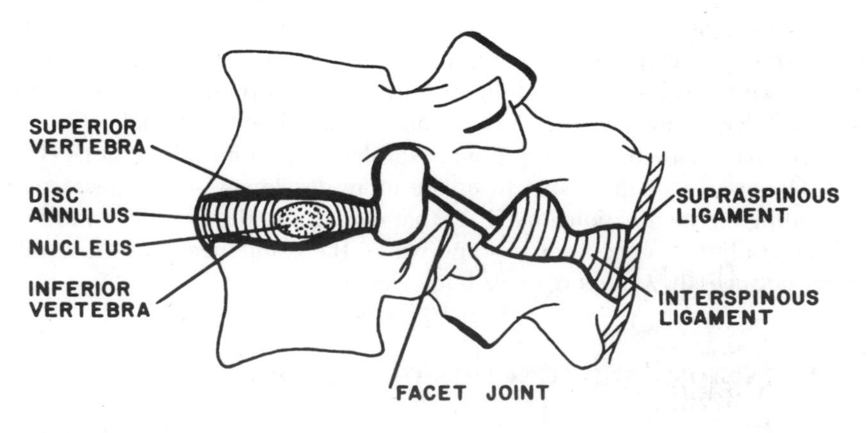

Figure 1.2
The basic structured unit of the spine.

as a framework for the attachment of internal organs (facilitated by the rib cage). It also supports the upper and lower extremities and the head. If it were only a support, the spine could be rigid, greatly simplifying its mechanical role. The mobility allows for the many physical tasks of daily living and work but complicates spine structure. Thus, instead of a single rigid column, the spine is a flexible stack of rigid blocks with flexible soft tissue in between. The rigid blocks are known as the vertebral bodies and the flexible soft tissue structure in between each vertebra is the intervertebral disc. The basic functional unit of the spine is termed the motion segment (see Fig. 1.3). The motions of the individual spinal segments, and the total motion of the spine and lumbar spine are shown in Figure 1.3. The largest motions of the lumbar spine are in forward bending (flexion) such as occurs when touching the toes, and backward bending (extension) as when leaning back to view the sky. Other important motions are twisting, which is termed axial rotation, and lateral bending. More complex motions involve combinations of flexing, bending to the side, and twisting.

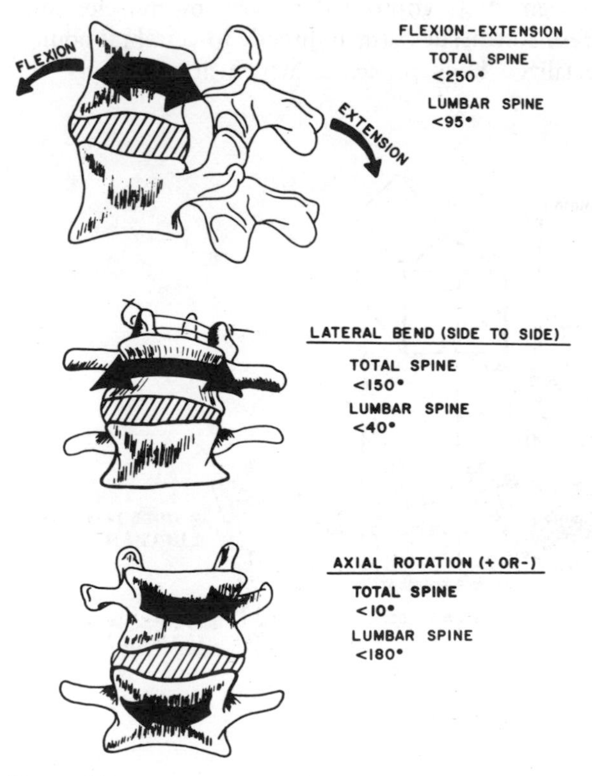

Figure 1.3
Motion segments and total motion of the lumbar spine.

The housing function of the spine protects the spinal cord and nerves as they pass from the head to their point of departure to either the upper or lower extremities. The spinal canal transverses vertically from the skull to the tailbone and functions as a conduit to protect the spinal cord and its nerve roots. Since the spinal cord terminates at the first or second lumbar vertebra, this housing also protects nerve roots caudal to the terminal part of the cord. This fluid-filled sac containing nerve roots within the lumbar spine is called the cauda equina. In cross-section, the spinal canal or conduit is similar in construction to a house (see Fig. 1.4). The back of the vertebral body and disc serves as the foundation, the pedicles on either side serve as walls, and the laminae are the laminae (roof) of this house. Once again the need for mobility and flexibility complicates the mechanical requirements for the house. The roof of one vertebra joins to the adjacent vertebrae through a pair of interlocking joints (termed facet or zygoapophyseal joints) illustrated in Figure 1.4. Although these joints are essential to normal kinematics, they also constrain motion dependent on their orientation. Thus, depending on the planes of these joints, only certain motions may be allowed.

The motion of each segment is controlled actively by muscles and passively by ligaments. These soft tissues attach directly to vertebral bodies and laminae, and to specialized bony processes which act as levers. As

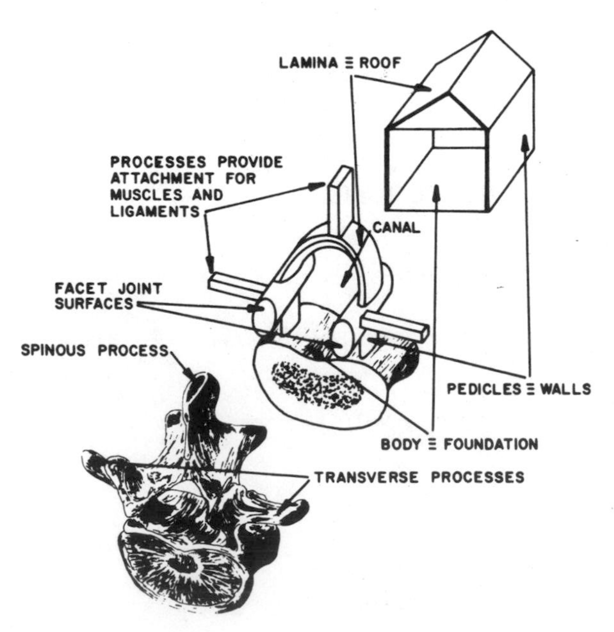

Figure 1.4
The housing function of the spine.

depicted in Figure 1.4, each vertebra has a dorsally located spinous process and a pair of laterally placed transverse processes. The ligaments and muscles that connect between these processes and the pelvis support the spine similar to the way cables or guy wires support a radio tower.

SPINAL LOADING

The application of an external load will result in deformations within the spinal structures and also produce movements between the structures in the manner previously described (forward, backward, and lateral bending, and twisting). Each spinal motion segment is subjected to a variety of loads and resultant forces as illustrated in Figure 1.5. In mechanical terms, these loads are divided by the area on which they act and are termed stresses. Application of a load results in deformations within the structure. These deformations are collectively termed strains.

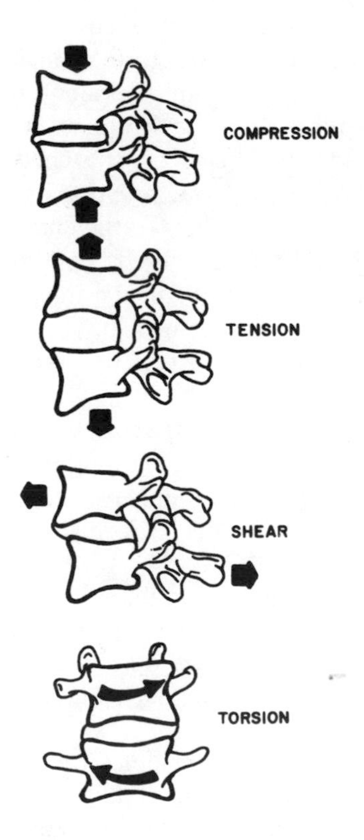

Figure 1.5
Loads and resultant forces on spinal motion segments.

Since the pattern of injury that one observes in the spine is a function of the type and magnitude of forces that were present, it is important to understand these forces, and the types of deformations and tissue failures they eventually produce.

A downward force (stress) perpendicular to the surface of the upper vertebra compresses the disc, and causes it to bulge. As the compressive load increases, pressure within the disc also increases. If the compressive force were increased, one would expect failure to occur at some point. In the laboratory, application of compressive load results in failure of the vertebral endplate, usually towards its center. This mechanism of injury is most common when a person falls, landing in the seated posture, producing the so-called compression fracture. This type of failure may relate to whether the spine was loaded in flexion or extension, with flexion tending to cause anterior collapse of the endplate, a point where the bone structure tends to be weakest. Forces sufficient to cause endplate failure may also occur through the act of lifting grossly excessive loads. This concept and prevention of endplate failure is explored further in Chapter 14.

A second type of loading is termed tension. Tensions tend to pull apart the structure being loaded, and the cross-sectional area of the structure decreases. Depending on the material that makes up the structure, tension loading may produce an elastic recoil. These principles are most evident in the highly elastic structure of the rubber band. In the spine, the ligaments are loaded in tension. In spinal extension, the anterior ligaments are stretched; while in flexion, the posterior ligaments are stretched. Overstretching any ligament may cause rupture of either individual ligament fibers or the entire ligament. Collectively, these failures are termed sprains. When there is major spinal trauma, such as might occur in a severe fall or vehicular injury, ligaments do rupture completely and may be accompanied by frank dislocation of a vertebra. The role of ligament strains in most occupational low back injuries is controversial, as discussed in Chapter 2.

Twisting motions of the spine produce strains in the tissues. In the normal lumbar spine, the facets tend to restrict torsion. In the laboratory, application of the torsional stresses produce fragmentation and disruption of the disc. This may be one of the earliest causes of acute low back episodes, as discussed in Chapter 3.

Shear loading is produced by translation motion of structures. For example, as the spine flexes, there is a tendency for a vertebra to slide forward relative to its next lowest neighbor. Shear stresses occur in the lumbar spine because of the lordotic curvature of this region, particularly at the lumbosacral junction (see Fig. 1.1).

In real life, we load our spines in many directions. Consider a worker transferring an object from a low pallet on his right to a higher table on his left. During the course of moving the object, the spine may be subjected to compressive loads (associated with the initial lift), lateral bending forces, torsional loads and shear. Each of these forces will occur in different phases of the lifting cycle and in different combinations. Unfortunately, these complex forces are difficult to analyze in the workplace and even more difficult to replicate in the laboratory. By separating specific activities into their component parts, one can obtain a fairly accurate assessment of the types of stresses involved and the resultant strains to which the vertebral structures have been subjected. This type of analysis involves the application of basic engineering principles coupled with an understanding of the spinal stresses and strains. This is the cornerstone of the concepts to be presented in subsequent chapters regarding education, prevention, and workplace design.

INDIVIDUAL SPINAL STRUCTURES: THE SUPPORT SYSTEM

Let us now consider the individual spinal structures and how they respond to motion and load.

The Vertebral Bodies

The vertebral bodies are short cylindrical bones with a kidney-shaped cross-section as shown in Figure 1.6. They are the key element in the load-bearing system of the spine. The core of the vertebral body is made primarily of cancellous bone which has a honeycomb-like construction. Since bone grows in a direction to withstand forces, the directions of trabeculae reveal the normal forces acting on the vertebra (see Fig. 1.7).

The vertically directed trabeculae support the compressive loads in the main vertebral body (see Fig. 1.7). At the upper and lower surfaces of the body there are oblique trabeculae sweeping up or down to aid in this compressive load bearing function. These trabeculae come together at the pedicles to resist the tensile forces there. The trabeculae sweep up and down to the superior and inferior facets and outward to the spinous process. The former trabeculae are oriented to support the compressive and shear forces in the facets whereas the latter trabeculae are oriented to withstand the tensile and bending forces applied to the spinous process.

The type of failure may relate to whether the spine was loaded in flexion or extension, with flexion tending to cause anterior collapse where the trabeculae are weaker (Fig. 1.7).

Krenz and Troup (1973) found that the pressure was higher in the center of the endplate than in the periphery during compressive loading. This is a common site for failure. The failure results from the nucleus (central part of the disc) rupturing the endplate. This may be a significant problem for those who have diminished bone strength such as occurs in

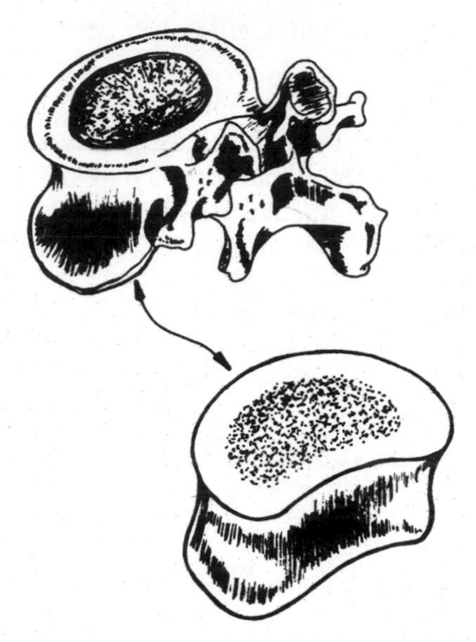

Figure 1.6
The vertebral body and cross-section.

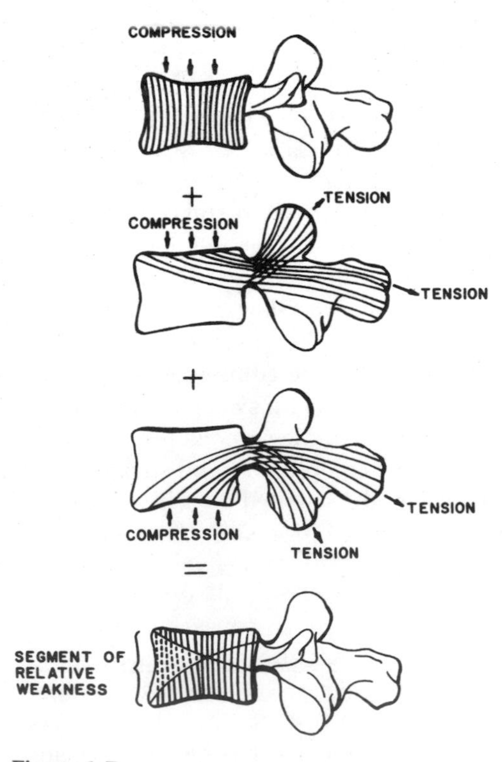

Figure 1.7
Trabeculae directions in the vertebrae.

osteoporosis. In osteoporosis there is a reduction of bone volume and an increased risk of fracture with smaller external loads, a concept discussed in Chapter 3. The vertebral bodies are about six times stiffer and three times thicker than the disc. Thus, the vertebrae deform about half as much as the disc under compression. Since the vertebrae are filled with blood, it is possible that they behave like hydraulically strengthened shock absorbers (Kazarian 1972).

The Posterior Elements and the Facet Joints (Apophyseal Joints)

Posteriorly the vertebral arch attaches to the vertebral body. The vertebral arch is made of the pedicles which are short and cylindrical, lamina, facet joints and spinous and transverse processes (Fig. 1.4). At the site where the lamina takes origin from the pedicles, the lamina is narrowed, an area referred to as the pars interarticularis or isthmus. Whereas pedicles rarely fracture, the pars is a frequent site of a peculiar fracture, apparently secondary to fatigue of bone rather than a sudden or acute fracture. Such fractures, when they occur, usually heal by fibrous union, a so-called spondylolysis occurs, which is a prerequisite for a condition termed spondylolisthesis, which will be discussed in Chapter 3. The articulations between one vertebral articular process and its neighbor are termed the facet joints. The facet joints are extensions of the lamina and are covered with hyaline cartilage on their articulating surfaces. The facets are particularly important in resisting torsion and shear, but also play a role in compression.

Normally, the facets and discs contribute about 80 percent of the torsional load resistance, with the facets contributing one-half of that amount (Farfan 1973). Hutton, Stott, and Cyron (1977) obtained similar findings in facet and disc load sharing when they studied the shear behavior of the lumbar spine.

Using pressure-sensitive film, Lorenz, Patwardhan, and Vanderby (1983) found that approximately 25 percent of the axial compressive load was transmitted through the facets. This load bearing was reduced markedly by excision of a single facet. This implies some kind of wedging mechanism in which both facet joints are responsible for force transmission. All of the load resistant characteristics can be altered when the facets are maloriented.

The amount of load bearing by the facet joints is related to whether the motion segment is loaded in flexion or extension. Thus, the difference in intradiscal pressures between erect sitting and erect standing discussed in Chapter 13 can be explained in part by load bearing of the facet joints while in extension or lordosis. Theoretically then, the disc would be protected from both torsional and compressive loads when the motion segment is in extension (or lordosis). However, excessive loading of the spine in extension may cause failure of this secondary load-bearing mechanism; that is, loads transmitted through the facet joints may produce high strains in the pars interarticularis leading to spondylolysis.

The Intervertebral Disc

The disc forms the primary articulation between the vertebral bodies. It apparently has a major role in weight bearing since the area increases as a direct function of body mass over the whole mammalian family (Wilder, Krag, and Pope 1984). The function of loadbearing in shear, compression, and torsion is shared between the discs and the facet joints which collectively form the so-called three joint complex.

The disc, which is avascular, is composed of two morphologically separate parts. The outer part (Fig. 1.8), called the annulus fibrosus, is made up of about 90 sheets laminated over each other. Each sheet is made of collagen fibers that are oriented vertically at the peripheral layer but become progressively more oblique with each underlying sheet. The fibers in adjacent sheets run somewhat at angles to each other. The lamination of the annular layers strengthens the annulus much like the plies strengthen an

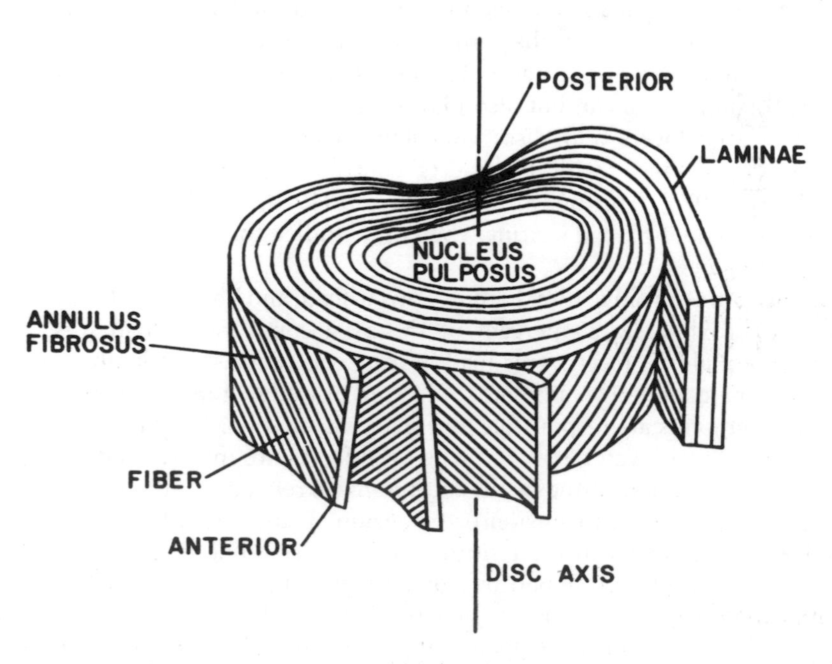

Figure 1.8
The intervertebral disc.

automobile tire. The central part of the disc is called the nucleus pulposus. In the younger individual, it is nearly 90 percent water, with the remaining structure being comprised of collagen and proteoglycans. In the young healthy disc, a positive pressure is present within the nucleus pulposus at rest and increases as the loads are applied to the spine. This pressure approximates to 1.5 times the mean applied pressure over the entire area of the endplate (Nachemson and Morris 1964). Disc pressures have been extensively studied in various postures and seating configurations, as discussed fully in Chapter 13.

In axial compression, the increased intradiscal pressure is counteracted by annular fiber tension and disc bulge, rather analogous to inflating a tire (see Fig. 1.9). Some disc space narrowing also occurs. In flexion, extension and lateral bending, the same process occurs, although usually focusing on a specific segment of the circumference of the disc. This process may be associated with a small displacement of the nucleus within the disc (Shah, Hampson, and Jayson 1978), but this view has been disputed more recently (Krag *et al.* 1983). Since the nucleus is roughly spherical, some have considered it as a ball that allows one vertebra to rotate over its neighbor (Fig. 1.10). The nucleus or ball is said to move backward during flexion and forward during extension. However, others have not confirmed this observation, and Krag *et al.* (1983) in specific, found little motion of the nucleus.

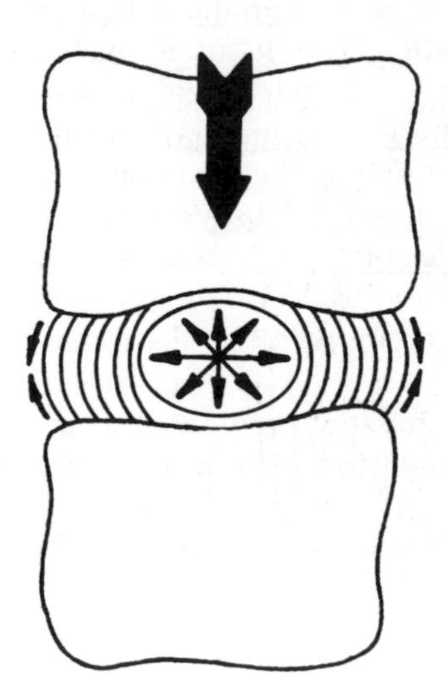

Figure 1.9
Tire analogy.

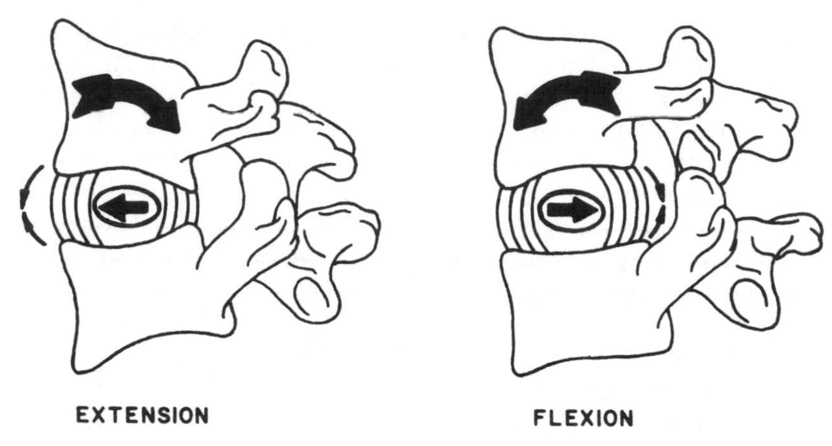

Figure 1.10
Nucleus pulposus in spine extension and flexion.

In axial rotation (i.e., torsion of the disc), the annular fibers in one direction are stretched significantly whereas those on the opposite side are shortened or crimped. Mathematical models based on the geometry of the lumbar disc demonstrate that torsion produces stress concentration at the region of the posterior lateral annulus which is a common site of disc herniation (Spilker 1980, Krause 1973, Broberg, and von Essen 1980) (see Chapter 2). Torsional loading of spinal segments *in vitro* produces fissures in the annulus in the same posterolateral location. This fissuring may be accompanied by tracking of the nuclear material, although a frank rupture has not been produced in torsion in the laboratory using motion segments from cadavera. These events are thought to be one of the common early causes of acute low back pain. However, acute ruptures of discs can only be produced in the laboratory by combining flexion and lateral bending motions. In flexion, the disc may be more susceptible to compression, torsion, and shear loads because the facet joints are relatively disengaged and therefore carry a lesser load.

Farfan (1973) believes that compression failures promote disc degeneration. The adult disc is avascular and endplate microfractures can result in vascular ingrowth and concomitant formation of granulation tissue. As a result, the chemistry of the disc and the mechanical behavior of the constituents may be altered.

The Effects of Surgery and other Treatments on the Mechanics of the Disc

Although the clinical role of spine surgery is introduced in Chapter 8, it is important to recognize some effects of common treatments and op erations on the mechanics of the disc and spine. Removal of the nucleus involves the creation of a surgical defect. The presence of such defects would produce abnormal stress concentrations and theoretically could predispose the disc to further herniation. Markolf (1972), however, observed than an annular defect after compressive loading did not lead to herniation and thus proposed a self-sealing mechanism in the annulus. Additional information came from Farfan *et al.* (1970) who found minimal changes in the torsional properties of motion segment specimens in which an experimental annular defect was produced. The above experiments do not take into account the continued wear, tear, and repair processes over time. It is yet to be determined if the recurrence rate of herniated discs will be less with chemonucleolysis (injection of chymopapain) which does not require additional surgical damage to the annulus.

Analysis of patients following disc excision or chymopapain injection reveals narrowing at the operated or injected interspace. This narrowing is typically accompanied by peripheral osteophyte formation. These observations imply altered compressive load-bearing capability of the motion segment, although the radiographic findings may not be of great clinical significance because there is little correlation between disc space narrowing and clinical reports of LBP. It is possible that over a lifetime the altered mechanics resulting from disc space narrowing may lead to spinal stenosis. Unfortunately, 40 to 50 year follow-ups would be required making a study of this potential association impractical. Recently, studies of chymopapain injected dogs reveal that the disc resumes its height after a period of narrowing (Bradford, Cooper, and Oegema 1983).

INDIVIDUAL SPINAL STRUCTURES: THE CONTROL SYSTEM

We have discussed those spinal structures which provide the basic functions of loadbearing, mobility, and housing. However, these structures, unlike most other joints, depend on other tissues for their stability. For example, the normal adult hip will not dislocate even if the surrounding muscles are paralyzed, unless there is severe trauma. In contrast, buckling of the spinal column occurs when its segmental muscles are paralyzed. The ligaments and muscles, as stated before, provide the inherent stability much like cables in a suspension bridge or cables supporting a radio tower.

Ligaments

Various investigators have shown that ligaments are vital for the structural stability of the spinal system. The ligaments are also the primary tensile load bearing elements. In general they act as check reins preventing excessive motion. It is important to realize that ligaments (unlike muscles) are passive structures so that the tension in them depends on their length. Since ligaments have a property known as viscoelasticity, their deformation and type of failure depends on their rate of loading. Ligament properties also depend upon the history of deformation of the ligament. It is known that, like all materials, ligaments fatigue and can fail with repetitive loading cycles (Weisman, Pope, and Johnson 1980).

The annulus of the intervertebral disc may also be considered a part of the ligamentous system. The posterior and anterior longitudinal ligaments both traverse the length of the spine, adding to the support of the vertebral body and disc (Fig. 1.11). They are interlinked at each level by the disc. These ligaments and the outer fibers of the annulus are richly supplied by nerve fibers that respond to painful stimuli. The remaining ligaments support and link the posterior elements. Of great functional importance is the ligamentum flavum which joins the lamina of adjacent vertebrae. The ligamentum flavum is highly elastic and strong compared to other ligaments, and its high content of elastic fibers gives it a yellow appearance (hence it is sometimes called the yellow ligament). Its elastic properties allow it to

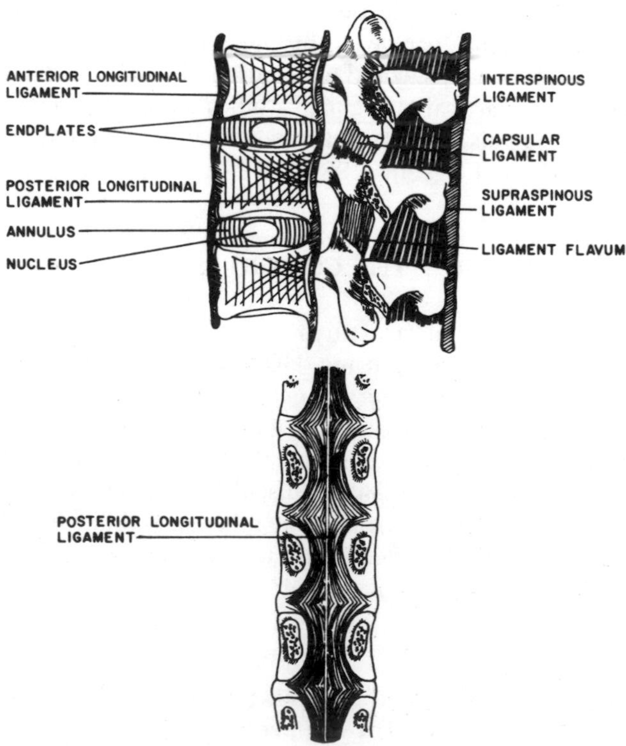

Figure 1.11
The ligaments of the spine

lengthen with spine flexion and shorten with extension. In patients with severe spine degeneration, it may be thickened, less elastic, and produce narrowing of the spinal canal in extension. This narrowing occurs because of the buckling of the ligament. Thus, at the time of surgery for spinal stenosis, its excision may be necessitated.

The tip and edges of the spinous process are joined by the interspinous and supraspinous ligaments. Because they are far from the disc and therefore act on long moment arms, these ligaments have an important functional role in resisting spine flexion. Similarly, intertransverse ligaments join the transverse processes of the vertebra.

The capsular structures around the apophyseal or facet joints are of functional importance in LBP. These capsules include separate thickenings which are at the front and back and have different functional roles. These capsular ligaments appear to limit the excursion of the facet joints, although their stabilizing functions have not yet been validated. They are richly supplied with pain-sensitive nerve endings but the facet joints themselves have no nerves. Therefore, pain from degenerated joints would have to be initiated by a mechanical strain or chemical irritation of these capsular structures. Facet joint pain has been estimated to be a common cause for chronic LBP and at times surgical procedures are employed to relieve this pain syndrome, as discussed in Chapter 8.

Overall tension stresses are far less common stresses caused by compression, shear or torsion. Tension occurs when the ligaments are resisting the effects of the compressive, shear, or torsional loads applied to other vertebral structures. One of the most common diagnoses given for acute LBP is spine sprain. It is likely that damage to these structures may occur and produce acute discomfort, but no precise diagnosis can yet be given. This concept is discussed in detail in Chapter 3.

The Musculature

The spinal musculature consists of flexors and extensors. The flexor muscles are made up of the psoas, which attaches directly to the vertebral bodies anterolaterally, and the abdominal muscles, which flex the spine through their attachments to the rib cage and pelvis.

The abdominal muscles are made up of the midline and anterior rectus abdominus muscles, and the anterior and laterally placed layers of the internal and external oblique, and transverse abdominal muscles.

The extensor muscles (the erector spinae which are all of the deep posterior muscles including the multifidi, and intertransversarii) attach to the outriggers (spinous processes and transverse processes) and other parts of the posterior elements. Their fibers may attach intrasegmentally (between adjacent vertebrae only) or intersegmentally (bridging multiple motion segments). When both sides contract symmetrically, they produce extension of the spine. When right and left sides contract and relax asymmetrically, lateral bending is produced. Both flexor and extensor muscles work together symmetrically.

From a preventive and therapeutic standpoint, the muscles are the most important structures of the spine. They are the motors that power the spine. Under voluntary control they position the spine and stabilize it during awkward work postures. Additionally, they provide the power necessary for lifting and carrying. Gregersen and Lucas (1967) showed how an excised spine with the ligaments intact, but without the muscles, buckles under very small compressive forces. Various animal models of the spine, when the mechanical function of the muscles has been disrupted, have shown how the spinal stability is delicately controlled by the musculature. Changes in muscular strength, or changes of muscle balance, may lead to an increased risk of LBP.

Chaffin (1974) has demonstrated that workers with inadequate lifting strength working in relatively stressful lifting tasks have higher low back injury rates than workers with less stressful lifting tasks. Apparently, when lifting near the strength limit of these muscles, excessive strain may be transmitted to other soft tissues (e.g., the ligaments and discs). It is possible that in the presence of muscle fatigue the probability of injury is increased. The ability to respond quickly to sudden loading may help prevent spine injury. Thus, when one of two workers carrying a load drops his share, a sudden load will be imparted to the other worker. When muscles are fatigued, the possibility of injury increases because of an alteration in the ability to respond to these conditions.

From a therapeutic standpoint, strengthening of these muscles and training them to hold the spine in comfortable postures may allow function even in the presence of disease. Muscles can be strengthened and trained under the voluntary control of the patient. In a rehabilitation program at The University of Iowa, there was a definite correlation between improvement in trunk muscle strength and overall improvement in the symptoms and functions of patients with LBP (Smidt, Amundsen, and Dostal 1980).

Farfan (1973) has suggested that the lumbodorsal fascia can change the length of the dorsal ligament by means of a laterally-directed force from the oblique musculature. This is said to shorten these ligaments. Farfan postulated that this is the mechanism by which intra-abdominal pressure creates extension moments about the spine and thus assists in compressive load bearing. Thus, the muscles act as controls of compressive load bearing by acting through the ligaments.

Other Support Structures

Both the abdominal and thoracic cavities have been shown to become pressurized during strenuous activity. The abdominal cavity can be pressurized by mechanical contraction of the muscles of the abdominal wall, together with the diaphragm. Concomitant closing of the glottis (Valsalva maneuver) results in further pressurization (Davis 1959). This increased pressure tends to force the pelvic wall (floor of abdominal cavity) and the lung's diaphragm (roof of the abdominal cavity) apart. Bartelink (1957) and Morris, Lucas, and Bresler (1961) suggest that this mechanism tends to extend the spine and thus reduce the contraction force required in the extensor muscles. In turn, this force rebalancing is said to reduce the compressive load bearing of the disc. This is controversial since the abdominal muscles must contract, producing a flexion moment in order to produce the intra-abdominal pressure. Morris, Lucas, and Bresler (1961) theorized that the extension forces (caused by intra-abdominal pressure) would greatly outweigh the flexion forces (caused by abdominal muscle contracting). However, other mathematical models are less optimistic (Gilbertson, Krag and Pope 1983).

The general concept of spinal support from the abdomen has led to a rationale for flexion exercises, the wearing of corsets, the protection of the spine by a belt for weight lifters. Davis (1981) has used this concept in the determination of safe loads to be used in industry. In lifting, there is a linear relationship between the amount of weight lifted and the intra-abdominal pressure that can be measured in the stomach or rectum (Grew 1980; Andersson, Ortengren, and Nachemson 1977). However, such relationships do not take into account the altered mechanics when the spine is located in a flexed or axially rotated position. Other measurements in those positions (Gilbertson, Krag, and Pope 1983) have shown that an increase in intra-abdominal pressure may not decrease the activity of the dorsal musculature and thus one would presume that the net axial loading of the disc is not reduced (Fig. 1.12). Thus, at the present time, the beneficial effects of abdominal pressurization are controversial.

McNeill *et al.* (1980), testing trunk strength in the standing position, found that LBP patients were relatively extensor-underpowered compared to persons without LBP. However, at The University of Iowa, when trunk strength was measured in the seated position, the flexor muscles were relatively weaker than the extensor muscles in chronic LBP patients (Smidt, Amundsen, and Dostal 1980). These differences are most likely a result of the position in which the patients are tested and the effect of muscles crossing the hip joint. In either case, chronic LBP patients have consistently been found to have weak trunk strength compared to normals. A trunk strengthening exercise program may be beneficial in the prophylaxis of recurrent injury to patients returning to work following injury. Poulsen and Jorgensen (1971) found that rehabilitation patients who had isometric trunk strengths equivalent to at least two-thirds of body weight had greater success at returning to work than weaker rehabilitation patients. All of these concepts will be discussed in detail in Chapters 7 and 11. Whether or not injury to these muscles is a common cause of LBP is still controversial.

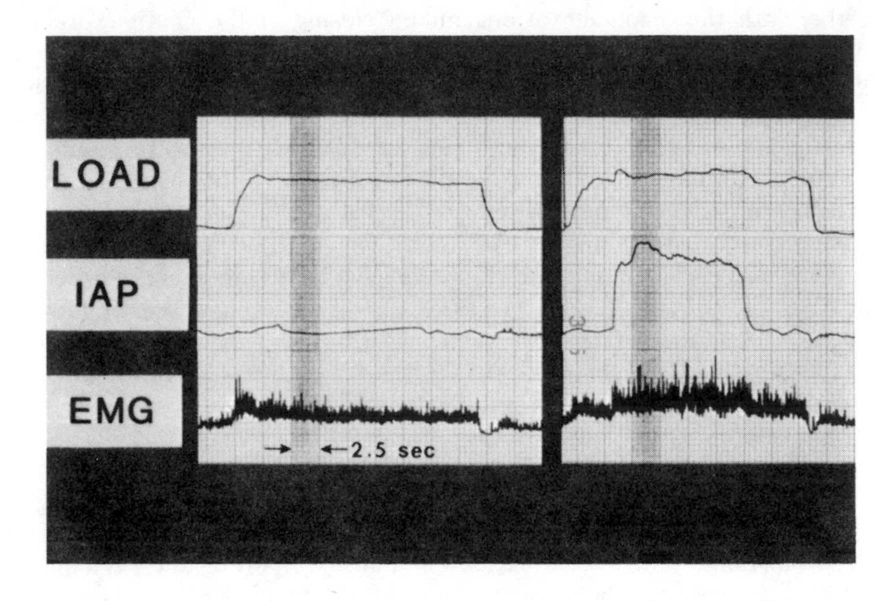

Figure 1.12
Chart paper tracing showing that abdominal pressurization during lifting does not necessarily reduce dorsal muscle tension.

PAIN AND THE NERVOUS SYSTEM

Low back pain can be defined as pain or discomfort presumed to be of musculoskeletal origin, constant or intermittent, located between the lower rib cage and posterior iliac crests, (in the lower back) commonly radiating to the buttocks and upper thighs. This definition of low back pain (low back symptoms) excludes nonmusculoskeletal disease that can cause pain with the same distribution (Fig. 1.13). Referred pain, for example, can result from numerous intra-abdominal, pelvic or retroperitoneal disease processes such as prostatitis, expanding abdominal aortic aneurysm, or intrapelvic disease in the female (Fig. 1.14).

The usual low back disease pain of musculoskeletal etiology may or may not be associated with leg symptoms suggestive of entrapment of one or more lumbar nerve roots. Typically, these referred leg symptoms are termed sciatica. A precise definition of sciatica is pain, in the distribution of one or more lumbar nerve roots (Foerster 1933), with the pain radiating below the level of the knee, and often into the foot, usually accompanied by subjective symptoms of numbness and/or weakness (Fig. 1.15). When the pain and symptoms are restricted to the distribution of one or more nerve roots (i.e. is not diffusely distributed), the term radiculopathy can be used.

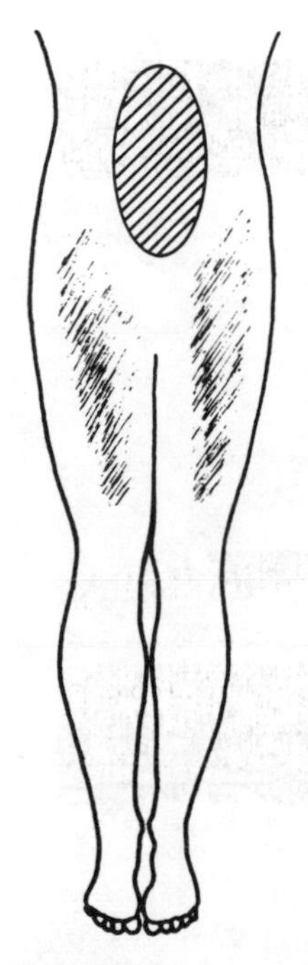

Figure 1.13
Typical pain referral in low back pain.

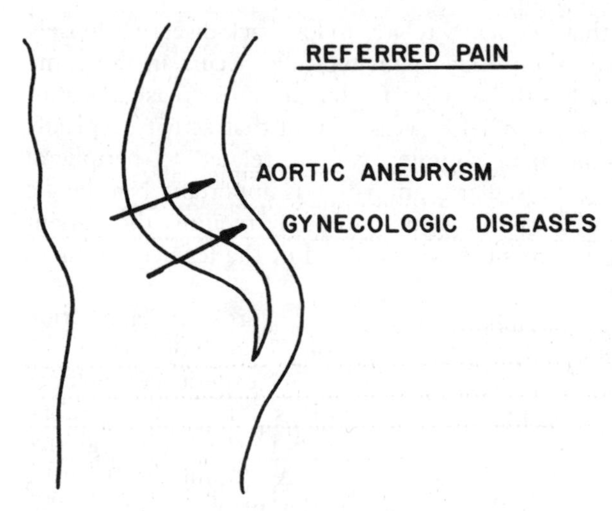

Figure 1.14
Referred pain from other causes.

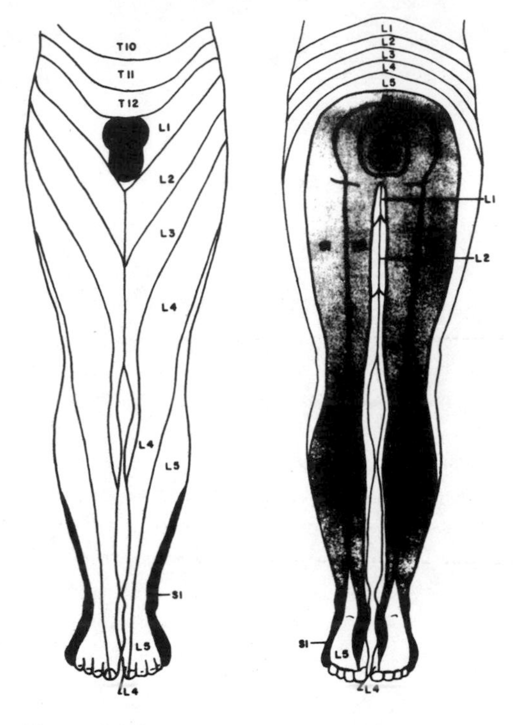

Figure 1.15
The dermatomes of the spinal nerves.

A third type of pain that frequently relates to low back disease is loosely termed "neurologic claudication." This pain typically occurs in the same distribution as sciatica but has the specific attribute of an increase in symptoms with physical activity and relief by cessation of that activity, typically walking. Because a similar pain complex can be related to peripheral vascular compromise, careful vascular examination is important. Neurologic claudication, unlike vascular claudication, commonly requires the patient not only to stop walking but to sit down for relief of the leg symptoms to occur.

LBP and the associated symptoms of sciatica or neurologic claudication involve complex neurophysiologic and psychologic phenomena which collectively form the 'pain experience.' The pain experience includes activation of specialized, peripheral nerve endings termed nociceptors, afferent transmission through spinal cord pathways to the brain, cognitive integration, and affective and behavioral outcomes. A voluminous literature has analyzed various aspects of this highly complex phenomenon. In this section, we have distilled the most useful aspects of the literature as a background of understanding for industrial LBP. As such, we will only touch upon the most relevant components of this highly complex subject for which there still remains a need for further investigation.

Peripheral Stimulation—LBP

A structure capable of initiating a pain experience must be innervated by specific nerve endings, which are collectively termed nociceptors (Wyke 1976, 1982, LaMotte 1982). These structures can be identified by a variety of microscopic methods. Additional criteria for a structure being pain-sensitive may include the production of pain by noxious stimulation and cessation of pain when the structure is selectively blocked with a local anesthetic.

The classic nociceptors are arborizing, unmyelinated, free nerve endings, although other more specialized nerve endings appear to initiate pain peripherally. Many of these are sensitive to mechanical stimuli and are termed mechanoreceptors. Wyke (1980, 1982) and others (Pedersen *et al.* 1956, Paris 1983) have identified nociceptor innervated structures in the spine which are as follows (Fig. 1.16):

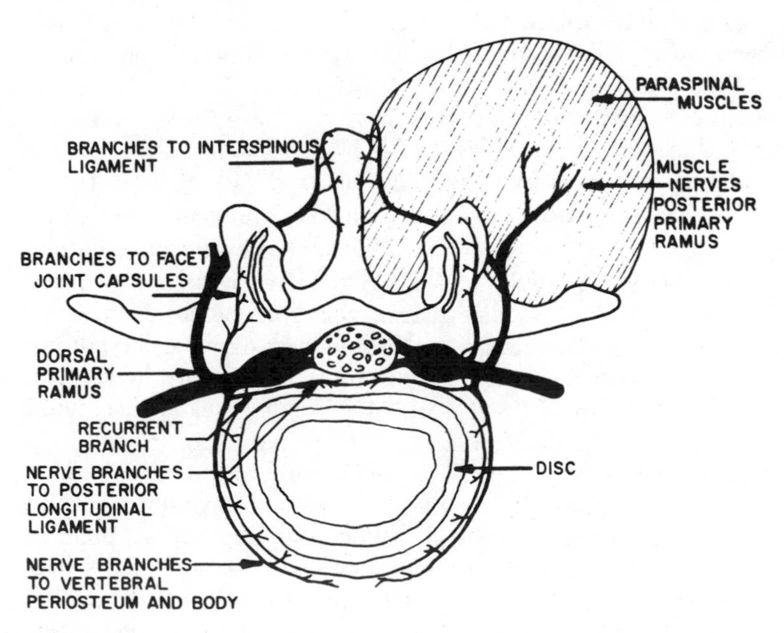

Figure 1.16
Innervated spinal structures.

1. Facet joint capsules, sacroiliac joint capsule.
2. Ligaments—posterior longitudinal, ligamentum flavum, interspinous.
3. Vertebrae—periosteum, nociceptors in association with intraosseous blood vessels.
4. Dura mater and epidural fat.
5. Muscles—nociceptors accompany blood vessels, and possibly may also be found within the muscle fibers.

Activation of these nociceptors may occur by a variety of mechanical and chemical stimuli. First, local pressure, or mechanical tension, appears to activate nociceptor signals. Alteration in the tensions of ligaments and joint capsules occurs under certain mechanical conditions. Increased pressure in the perivascular nociceptor sites probably occurs also. For example, patients with chronic spinal degeneration appear to have an increased intraosseous pressure, and venous dilitation, which might be one mechanism for nociceptor activation (Arnoldi 1972).

Second, alterations in the chemical environment may stimulate nociceptors. Some of the chemicals identified as activators include alterations in acidity (decreased pH), lactic acid, histamines, bradykinens, and prostaglandins, many of which are associated with the inflammatory response.

Third, decreased oxygen tension (ischemia) appears to be a potent innervator of pain. Muscular back pain is often presumed to result from decreased oxygen, and increased lactic acid which accompanies increased and prolonged muscle contractions or muscle overuse, although this may not be the case (Jokl 1982).

The nociceptor sites within the spine have also been identified further by other clinical experiments. Classic studies by Lewis and Kellgren (1939), later duplicated by Hirsch *et al.* (1963), McCall *et al.* (1979), and Mooney (1976) have involved selective injections of spinal structures with hypertonic saline solutions which activate nociceptors. Injection of hypertonic solutions into facets and ligaments produce a rather typical low back syndrome, with pain distribution over the sacroiliac joints, and often radiating in to the posterior aspects of the upper thighs. This noxiously-produced syndrome sometimes can be reversed by selective injection of local anesthetics. This latter observation has formed the basis for the diagnostic and therapeutic use of facet blocks.

It is of particular note that the list of pain-sensitive structures does not include the nucleus pulposus or annulus. Most studies have not identified nociceptors within the disc, although some studies suggest there is innervation of the outermost fibers of the annulus fibrosus. Others have reported that nociceptors are found in discs in material taken at the time of surgery. It is possible that disc injury, and the subsequent repair mechanism, includes invasion of the disc by blood vessels and their nociceptors. Although the annulus is not well innervated, except at its periphery, the adjacent epidural vessels and fat, and the posterolongitudinal ligaments' nociceptors may be activated by bulging of the disc under mechanical loading or by chemical changes that accompany disc degeneration.

Peripheral Stimulation—Sciatica and Neurologic Claudication

The production of sciatica involves the direct compression or increased tension on a nerve root, and dorsal sensory ganglion (Fig. 1.17). Smyth and Wright (1958) have noted that stimulation of a normal peripheral nerve does not produce sciatic pain unless that nerve has been compressed previously by a herniated nucleus pulposus (HNP), or some other pathologic condition. This observation suggests that sciatic pain requires an intermediary step beyond simple compression, which is most likely some alteration in the local chemical environment surrounding or within the nerve (Murphy 1977). Increased lactic acid and decreased pH have been shown to be present in the local environment of nerve roots compressed by HNP. Another explanation is that the blood supply to the nerve root is diminished (Murphy 1977, Lindahl and Rexed 1951). This latter explanation is particularly attractive in the understanding of neurologic claudication. It has been shown that the caudal blood supply is tenuous (Crock and Youshizawa 1977), and chronic

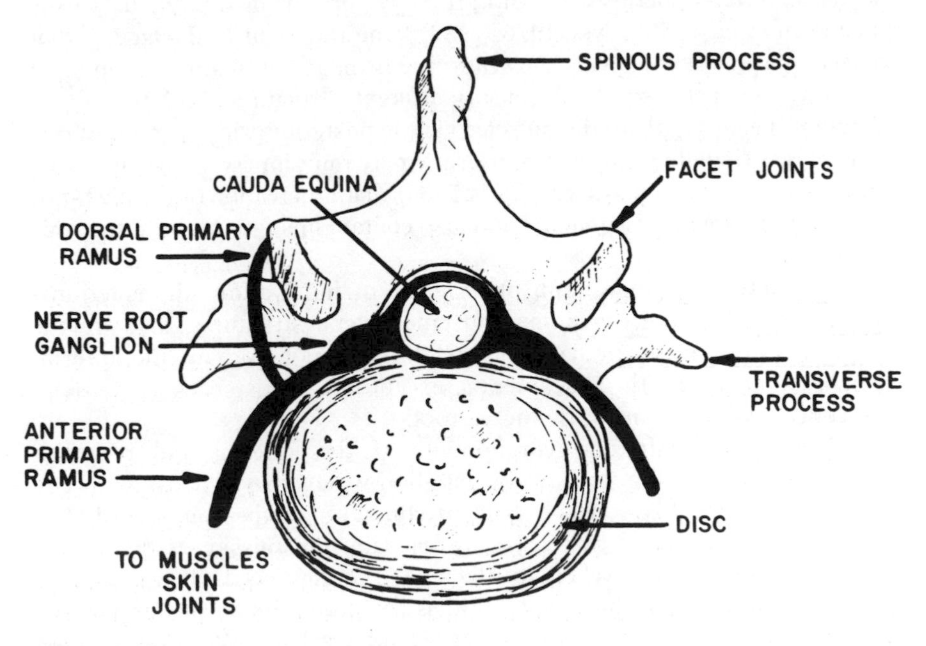

Figure 1.17
Dorsal and anterior primary rami.

compression of this tenuous blood supply from narrowing of the spinal canal might well occur in stenotic conditions. Recently, Panjabi has also evaluated how the nerves might become compressed within the foramina, as a function of the biomechanics of the spine and the foramina. Although the pathology of chronic nerve conditions is only partially known, some of the demonstrated changes include scarring between nerve fibers, as well as extensive chronic inflammatory cells (Lindahl and Rexed 1951, Murphy 1977). It is not known how these pathologic changes affect the sensitivity of the nerve to mechanical stimuli, nor is it well established whether these changes in and of themselves may be pain productive.

Transmission

Those structures capable of initiating the LBP experience are largely innervated by branching nerves derived from the posterior primary rami, whereas sciatic pain is transmitted via fibers contained within the anterior primary rami (Fig. 1.17). Figure 1.17 outlines the overall distribution of these two pathways. Branches of the posterior (dorsal) primary rami supply the facet joints, ligaments, and skin adjacent to the spine, as well as containing efferent (motor) fibers to the paraspinal musculature. A recurrent sinovertebral nerve of Luschka supplies the posterolongitudinal ligaments, posterior vertebral periosteum, and probably portions of the venous plexus (Pedersen *et al.* 1956, Wyke 1976, 1982). One theory of LBP suggests that muscle spasm accompanying the syndrome is the result of stimulation of an unknown arc between these sensory-afferent structures, and the motor (efferent) nerve supply to the muscles via the posterior primary rami. At one time, it was thought that the posterior primary rami innervated segmentally only at one level. It is now clear (Paris 1983) that innervation of the posterior primary rami are multisegmental and also contain linkages to the autonomic nervous system.

The anterior primary rami provide motor innervation, afferent transmission to all of the remaining musculoskeletal structures, and are the nerves activated in sciatic pain. Unlike the posterior primary, the anterior primary rami are strictly unisegmental, except in circumstances where there are congenital variations in the nerve roots.

The sensory (afferent) components of the anterior and posterior primary rami enter the dorsal root ganglion where they synapse with cell bodies containing fibers that extend proximally into the spinal cord (Fig. 1.18). The classic spinal cord pathway for pain transmission is the spinal-thalamic track, which is schematically shown in Figure 1.18 It is now clear that this pathway for pain conduction is far more complex (Casey 1982, Dennis and Melzack 1977, Melzack 1973, Melzack and Wall, 1968). Not only are there other less well understood pathways, but there are synaptic connections between pain conductive pathways, and both afferent and

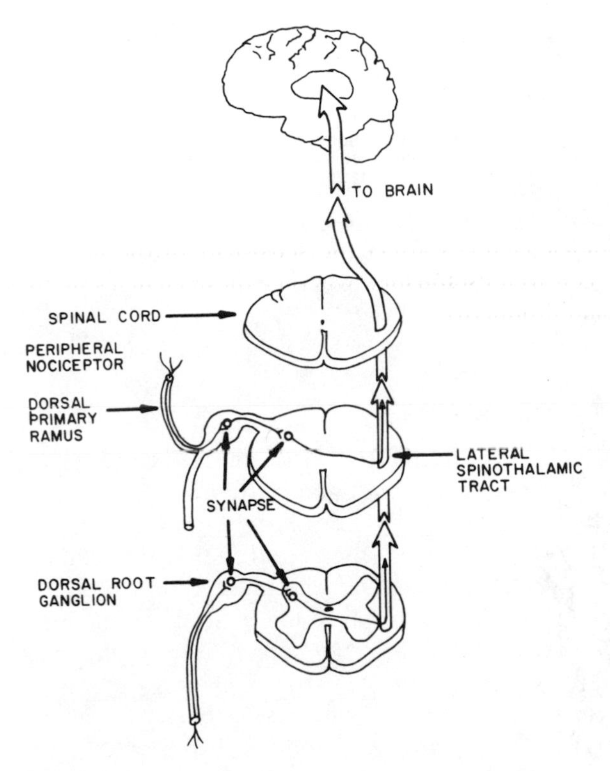

Figure 1.18
Pain transmission pathway.

efferent pathways which probably modulate pain transmission. This oversimplification of a highly complex subject underlies one of the most attractive explanations for pain modulation which has been termed "the Gate theory" (Melzack 1973, Melzack and Wall 1968). Figure 1.19 schematically demonstrates how nerve conduction may be "gated" by other afferent and efferent stimuli. This theory provides an attractive explanation for the sometimes beneficial effects of acupuncture, transcutaneous nerve stimulation, and counterirritation. This complexity of nerve transmission also may explain the frequent ineffectiveness of selective transsection of the spinal-thalamic track which has sometimes been used in the surgical management of intractable pain. (This operation is rarely indicated in low back disease.) (White and Sweet 1969).

Another newly-discovered, exciting modulator of pain transmission is chemical (Fields 1982). Within spinal fluid, chemical substances are identified which have morphine-like analgesic action. These substances, collectively termed endorphins, appear to be excreted by cells, most prominently located in the periaqueductal grey matter. Elevations of the

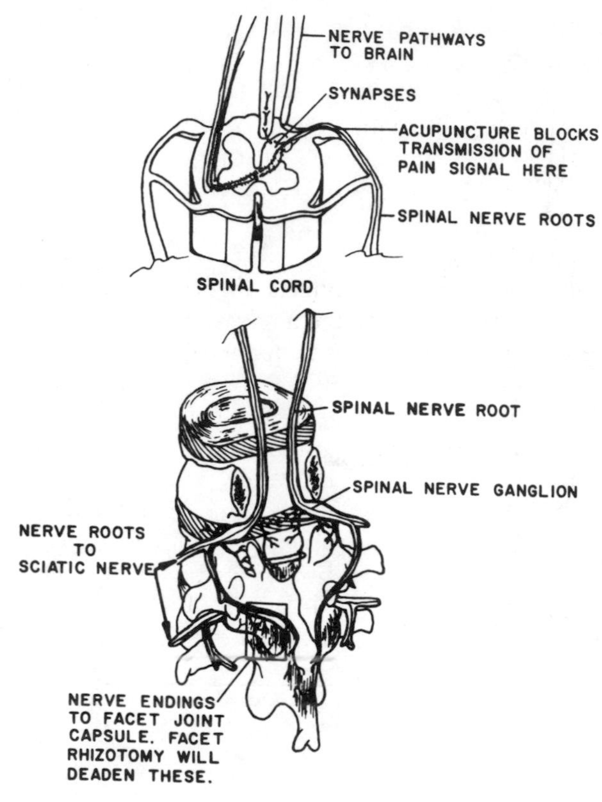

Figure 1.19
Pain modulation.

endorphins are demonstrable in a variety of conditions, for example after sustained aerobic exercise. Administration of morphine into the spinal fluid also appears to be a far more potent inhibitor of pain, than when the drug is given systemically as in intramuscular injection. Currently, this effect of morphine is being used to relieve pain after spinal surgery, and in selected cases the implanted resevoirs permit the chronic instillation of morphine into the spinal fluid. (Again, this latter technique is currently applicable to patients with malignancies, and not those with LBP.)

Pain Cognition

The pathways and areas of the brain involved in pain cognition are beyond the scope of this book. However, the cognitive experience of pain is clearly modified by many pharmacologic and psychologic phenomena. Some of the relevant modifiers of pain cognition include the following: 1) the setting of the pain experience; 2) prior pain experience; 3) age; 4) ethnic and social background; 5) psychologic status.

Beecher (1956) compared chronic pain in soldiers and civilians. He observed that soldiers injured on the battlefield frequently had few complaints of pain and often did not require morphine injections for relief. Civilians with comparable injuries almost always complained of severe pain and usually required morphine. Based on these observations, he suggested the setting in which pain occurs markedly affects the cognition of the experience. For the soldier, the injury represented a release from intolerable battlefield conditions, thus favorably modifying the pain experience.

Similarly, prior experiences of pain can modify favorably or unfavorably the current pain episode. In low back disease, as in other diseases, it is observed that the pain experience tends to increase as a function of repeated episodes. Pain is also modified by age and sex (Woodrow 1972, Notermans and Topf 1974). In general, diminished tolerance to certain painful stimuli can be demonstrated as a function of aging and sex, with males generally more tolerant than females. Whether these differences are the result of anatomic, hormonal, or other factors is unclear. It has also been shown that specific ethnic groups have greater or lesser tolerance to pain (Wolff and Langley 1975).

Affect

Pain, particularly when chronic, evokes a variety of emotions, which in turn may alter pain cognition. The most common affective states are anxiety, depression, focusing on bodily symptoms (somatization), and feelings of hopelessness and helplessness. These emotions can be quantified, usually by psychologic tests such as the Minnesota Multiphasic Personality Inventory (MMPI) and the Cornell Medical Inventory. For example, in a study of males aged 18 to 55, comparisons were made between those who had never experienced LBP and those who had chronic disabling pain (Frymoyer *et al.* 1983). Table 1.1 compares the responses given by these three groups expressed as the percentage who responded affirmatively to each question. Similarly, Beals and Hickman (1972) have shown that patients with chronic disabling LBP typically have elevations in the hysteria, hyperchondriasis, somatization, and depression scales of the MMPI. These same attributes are associated with poor outcomes from the medical and surgical treatment of low back diseases. These emotional states also modify the cognition of pain. For example, studies of patients with arthritis demonstrate that complaints of pain and, in fact, objective signs of disease often wax and wane, in part as a function of their emotional state. During periods of depression and anxiety, symptoms, and sometimes signs, worsen.

TABLE 1.1
Comparison of Asymptomatic, Non-Disabling and Disabling Low Back Pain Patients

	Asymptomatic	*LBP Non-disabling**	*LBP Disabling***
	n = 106	n = 193	n = 20
Unhappy	5.2	10.1	30.0
Hopeless	8.0	12.4	33.0
Worried	15.6	32.8	55.0
Scared	7.4	18.3	30.0
Nervous	11.2	26.5	36.8
Annoyed	32.3	54.3	55.0
Temper Outbursts	8.0	18.7	31.6
Lonely	21.1	23.5	35.0
Touchy	7.8	14.8	15.0
Hurt	4.2	8.6	15.0
Unsympathetic	9.1	13.9	50.0
Headaches	6.1	20.0	30.0
Sleep Disturbances	28.3	52.1	80.0
Dissatisfied with Medical Care	79.1	78.9	60.0
Feels Handicapped	0.0	5.3	50.0
Feels Miserable	6.1	23.1	60.0
Alcohol & Drug Abuse	7.1	7.0	15.0
Has Had Nervous Breakdown	3.0	1.1	10.5
Requires Psychiatric Help	8.9	8.2	15.8
Requires Counselling	8.6	12.8	35.0

*Non-disabling pain is defined as less than 7 days of work loss in the previous year because of back pain (many of the 193 had lost no time from work whatsoever.)

**Disabling is defined as greater than 7 days work loss in the past year.

Source: Frymoyer, J.W.; Rosen, J.C.; Clements, J.; and Pope, M.H. Psychologic factors in low back pain. Submitted to *Clin. Orthop.*

In all of these studies, it remains unclear whether the emotional state is the cause or the result of the pain experience. It is frequently suggested that there is a pain-prone personality, but to date there has been no study that either proves or disproves this theory with respect to low back disease. There is no doubt that successful rehabilitation of patients with low back diseases requires careful attention to the emotional components of the pain experience. This general belief forms the basis for much of the management programs for dealing with chronic LBP discussed in Chapter 9.

Behavior

Outward manifestations of the pain experience are behavioral. These include both physical and emotional expressions. The physical manifestations may vary, usually including stiff, guarded movements, and postural aberrations. Alterations in facial expression, postural abnormalities and sweating are other signs. In LBP patients, exaggeration of these physical signs forms the basis for a screening physical examination, described by Waddell (1980) and detailed in Chapter 8.

The psychologic manifestations of pain include the signs of depression and anxiety, and if the syndrome is chronic, usually a large element of anger. Patients with chronic low back symptoms are also more likely to manifest other pain complaints such as headache (Frymoyer *et al.* 1983, Svensson and Andersson 1983). Not surprisingly, greater reliance on doctors and other health professionals is demonstrated. Alcohol and drug abuse, disrupted marital and other social relationships are common, particularly when there is associated disability. Again, the ultimate management of chronic disabling pain syndromes, LBP or otherwise, depends upon the recognition and modification of these behaviors, a point emphasized in Chapter 9.

SUMMARY

The spine is a flexible multicurved column which has four primary functions. These are load bearing, mobility, control, and protective housing of the spinal cord. The dual roles of motion and support complicate the structure. The vertebral body and the disc have the primary role in compressive load bearing. The disc is of a fibrous laminated construction filled with a gel. Compressive forces increase the disc pressure and cause the disc to bulge. The vertebral bodies are made of trabecular bone that is oriented so as to provide support against the normal forces on the vertebrae. The vertebrae are filled with blood and may act as hydraulic shock absorbers. The ligaments have a primary role in tensile load support and such tensile load support can occur in many of the ligaments' movements in the industrial environment. Mechanical overload may result in ligament sprain. The facet joints combined have a load bearing function in compression and shear. The muscles have a role in control of posture and position and act as the motors to lift and move weights. They also provide the major stabilizing effect of the spine structures. The nervous system has a primary role in giving signals for the control of muscles, but has a secondary role and important role in sensory and pain feedback.

In low back disease, pain can be classified as LBP and radiculopathies (sciatica and neurologic claudication) which are produced by different branches of the nervous system. A complex neurophysiologic and emotional structuring creates the pain experience which involves stimulation of peripheral sensory receptors (nociceptors), transmission, cognition, and affective and behavioral states.

REFERENCES

Andersson, Gunnar B.J.; Ortengren, Roland; and Nachemson, Alf. 1977. Intradiskal pressure, intra-abdominal pressure and myoelectric back muscle activity related to posture and loading. *Clinical Orthopaedics and Related Research* 129:156–164.

Arnoldi, Carl C. 1972. Intravertebral pressures in patients with lumbar pain. A preliminary communication. *Acta Orthop. Scand.* 43 (2):109–17.

Bartelink, D.L. 1957. The role of abdominal pressure in relieving the pressure on the lumbar intervertebral discs. *J Bone Joint Surg.* 39-B(4):718–25.

Beals, Rodney K.; and Hickman, Norman W. 1972. Industrial injuries of the back and extremities. Comprehensive evaluation—an aid in prognosis and management: A study of one hundred and eighty patients. *J. Bone Joint Surg.* 54A (8):1593–1611.

Beecher, Henry K. 1956. Relationship of significance of wound to pain experienced. *JAMA* 161:1609–1613.

Bradford, D.S.; Cooper, K.M.; and Oegema, T.R., Jr. 1983. Chymopapain chemonucleolysis and nucleus pulposus regeneration. *Trans. Orthopaedic Research Society* 8:179.

Broberg, K.B.; and von Essen, H.O. 1980. Modeling of intervertebral discs. *Spine* 5(2):155–67.

Casey, Kenneth L. 1982. Pain pathways and the mechanisms in the central nervous system with special reference to low back pain. A.A.O.S. *Symposium on idiopathic low back pain.* White, Augustus A. III; and Gordon, S (eds.) St. Louis: C.V. Mosby, pp. 444–55.

Chaffin, Don B. 1974. Human strength capability and low-back pain. *Journal of Occupational Medicine* 16(4):248–54.

Crock, Henry V.; and Yoshizawa, H. 1977. *The blood supply of the vertebral column and spinal cord in man.* New York: Springer-Verlag.

Davis, Peter R. 1959. The causation of herniae by weight-lifting. *Lancet* 2:155–57.

Davis, Peter R. 1981. The use of intra-abdominal pressure in evaluating stress on the lumbar spine. *Spine* 6(1):90–92.

Dennis, Stephen G.; and Melzack, Ronald 1977. Pain signalling systems in the dorsal and ventral spinal cord. *Pain* 4 (2):97–132.

Farfan, H.F.; Cossette, J.W.; Robertson, G.H.; Wells, R.V.; and Kraus, H. 1970. The effects of torsion on the lumbar intervertebral joints: The role of torsion in the production of disc degeneration. *Journal of Bone and Joint Surgery* 52A(3):468–197.

Farfan, H.F. 1973. *Mechanical disorders of the low back.* Philadelphia: Lea & Febiger.

Fields, Howard. 1982. Endogenous mechanisms of pain modulation. A.A.O.S. *Symposium on idiopathic low back pain.* White, Augustus A. III; and Gordon, Stephen (eds.) St. Louis: C.V. Mosby Co., pp. 463–475.

Foerster, O. 1933. The dermatomes in man. *Brain* 56:1–39.

Frymoyer, J.W.; Rosen, J.C.; Clements, J.; and Pope, M.H. 1983. Psychologic factors in low back pain disability. Submitted to *Clin. Orthop.* 11/83.

Gilbertson, L.G.; Krag, M.H.; and Pope, M.H. 1983. Investigation of the effect of intra-abdominal pressure on the load bearing of the spine. *Trans. Orthop. Research Soc.* 8:177.

Gill, Gerald G.; Manning, John G.; and White, Hugh L. 1955. Surgical treatment of spondylolisthesis without spine fusion. Excision of the loose lamina with decompression of the nerve roots. *Journal of Bone and Joint Surgery* 37A(3):493–520.

Gregersen, Gary G.; and Lucas, Donald B. 1967. An *in vivo* study of the axial rotation of the human thoracolumbar spine. *Journal of Bone and Joint Surgery* 49A(2): 247–62.

Grew, N.D. 1980. Intraabdominal pressure response to loads applied to the torso in normal subjects. *Spine* 5(2):149–54.

Hagelstam, L. 1949. Retroposition of lumbar vertebrae. *Acta Chir. Scand.* (Suppl) 143.

Hirsch, C.; Ingelmark, B.E.; and Miller, M. 1963. The anatomical basis for low back pain. Studies on the presence of sensory nerve endings in ligamentous, capsular, and intervertebral disc structures in the human lumbar spine. *Acta Orthop. Scand.* 33:1–17.

Howe, John F.; Loeser, John D.; and Calvin, William H. 1977. Mechanosensitivity of dorsal root ganglia and chronically injured axons: a physiological basis for the radicular pain of nerve root compression. *Pain* 3:25–41.

Hutton, W.C.; Stott, J.R.R.; and Cyron, B.M. 1977. Is spondylolysis a fatigue fracture? *Spine* 2 (3):202–209.

Jackson, Henry C.; Winkelmann, R.K.; and Bickel, William H. 1966. Nerve endings in the human lumbar spinal column and related structures. *J. Bone Joint Surg.* 48A (7):1272–1281.

Jokl, Peter. 1982. Muscle and low back pain. A.A.O.S. *Symposium on idiopathic low back pain.* White, Augustus A. III; and Gordon, S (eds.) St. Louis: C.V. Mosby Co., pp. 456–62.

Kazarian, Leon. 1972. Dynamic response characteristics of the human vertebral column: An experimental study of human autopsy specimens. *Acta Orthop. Scand.* (Suppl) 146:1–186.

Kazarian, Leon E. 1975. Creep characteristics of the human spinal column. *Orthopaedic Clinics of North America* 6(1):3–18.

Kelsey, Jennifer L.; Pastides, Harris; and Bisbee, Gerald E., Jr. 1978. *Musculo-skeletal disorders: Their frequency of occurrence and their impact on the population of the United States.* New York: Prodist.

Knutsson, F. 1944. The instability associated with disk degeneration in the lumbar spine. *Acta Radiol.* 25:593–609.

Krag, Martin H.; Wilder, David G.; and Pope, Malcolm H. 1983. Internal strain and nuclear movements of the intervertebral disc. Presented at and *In Proceedings of Int'l Soc for Study of the Lumbar Spine*, 8 April 1983, Cambridge, England.

Kraus, H. 1973. Stress Analysis. *In Mechanical disorders of the low back*. pp. 112–33, H.F. Farfan (ed.), Philadelphia: Lea & Febiger.

Krenz, J.; and Troup, J.D.G. 1973. The structure of the pars inter-articularis of the lower lumbar vertebrae and its relation to the etiology of spondylolysis with a report of a healing fracture in the neural arch of a fourth lumbar vertebra. *Journal of Bone and Joint Surgery* 55B(4):735–41.

LaMotte, Robert H. 1982. Nociceptors in skin, joint, muscle and bone. A.A.O.S. *Symposium on idiopathic low back pain*. White, Augustus A. III; and Gordon, S. (eds.) St. Louis: C.V. Mosby Co., pp. 417–32.

Lane, Joseph M.; and Vigorita, Vincent J. 1983. Osteoporosis. *Journal of Bone and Joint Surgery* 65A(2):274-78.

Lewis, T.; and Kellgren, J.H. 1939. Observations relating to referred pain, visceromotor reflexes and other associated phenomena. *Clin. Sci.* 4 (1):47–71.

Lindahl, O.; and Rexed, B. 1951. Histologic changes in spinal nerve roots of operated cases of sciatica. *Acta Orthop. Scand.* 20:215–25.

Lorenz, M.; Patwardhan, A.; and Vanderby, R. 1983. Load-bearing characteristics of lumbar facets in normal and surgically altered spinal segments. *Spine* 8 (2):122–30.

Markolf, Keith L 1972. Deformation of the thoracolumbar intervertebral joint in response to external loads. A biomechanical study using autopsy material. *Journal of Bone and Joint Surgery* 54A(3):511–33.

McCall, I.W.; Park, W.M.; and O'Brien, J.P. 1979. Induced pain referral from posterior lumbar elements in normal subjects. *Spine* 4 (5):441–46.

McNeill, Thomas; Warwick, David; Andersson, Gunnar; and Schultz, Albert. 1980. Trunk and strengths in attempted flexion, extension, and lateral bending in healthy subjects and patients with low-back disorders. *Spine* 5(6):529–538.

Melzack, R. 1973. The puzzle of pain. New York: Basic Books.

Melzack, Ronald; and Wall, Patrick D. 1965. Pain mechanisms: A new theory. *Science* 150 (3699):971–79.

Melzack, R.; and Wall, P.D. 1968. Gate control theory of pain. In *Pain*, A. Soulairac, J. Cahn, and J. Charpentier (ed.). New York: Academic Press, pp. 11–31

Melzack, R.; and Wall, P.D. 1970. Psychophysiology of pain. *Int. Anesthesiol. Clin.* 8:3–34.

Mooney, Vert; and Robertson, James. 1976. The facet syndrome. *Clin. Orthop.* 115:149–56.

Morris, J.M.; Lucas, D.B.; and Bresler, B. 1961. Role of the trunk in stability of the spine. *Journal of Bone and Joint Surgery* 43-A(3):327–51.

Mountcastle, Vernon B. (Ed.) 1968 *Medical Physiology* 12th Ed. V.1 and V.2., St. Louis: Mosby.

Murphy, Robert W. 1977. Nerve roots and spinal nerves in degenerative disk disease. *Clin. Orthop.* 129:46–60.

Nachemson, Alf; and Morris, James M. 1964. *In vivo* measurements of intradiscal pressure. Discometry, a method for the determination of pressure in the lower lumbar discs. *Journal of Bone and Joint Surgery* 46A(5):1077–92.

Newman, P.H. 1952 Sprung back. *Journal of Bone and Joint Surgery* 34B(1):30–37.

Notermans, S.L.H.; and Tophoff, M.M.W.A. 1967. Sex differences in pain tolerance and pain appreciation. *Psych., Neurol. Neurochir.* 70:23–29.

Panjabi, M.M.; Takata, K.; and Goel, V.K. 1983. Kinematics of lumbar intervertebral foramina. *Spine* 8 (4):348–64.

Paris, Stanley V. 1983. Anatomy as related to function and pain. *Orthop. Clin. North Am.* 14 (3):475–90.

Pedersen, Herbert E.; Blunck, Conrad F.J.; and Gardner, Ernest. 1956. The anatomy of the lumbosacral posterior rami and meningeal branches of the spinal nerves (sinuvertebral nerves) with an experimental study of their functions. *Journal of Bone and Joint Surgery* 38A (2):377–91.

Perey, Olof. 1957. Fracture of the vertebral end-plate in the lumbar spine: An experimental biomechanical investigation. *Acta Orthop. Scand.* (Suppl) 25:1–101.

Poulsen, E.; and Jorgensen, K. 1971. Back muscle strength, lifting and stooped working postures. *Applied Ergon.* 2:133–37.

Shah, J.S.; Hampson, W.G.J.; and Jayson, M.I.V. 1978. The distribution of surface strain in the cadaveric lumbar spine. *Journal of Bone and Joint Surgery* 60B(2):246–51.

Smidt, G.L.; Amundsen, L.R.; and Dostal, W.F. 1980. Muscle strength at the trunk. *J. Orthop. and Sports Physical Therapy* 1(3): 165–70.

Smyth, M.J.; and Wright, V. 1958. Sciatica and the intervertebral disc: An experimental study. *Journal of Bone and Joint Surgery* 40A(6):1401–18.

Spilker, Robert L. 1980. Mechanical behavior of a simple model of an intervertebral disk under compressive loading. *Journal of Biomechanics* 13(10):895–901.

Stokes, I.A.F.; Wilder, D.G. Frymoyer, J.W.; and Pope, M.H. 1981. 1980 Volvo Award in clinical sciences. Assessment of patients with low back pain by biplanar radiographic measurement of intervertebral motion. *Spine* 6(3):233–40.

Svensson, Hans-Olof; and Andersson, G.B.J. 1983. Low back pain in forty to forty-seven year old men. Work history and work environment factors. *Spine*, 8 (3):272-76.

Waddell, G.; McCulloch, J.A.; Kummel, E.; and Venner, R.M. 1980. Nonorganic physical signs in low back pain. *Spine* 5 (2):117–25.

Weisman, Gerald; Pope, Malcolm H.; and Johnson, Robert J. 1980. Cyclic loading in knee ligament injuries. *American Journal of Sports Medicine* 8(1):24–30.

White, James C.; and Sweet, William H. 1969. *Pain and the neurosurgeon. A forty-year experience.* Springfield, Il.: Charles C. Thomas.

Wilder, David G.; Pope, Malcolm H.; and Frymoyer, John W. 1982 Cyclic loading of the intervertebral motion segment. 1982. *In Proceedings Tenth Northeast Bioengineering Conference.* March 15-16, 1982. Eric W. Hansen (ed.), IEEE catalog #82Ch1747-5, p. 9–11.

Wilder, D.G.; Krag, M.H.; and Pope, M.H. 1984. *Atlas of mammalian—lumbar vertebrae—pictorial and dimensional information.* In preparation. Springfield, Il.: Charles C. Thomas.

Wolff, B.B.; and Langley, S. 1968. Cultural factors and the response to pain. A review. *Am. Anthropol.* 70:494–501.

Wolff, B.B.; and Langley, S. 1975. Cultural factors and the response to pain. A review. In M. Weisenberg (ed.) *Pain: Clinical and experimental perspectives.* St. Louis: C.V. Mosby.

Woodrow, K.M.; Friedman, G.D.; Seiglanb, A.B.; and Collen, M.F. 1972. Pain tolerances: Differences according to age, sex and race. *Psychosom. Med.* 34:548–56.

Wyke, Barry D. 1976. Neurological aspects of low back pain. ***The lumbar spine and back pain.*** 1st. ed. Jayson Malcolm I.V. (ed.) London: Pitman Medical Publishing Co., pp. 189–256.

Wyke, B.D. 1982. Receptor systems in lumbosacral tissues in relation to the production of low back pain. A.A.O.S. *Symposium on idiopathic low back pain.* White, Augustus A. III; and Gordon, S. (eds.) St. Louis: C.V. Mosby Co. pp. 97–107.

2 Occupational Biomechanics of the Lumbar Spine

Gunnar B.J. Andersson, Don B. Chaffin, and Malcolm H. Pope

The preceding chapter has dealt with the biomechanics of the spine and its motion segments in broad context. Additional data are important to understand the biomechanics of the spine as it relates to work and the work environment. Occupational biomechanics of the spine, as presented here, form the basis for subsequent discussion on workplace evaluation (Chapter 13) and workplace design (Chapter 14). As noted in the Introduction, this chapter is necessarily technical and dependent on engineering concepts. The basic message, however, should be of interest to all readers.

The forces and moments acting on the lumbar spine come from body segment weights, movements of the trunk and extremities, and any external loads being handled or applied. These forces and moments must be equilibrated by internal forces, mainly contractions of the trunk muscles and resistances of the soft tissues. Pressures within the trunk cavities may also aid in this respect, as discussed in Chapter 1.

In the following pages we discuss the forces and moments on the spine as they relate to body posture, and to the handling of materials and tools. We also discuss how externally applied forces, such as vibration, can influence the spine and its substructures. The chapter as such is introductory to the subject. Although we cover this topic in some detail here, the interested reader is referred to Chaffin and Andersson (1984) for all aspects of occupational biomechanics.

BIOMECHANICS OF POSTURE

When considering the influence of postural factors on the load on the spine there are three issues to consider: 1) basic posture, 2) postural symmetry, and 3) postural constraint.

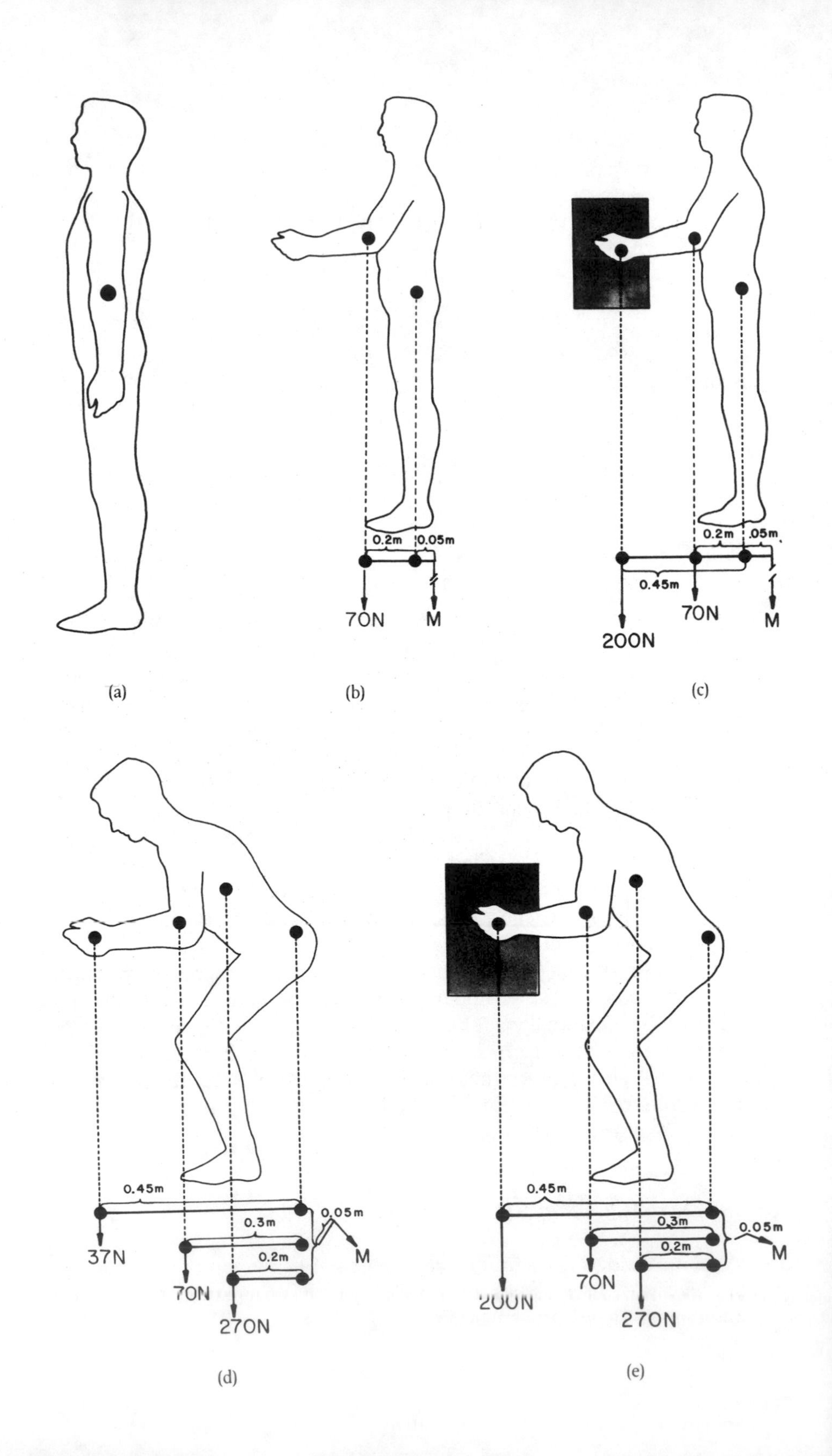

0.2m
0.05m
70N
M
0.2m
.05m
0.45m
70N
M
200N
0.45m
0.05m
0.3m
37N
M
0.2m
70N
270N
0.45m
0.3m
0.05m
0.2m
M
70N
200N
270N

(a) (b) (c) (d) (e)

At the workplace, standing and sitting are the two forms of basic posture. The main advantages of standing are mobility, reach, and exertion of force, while sitting is less stressful for the legs, less energy-consuming, less demanding on the blood circulation to the legs, and more stable for precision work. As will be discussed subsequently, there are also disadvantages to sitting work postures, especially with respect to the back. Ergonomic problems cannot be solved by simply providing a chair.

Even the most comfortable posture should not be maintained for long periods. Static loading of muscles and joint tissues will occur and lead to discomfort.

Standing Postures

Muscular activity is required to maintain an upright posture. As long as the body segments are well aligned with respect to the center of gravity, the activity is small (Andersson 1974, Andersson *et al.* 1974a-f, 1977; Andersson and Ortengren 1974a; Carlsoo 1961; Donish and Basmajian 1972; Morris *et al.* 1962). Any shift in the center of gravity of the trunk requires active counterbalancing by muscle force to maintain equilibrium (Klausen 1965; Andersson *et al.* 1977). Muscle forces are also required to counterbalance the moment caused by an outstretched arm, an external weight or any other force applied to the trunk, head and upper extremities (Andersson *et al.* 1980; Schultz *et al.* 1982). The combined effect of all these forces upon the lumbar spine produces a moment that must be counterbalanced by the spinal muscles to maintain equilibrium. This moment is referred to as the load moment. While the load moment is illustrated in Figure 2.1 for sagittally symmetric situations, a much more complex situation occurs when

Figure 2.1
(a) In upright standing the body segments are well aligned with respect to the center of gravity, and little muscular effort is required to maintain equilibrium. (b) An elevated arm causes a load moment of about 70 N × 0.2 m (14 Nm) which must be equilibrated by the back muscles acting with an average moment arm of 0.05 m (muscle tension = 620 N) plus moment due to the hand (no shown) of 17 Nm. An additional object with a weight of 200 N held at 0.45 M (c) causes an additional moment of 70 Nm (total muscle tension 2080 N). When leaning forward holding no weight (d) the trunk moment (270 × 0.2 54 Nm), arm moment (70 × 0.3 = 21 Nm), and hand moment (37 × 0.45 = 17 Nm) act together causing a total load moment of 92 Nm (muscle tension = 1840 N), and an additional weight together with the weight of the hand (e) causes an additional moment of 200 × 0.45 = 90 NM, all together 165 Nm (muscle tension = 3300 N).

asymmetry prevails. In such postures–lateral flexion, rotation, and combinations thereof–other appropriate muscles will contract. The asymmetry in muscle force is illustrated in Figure 2.2, from a study by Andersson *et al.* (1977). In rotation and lateral bending, high levels of activity occur on the contralateral side, while the activities on the ipsilateral side are small. This asymmetry in muscle activity can lead to unequal stress concentrations on the different component structures of the spine. Thus, to maintain low muscle forces and consequently low stresses on the spine structures, an upright symmetric posture should always be advocated, and all material should be handled as close to the body as possible. Disc pressure measurements obtained *in vivo* when standing indicate that the load on the L3 disc in a person weighing 150 lbs (70 kg) is about 107 lbs (500N) (Nachemson and Elfstrom 1970; Andersson *et al.* 1974a-f). When leaning forward, an increase in pressure occurs in parallel to the increase in load moment (Andersson *et al.* 1977; Schultz *et al.* 1982).

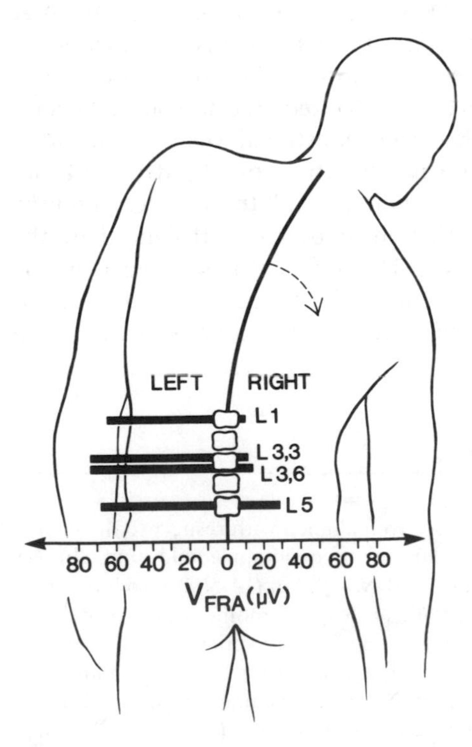

Figure 2.2
Myoelectric activity (EMG) recorded in lateral bending. A weight was held in the right arm. Note high activity levels on the convex side (contralateral to the load), and low levels on ipisilateral side. (Adapted from Andersson *et al.* 1977).

Sitting Postures

Sitting is a position in which the weight of the body is transferred to the supporting area mainly by the ischial tuberosities of the pelvis and their surrounding soft tissues. Depending on the chair and posture, some proportion of the total body weight will also be transferred to the floor, as well as to the backrest and armrests.

Three different types of sitting can be distinguished and will be referred to as anterior, middle, or posterior sitting postures. In the middle posture, the center of mass of the trunk is directly above the ischial tuberosities. This is the balanced position used on a stool. When relaxed in a middle posture, the lumbar spine is either straight or in slight kyphosis. The anterior (forward leaning) posture can be reached from the middle posture either by a forward rotation of the pelvis or by inducing a large induced kyphosis of the lumbar spine. Anterior sitting postures are adopted most often when desk work is performed. In the anterior posture, the center of mass is in front of the ischial tuberosities while in the posterior (backward leaning) posture, the center of mass is behind the ischial tuberosities. That posture is obtained by a backward rotation of the pelvis. It is a posture assumed in rest chairs and in chairs with reclining backrests such as lounge chairs.

In general, the posture of a seated person depends not only on the design of the chair, but on sitting habits and the task to be performed. The height and inclination of the seat of the chair, the position, shape and inclination of the backrest, and the presence of other types of support all

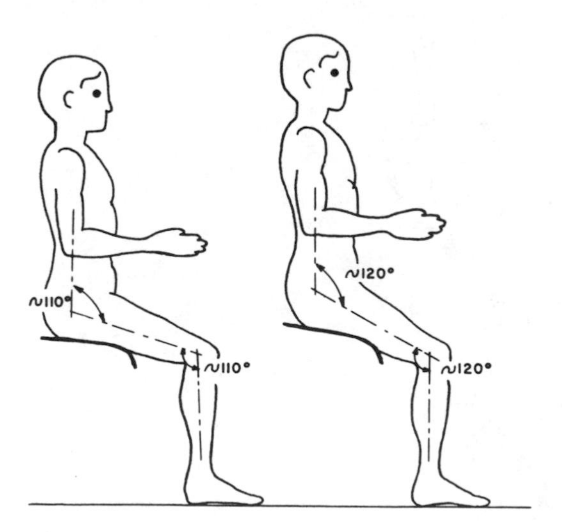

Figure 2.3
A semi-sitting posture allows for rapid changes from sitting to standing. (Adapted from Engdal 1983).

influence the resulting posture. The chair should also permit regular and easy alterations in posture, since continuous sitting in one position is a risk factor for LBP. To facilitate transitions between certain tasks, an intermediate posture, *semisitting*, is often desirable. In *semisitting*, a higher than normal chair is used, usually with a forward sloping seat such that a person leans on it, dividing the weight between the buttocks and feet (Laurig 1969) (Fig. 2.4).

Information on the biomechanics of sitting comes from radiographic studies, from studies of the myoelectric activity of muscles (EMG), and from disc pressure measurements.

Radiographic studies show that the pelvis rotates backward and the lumbar spine flattens when sitting (Akerblom 1948, Burandt 1969, Carlsoo 1972, Keegan 1953, Schoberth 1962, Umezawa 1971, Rosemeyer 1972, Andersson *et al.* 1979). Andersson *et al.* studied the influence of different types of backrests on pelvic rotation and on the lumbar lordosis angle. Changes in pelvic rotation influenced the shape of the lumbar spine because the sacral-horizontal angle changed, i.e., the foundation of the lumbar spine changed (Fig. 2.4). There must be compensation for this change in angle to keep the trunk upright.

To influence pelvic rotation a forward tilted seat was proposed by Burandt (1969), Carlsoo (1963) and Mandal (1975), while Rosemeyer (1972) suggested fixation of the pelvis on the seat by a pillow support. These measures all influence spine configuration. The shape of the lumbar spine

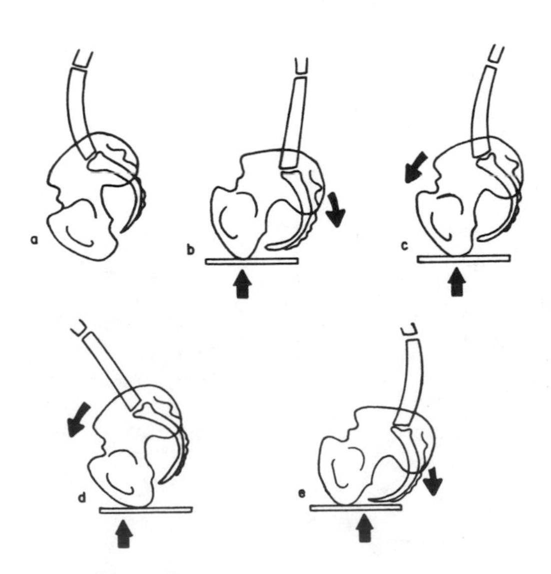

Figure 2.4
Changes in pelvic rotation influence the shape of the lumbar spine.

(the flattening) can also be influenced by a well-designed low-back support (Akerblom 1948, Keegan 1953, and Schoberth 1962).

The knee flexion angle also is important to the spine curvatures during sitting and the same is true for the hip flexion angle. Keegan (1963) obtained radiographs of a subject and found that the lumbar curve flattened when a 90° trunk-thigh angle was obtained, compared to a 135° angle. He attributed this to the combined actions of the various muscles that rotate the pelvis and thus influence the lumbar curve.

In vivo pressures measured within a lumbar disc when sitting without a support have been found to be about 35 percent higher than those when standing (Okushima 1970; Nachemson and Elfstrom 1970). To study the influence of supports, disc pressures have also been measured when sitting in different chairs with different back supports (Andersson 1974; Andersson *et al.* 1974b-f). These studies confirmed that the disc pressure is considerably lower in standing than in unsupported sitting (Fig. 2.4). Of different unsupported sitting postures, the lowest pressure was found when sitting with the back straight. The reasons for this increased pressure in sitting postures are: 1) an increase in the trunk load moment when the pelvis is rotated backward and the lumbar spine and torso are rotated forward; and 2) the deformation of the disc itself caused by the lumbar spine flattening.

Figure 2.5
Disc pressure measurements in standing, unsupported sitting, office work activities, and with and without armrests (adapted from Andersson *et al.* 1974).

When supports were added to the chair, disc pressure was found to be influenced by several factors (Fig. 2.6). Inclination of the backrest from vertical backward resulted in a decrease in disc pressure. An increase in lumbar support resulted also in a decrease in disc pressure. The decrease was generally larger when the backrest-seat angle was small. Studies performed in an office chair placing the back support at different lumbar levels showed a slightly lower pressure when the support was at the level of the fourth and fifth lumbar vertebrae rather than at the first and second vertebrae. The use of arm rests always resulted in a decrease in disc pressure, but was less pronounced when the backrest-seat angle was large.

The disc pressure measurements can be interpreted as follows: 1) when backrests are used, part of the body weight is transferred to them when a person leans back, reducing the load on the lumbar spine caused by the upper body weight; 2) an increase in backrest inclination (leaning the backrest more backward) means an increase in load transfer to the backrest and results in a reduced disc pressure; 3) the use of arm rests supports the weight of the arms, reducing the disc pressure; and 4) the use of a lumbar support changes the posture of the lumbar spine toward lordosis and hence reduces the deformation of the lumbar spine and corresponding disc pressure (Figure 2.7).

Studies were also made of typical seated office work. When writing at a desk, a decrease in disc pressure was noted (Fig. 2.5) compared to other

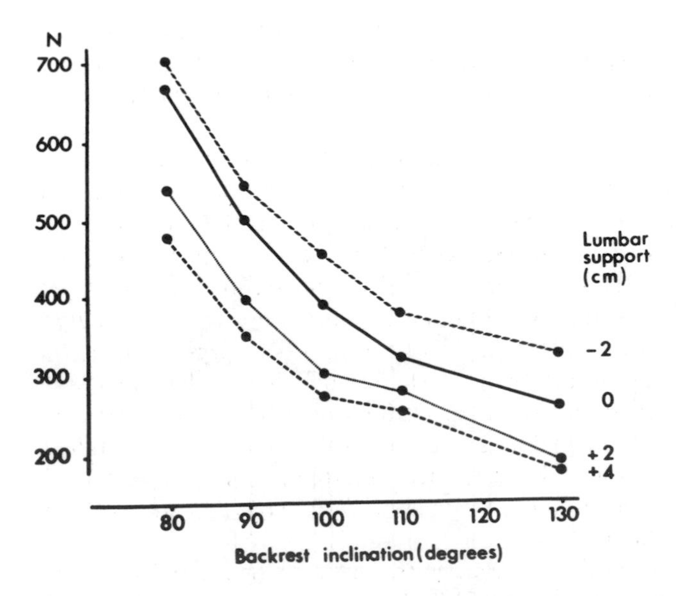

Figure 2.6
The disc pressure decreases when the backrest inclination is increased and when a lumbar support is used (adapted from Andersson *et al.* 1974).

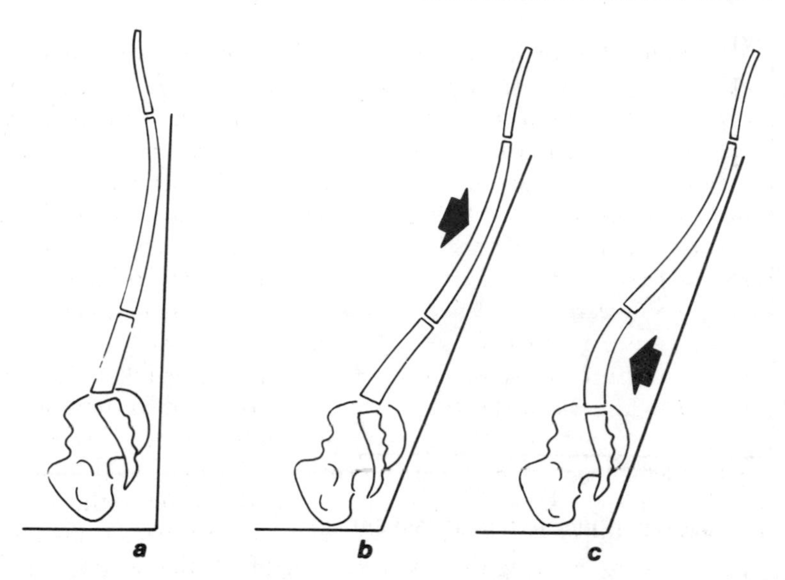

Figure 2.7
When backrests are used an increasing part of the body weight is transferred to the backrest (b). The use of a lumbar support changes the posture of the lumbar spine towards lordosis (c).

tasks. This was expected since the arms can be well supported by the desk. Other office activities, such as typing and lifting a phone at arm's length, increased the pressure because of larger external load moments imparted to the spine during such tasks.

Electromyography has also been used to study the activity of back muscles when sitting. The EMG activity levels are important because high activities indicate muscle contractions. Generally, similar activity levels have been recorded when standing and sitting without a back support (Floyd and Silver 1955; Carlsoo 1963; Rosemeyer 1971; Andersson and Ortengren 1974a; Andersson, Jonsson and Ortengren 1974b). There is general agreement that, in sitting, the myoelectric activity (EMG) decreases when: 1) the back is slumped forward in full flexion (Akerblom 1948; Lundervold 1951a, b; Floyd and Silver 1955; Schoberth 1962; Jonsson 1970); 2) the arms are supported (Carlsoo 1963; Floyd and Ward 1969; Rosemeyer 1971; Andersson *et al.* 1974a,b); or 3) a backrest is used (Akerblom 1948; Carlsoo 1963; Knutsson, Lindh and Telhag 1966; Floyd and Ward 1969; Rosemeyer 1971; Andersson *et al.* 1974a,b).

Of these different support parameters, backrest inclination has been found to be very important, with the EMG levels decreasing as the backrest-seat angle is increased (Knutsson, Lindh and Telhag 1966, Rosemeyer 1971) and is illustrated in Figure 2.8 (Andersson *et al.* 1974a). Andersson *et al.*

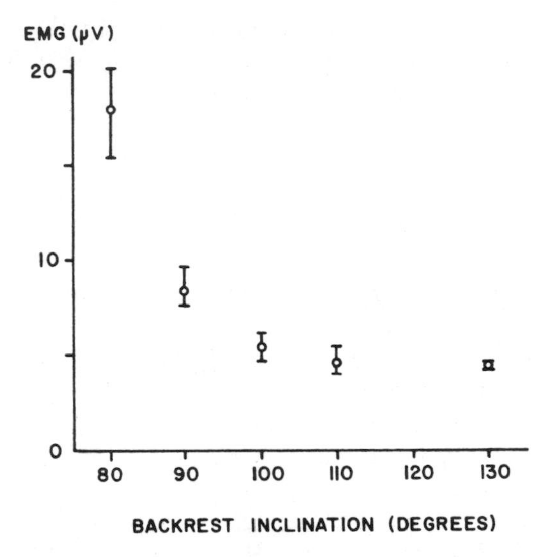

Figure 2.8
The myoelectric (EMG) activity decreases when the backrest inclination is increased (adapted from Andersson and Ortengren 1974).

(1974a-f) also found that the myoelectric activity not only decreased in the lumbar region but also in the thoracic and cervical areas of the spine when backrest inclination was increased. When the angle was greater than 110° there was little further effect on the EMG levels. The influence of a lumbar support was small on the muscle activity, whereas it was considerable on the disc pressure.

Yamaguchi, Umezawa and Ishinada (1972) also found that muscle activity decreased when the seat inclination was increased backward. The effect of different locations of the backrest was studied by Lundervold (1951a,b,1958) and Floyd and Roberts (1958). They found that the myoelectric activity was less when the back support was located in the lumbar region than in the thoracic, thus confirming Akerblom's (1948) observation that a support in the lumbar region was as effective as a full back support.

The effect of the height of the seat and of the table has also been investigated. Too high or too low a seat has been found to increase muscle activity (Lundervold 1951a, 1951b, 1958; Laurig 1969). The vertical distance between the seat and the table appears also to be an important factor (Lundervold 1951a, 1951b, 1958; Laurig 1969; Chaffin 1973; Andersson *et al.* 1974a-f).

Leg support is critical to distribute better and reduce the load on the buttocks and the back of thighs. To do so, the feet should rest firmly on the floor or foot support so that the weight of the lower legs is not supported by the front part of the thighs resting on the seat. When a chair is so high that the feet do not reach the floor, there may be uncomfortable pressures on the back of the thighs (Akerblom 1948, 1969; Schoberth 1962; Bush 1969). All of these concepts provide the basis for optimal seat design discussed in Chapter 14.

Recumbent Posture

Most people's occupations do not involve the recumbent position, although the reader can undoubtedly think of exceptions. However, a significant portion of each day is spent in the recumbent position. There is much opinion, but only scant scientific data that support optimal sleeping postures and bedding. The usual teaching has stressed the importance of lying supine on a relatively firm bedding surface, thus giving overall support to the spine while avoiding areas of excessive focal pressure on bony prominences. Flexion at the knees and hips is thought to decrease further stresses on the spine and supportive musculature. The only real direct evidence favoring this position comes from measurements of interdiscal pressure (usually at the L3–4 level) which show the lowest measurements in the supine, hip-knee flexed posture. In this position, the solus muscles are also relaxed. Patients with acute and chronic low back conditions often observe that this is the position of maximal comfort. In particular, individuals with advanced spinal stenosis and claudicatory leg pain often can only find comfort in this position, presumably because the canal dimensions are greatest. The position also reduces lumbar lordosis and theoretically decreases stresses on the facet joints and their capsules. However, not all individuals or patients with low back complaints find this posture comfortable. Indeed, some patients are more comfortable in the prone position, an observation emphasized by the proponents of extension exercise programs for the relief of certain low back conditions. The theoretical reason that the extended posture is comfortable for some patients is based on the assumption that a posteriorly bulging annulus becomes less bulged in extension as the nucleus pulposus moves into a more forward position.

In the side-lying position, particularly when bedding is soft, there are increases observed in the intradiscal pressure. Typically, the side-lying posture is associated with some element of twisting, thus placing abnormal stresses upon the facet capsules, as well as producing torsional stresses on the annulus fibrosus. Despite these theoretical mechanical consequences, many people, with or without low back pain, adopt this sleeping posture. Possibly one reason for this being a comfortable position is the associated hip and knee flexion. Based on the limited information available, we may conclude that there is insufficient evidence to justify insistence on a single recumbent body posture, because there is no evidence to indicate which posture might or might not reduce the risk of low back episodes, and it is not practical to alter many individuals' normal sleeping habits. It does seem reasonable for individuals whose complaint is sharp spinal stiffness or pain on arising from bed to try a variety of resting postures to determine which one is most comfortable. Based on the limited scientific evidence, the knee-hip flexed supine posture would seem to be the most useful starting point.

The question of optimal bedding design is equally elusive. There is clearly a great deal of commercial interest in promoting one or another alternative in bed design, both for the prevention and for the relief of low back diseases. Among the most popular current fads is the waterbed. Again, the evidence is scant and largely based on uncontrolled consumer reports. It seems logical that supportive mattresses and springs would have adequate structural integrity to give uniform body support and would not force the individual into an abnormal posture. For example, a stomach sleeper lying on inadequate bed support will be forced into a position of hyperextension. Similarly, side-lying sleeping postures under these circumstances would tend to accentuate the lateral bend and axial rotation. Perhaps, the simplest and most efficient sleeping surface is that advocated by the Scandinavians: thick plywood support with a foam rubber surface of six to eight inches.

BIOMECHANICS OF MANUAL MATERIALS HANDLING

Contrary to popular opinion, manual material handling has not been replaced by automation in modern industries. Heavy loads are still being lifted by people both in thei employment and leisure time activities. Often the body position allowed by the immediate physical environment or by the size of the object being lifted does not allocate the stresses on the body to those components that are best capable of coping with them. One of the parts of the body that is often the most highly stressed during a lifting act is the low back, specifically the lower lumbar segments of the spinal column and their associated muscles and ligaments.

What follows is an introduction to modelling the L5/S1 motion segment during lifting loads in industry. Through the use of such models one can begin to understand the complex interrelated mechanical factors that cause LBP. More detailed discussions of the mechanics of the spine are presented in Schultz *et al.* (1982a,b) and Chaffin and Andersson (1984).

Biomechanical modeling of load-lifting activities

It is evident from papers by Tichauer (1965), Roaf (1960), and Davis, Troup, and Burnard (1965) that the estimation of stresses on various parts of the musculoskeletal system during lifting activities will require a complex model that takes account of such factors as 1) instantaneous positions and accelerations of the extremities and trunk; 2) changes in spinal geometry; 3) strength variations within different muscle groups and people; and 4) effects of the abdominal pressure reflex during lifting.

Recent efforts have been made to extend an earlier biomechanical model of the human body created by Plagenhoef (1966) for sports activities to include 1) an estimate of the stresses in the lumbosacral disc; 2) the addition of external loads on the hands (e.g., in a materials handling task); and 3) an evaluation of the effects of various muscle group strengths on the performance of the person being studied. In accomplishing this, several different computerized biomechanical models have evolved. The following is a description of one of these models, and a presentation of some of the more interesting results that have been obtained from using the model to study various whole-body lifting activities.

The model to be described has been referred to as the *Static Sagittal Plane* (SSP) model developed by the research group at the University of Michigan which has used it over the last 15 years. As the name implies, this particular model has been developed to evaluate various static situations, such as when one is holding a weight or pushing or pulling on a non-moving object with both hands. In addition to these applications, the model can be used to analyze the more normal slow moves used in heavy materials handling by formulating the input data to describe a sequence of static positions with very small changes in each successive position during a movement. In making this type of pseudo-dynamic analysis it must be assumed that the effects of acceleration and momentum are negligible.

The SSP model is also restricted to symmetric sagittal plane activities; thus, a rotation or lateral deviation of the body cannot be analyzed. Three-dimensional models do allow such analysis (Schanne 1972, Schultz *et al.* 1982, 1982a,b).

The SSP model develops estimates of the forces and load moments at each of the major articulations of the extremities and back by treating the body as a series of seven solid links which are articulated at the ankles, knees, hips, L5/S1 disc, shoulders, elbows, and wrists as shown in Figure 2.9. Each of the links in the model is considered to have a mass whose estimate is based on the body weight proportionality constants presented by both Dempster and Gaughran (1967) and Clauser, McConville, and Young (1970). The distribution of the mass within each link is based on the data of Dempster (1955). The length of each link is established from over-the-body measurements with reference landmarks described by Dempster (1955). Specifically, the body measurements needed as input data are body stature, body weight, center of gravity of the hand to wrist distance, lower arm length, lower leg length, foot length, and elbow height when standing erect. From these the link lengths (i.e., the straight line distances between the articulation points of rotation) are estimated based on the empirical relationships developed by Dempster and co-workers (1964, 1967).

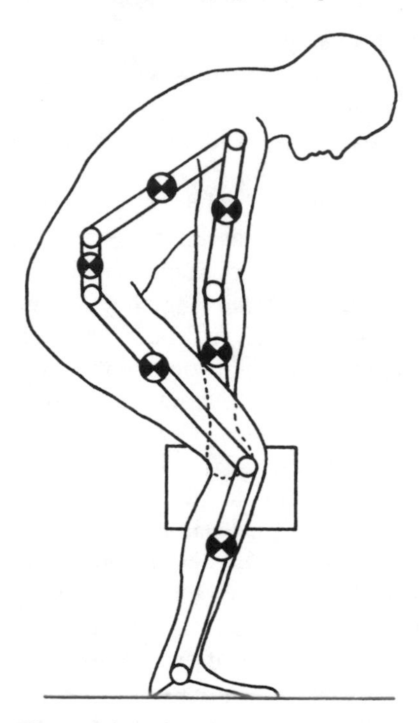

Figure 2.9
The static sagittal plane (SSP) model. Wrist joint not shown.

A task being analyzed with the SSP model is described by two types of data. First, any external force that may be exerted against the hands is measured and considered in the model as a vector acting at the center of gravity of the hands. For example, if a person is holding a 21 lb box (10 kg) it is entered into the program as a 21 lb force acting downward (i.e., a force acting in respect to some defined reference axis). The second type of information required by the model is a description of the posture of interest. These data are obtained by measuring the articulation angles from either a lateral photograph (or similar) of a person who is in the position of interest or from an articulated drawing-board body template placed in the "Task" position.

The above data are sufficient for the model to compute the load moments and forces at each of the six major articulations of the body. The remainder of the discussion focuses on the development and use of a model that transforms the externally-produced forces and load moments acting on the L5/S1 disc into predictions of compression and shear forces acting on the disc. Specifically, the compression force on the disc will be estimated as a function of: 1) posterior muscle actions; 2) body weight and load effects; and 3) abdominal pressure against the diaphragm.

Compressive forces on the lumbar spine caused by the abdominal muscles *per se* are assumed to be negligible since Bartelink (1957) disclosed that the rectus abdominis which could mechanically cause a spinal compressive force was relatively inactive during lifting activities. The abdominal pressure was therefore attributed to the oblique and transverse muscles which are not well positioned to assist or hinder directly in sagittal plane flexion or extension of the trunk, but rather act to create pressure against the diaphragm which causes erection of the torso.

The line of action of the muscles of the lower lumbar spine are assumed to act parallel to the force of compression on the vertebral discs with a moment arm of 2.0″ (5.0 cm) as illustrated in Figure 2.10. The estimate of the magnitude of the muscle force required to maintain a particular trunk position is accomplished by dividing the estimated L5/S1 load moment caused by the body weight above the L5/S1 disc and load in the hands (after the abdominal pressure effect has been subtracted) by the 2.0″ moment assumed for the back muscles. This logic assumes that the bending strength of the column is only caused by the spinal muscles and that the articulating facets do not resist compression in the lumbar column during a lifting act.

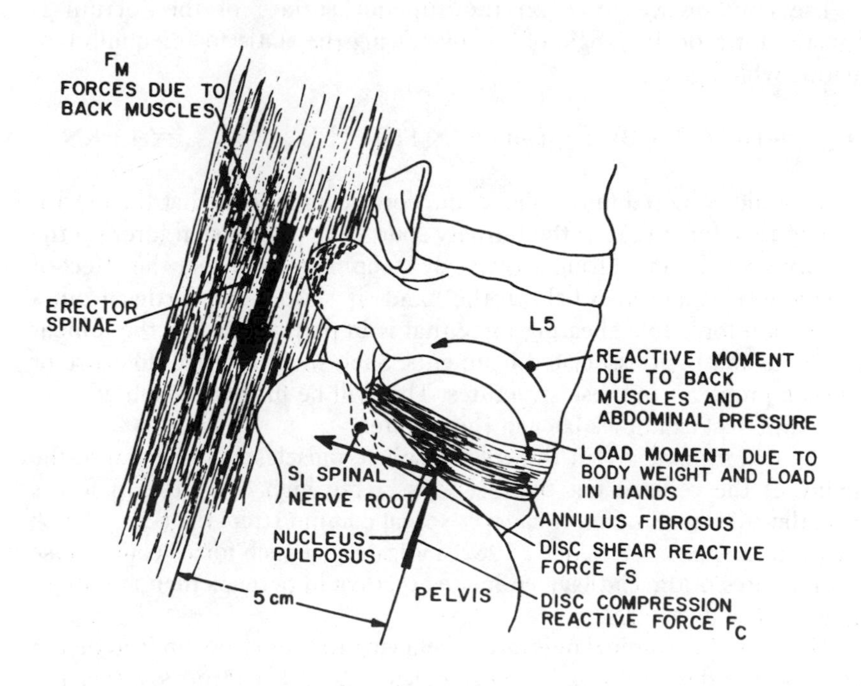

Figure 2.10
Forces acting around disc.

Some Numerical Examples

To assist in understanding the biomechanics of the spine, the SSP model has been used to study various lifting situations. The results of these simulations are now described.

Case 1. Lifting a 500 N load close to the body

For this case we assume a common lifting posture as illustrated in Figure 2.11. Here the trunk is in forward flexion (30° from vertical). In an average sized male, the resulting load moment at the hips is 200 Nm because of the weight BW of the trunk, neck, head and arms and the 500 N load FH held in the hands. The abdominal pressure P_{abdom} is predicted based on the hip load moment and the angle between the trunk and thighs assumed to be about 160 N.

The static moment equilibrium equation results in an estimate of the erector spinae forces as follows:

$$F_{musc} = \frac{(BW \times B) + (FH \times H) - F_{abdom} \times A)}{M} = 3.5 \text{ KN}$$

The compressive force on the superior surface of the sacrum is estimated for a flexion angle of 66° by solving the static force equilibrium equation which gives:

$$F_{comp} = (\sin 66° \times BW) + (\sin 66° \times FH) - F_{abdom} + F_{musc} = 4.1 \text{ KN}$$

It should be noted in the above four force components that the erector spinae muscle force F_{musc} is the primary source of compression force on the disc during a volitional lifting activity. By stooping over further the effect of the body weight and load held in the hands is shifted still further from a compression force to a shearing force that is in part resisted by the lumbar facet joints and their associated ligaments. Such shear force could cause or aggravate problems in these structures. This will be further illustrated with other numerical examples later in the chapter.

The major point is that the erector spinae muscles are essential to the stability of the column but their actions create high compression forces within the column. Studies of cadaver spinal column strengths (Perey 1957; Evans and Lissner 1954; Sonoda 1962) indicate that such forces could cause microfractures of the cartilage endplates that could perhaps then initiate or aggravate disc degeneration.

The role of abdominal pressure in relieving the compression load on the spinal column during lifting has been calculated earlier (Morris, Lucas and Bresler 1961) as a major source of column support. Data obtained by Asmussen and Poulsen (1968) would appear to limit the general effectiveness of this reflex mechanism. A maximum limit of 150 mmHg appears to be possible with highly trained individuals although 90 mm is probably a more reasonable limit for the normal population. It should be noted though that this spinal relief may not be possible when a person is carrying a load for a sustained period. If assumed to be a well-developed reflex in an individual, abdominal pressurization probably can relieve approximately 15 to 20 percent of the lumbar spinal column compression.

Case 2. Lifting variable loads close to body

As a more general case let us assume the body position shown earlier in Figure 2.11 but vary the load from zero to 600 N. The predicted L5/S1 compression forces for each load resulting from this procedure are shown in Figure 2.12. Even with moderate loads held reasonably close to the body, high compression forces occur at the lumbosacral level. Perhaps this is why one epidemiological study (Chaffin and Park 1973) has shown that some people had an increased incidence of low-back pain when lifting weights not exceeding 200 N in their jobs.

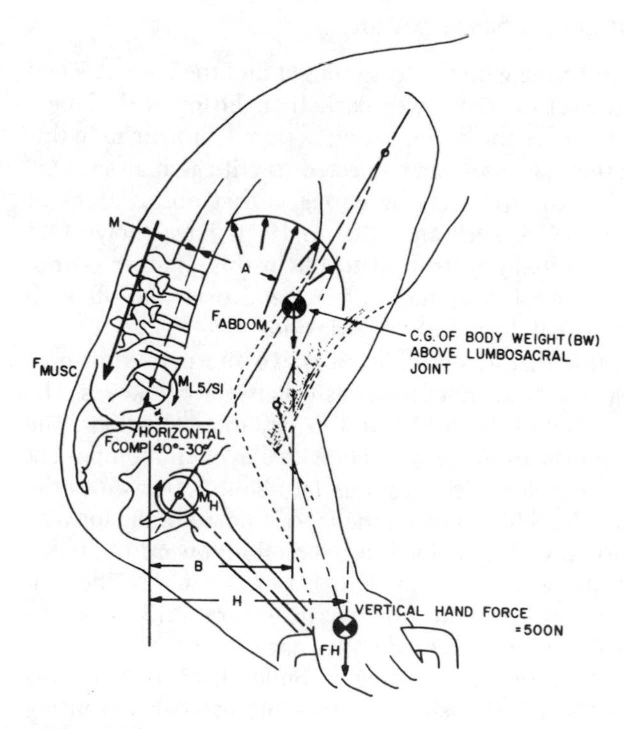

Figure 2.11
Case 1. Lifting close to body.

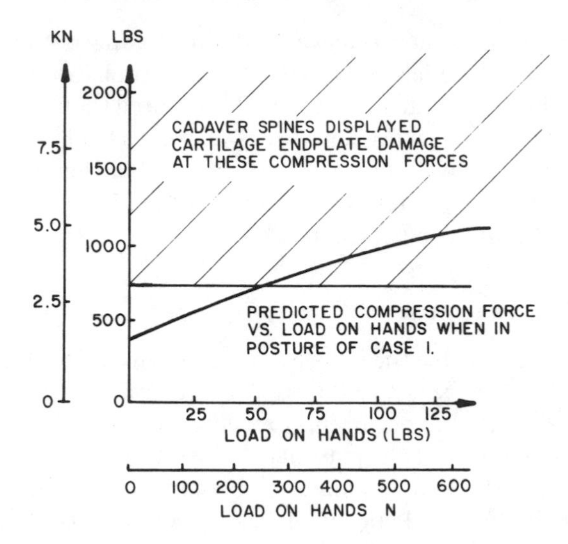

Figure 2.12
Predicted forces compared to endplate damage.

Case 3. Lifting loads with two different postures

It is often stated that lifting with the legs straight and the back stooped over is much more stressful on the lower back than lifting with a near vertical back, using the legs for the major lifting action. Unfortunately this rule, though widely quoted, has not been subjected to critical analysis. One reason for suspecting that the rule may be wrong is that observations of people lifting loads (Park 1973; Park and Chaffin 1974) have shown that people who have been repeatedly instructed to lift in this manner do not follow these instructions; rather, they more often bend over and lift with their backs, using only a small degree of knee flexion.

The numerical example that follows illustrates two alternative postures when lifting a bulky object which cannot pass easily between the knees. The load is located 37 cm in front of the ankles and 37 cm above the floor. The static load on the hands for the example is 130 N with a dynamic component added to represent the initial load accelerations (at about 100 ms into the lifting action). In dynamic load lifting when the load is horizontally forward of the feet, a person normally lifts the load in a direction toward the torso, rather than only vertically as was assumed in Cases 1 and 2. The two postures assumed in the example are shown in Figure 2.13 with the predicted L5/S1 compression forces shown under each figure.

The back-stooped posture when lifting a bulky load reduces the compressive loading on the L5/S1 disc below that predicted when lifting with a more erect back. The reasons for the larger compression forces when lifting with the back near vertical are that 1) the load moment arm is

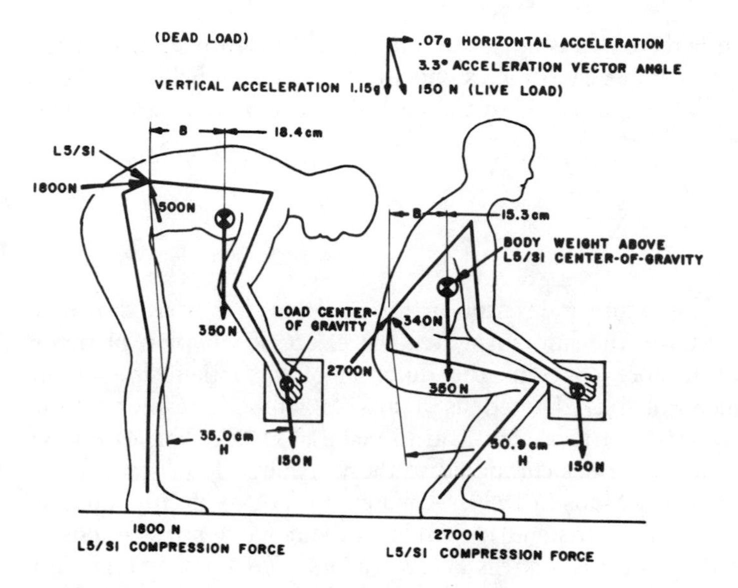

Figure 2.13
Comparison of straight and bent leg lifting.

increased and 2) the vertical component of both the body weight and hand forces add more directly to the compressive forces on the more vertical spine. Of course the shearing forces are greater when lifting with the back flexed. Also, depending on the precise curvature of the lumbar spine during the lift, the articular facet capsules and the posterior ligaments may be overstrained, especially with the more flexed torso posture required when performing a stoop type of lift.

In short, it appears that people lift loads with their backs rather than their legs. In so doing, they minimize the energy necessary to move the body-load mass combination as described by both Brown (1972) and Park (1973). Although it was commonly believed that "back lifting" was more stressful than "leg lifting," this does not appear to be so in regard to compression loading of the column, especially when lifting loads that are larger than can be brought between the knees or loads which are located horizontally away from the feet. Disc pressure measurements (Andersson *et al.* 1976) have shown the pressure to be similar when "back lift" and "leg lift" methods were compared provided the moment arm to the load was kept constant.

If a person can maintain the torso in a completely erect position while lifting, then the compressive force is minimized on the L5/S1 disc. Because of limited shoulder strength and arm reach, this normally requires a small object held in close to the body.

The most important principle when lifting loads is to minimize the distance the load is from the L5/S1 disc, especially if the load is heavy. The effect of varied H distances is illustrated in Figure 2.14. Also shown are the NIOSH limits to compression forces. Clearly a heavy load held in close to the body is much less hazardous to the back than when lifted further away from the body.

Cyclic Loading

Cyclic loading, another description for vibration, has been shown to have many effects on the human being. The effect of vibration of major mechanical significance concerns the duration of the fatigue life of the tissues. Fatigue, which is defined as a loss of strength resulting from intermittent stresses over time, can lead to material failure. This can occur with relatively small stresses compared to those required for a static stress failure (Hertzberg and Manson 1980). A material's fatigue life (the number of loading cycles it can withstand) depends on the range of stress imposed. For most materials, there is a stress below which the material's fatigue life can be considered infinite. Even a small increase in that stress over that limit can cause a significant decrease in the material's fatigue life. For instance, in

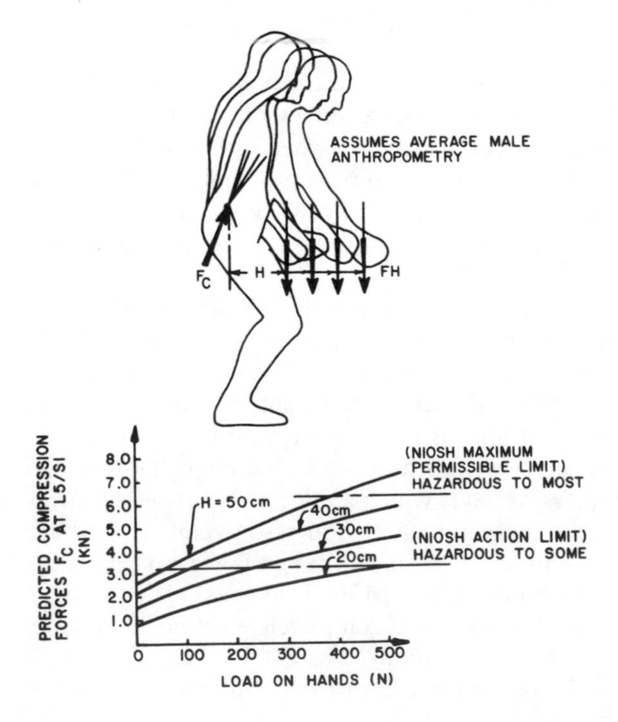

Figure 2.14
Effect of moment arm on disc load.

one polymer of polytetrafluorethylene, a 10 percent increase in stress range causes a 73.2 percent decrease in its fatigue life (Riddell, Koo, and O'Toole 1966). There are two possible modes of failure caused by specimen fatigue. One is the result of crack nucleation and growth; the other occurs as hysteretic heating produces an accelerating decrease in elastic modulus until failure results (Hertzberg and Manson 1980).

Mechanical softening in polymers caused by cyclic loading has also been demonstrated in ligaments by Weisman, Pope and Johnson (1980). A ligament is a biological connective tissue and consists of collagen (a polymer) composed of polypeptide chains (Fung 1981). After cyclic loading, the ultimate strength of the ligament was significantly less than its contralateral, unstressed fellow. The amount of softening in the ligament was related to the cyclic stress.

Vibration of the spine induces so-called "vibrocreep," which has been defined as the acceleration of creep under vibration (Kazarian 1972). Vibrocreep was demonstrated in cadaveric specimens, and it may be assumed that it occurs in living subjects. This will change the shape and configuration of the spine.

Mechanical studies have also been performed to evaluate the effect of vibrating the whole human being in various postures, in single or multiple directions. With epidemiologic evidence associating low back pain with vibration environments (Chapter 6), there has been an increasing focus on

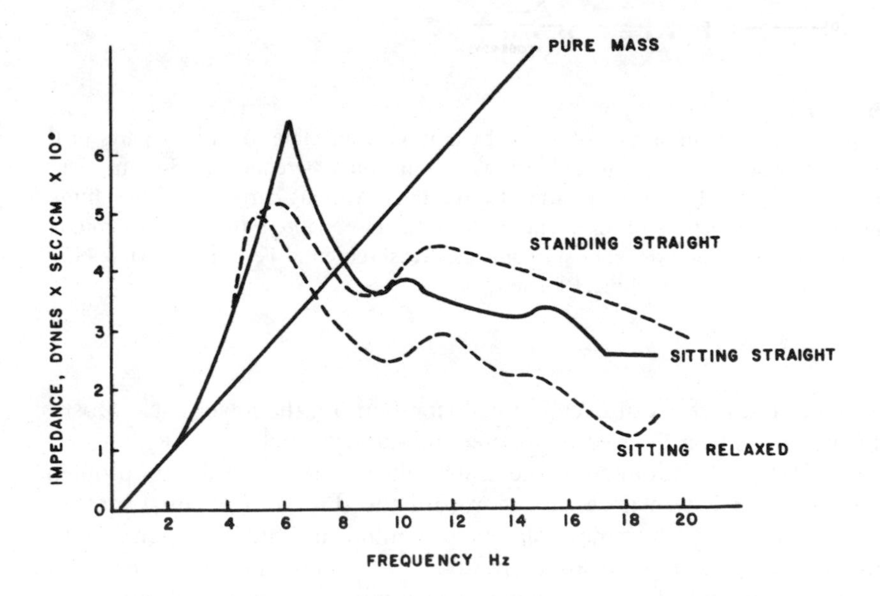

Figure 2.15
Mechanical impedance of the body as a function of frequency and body posture (Rao & Ashley 1976).

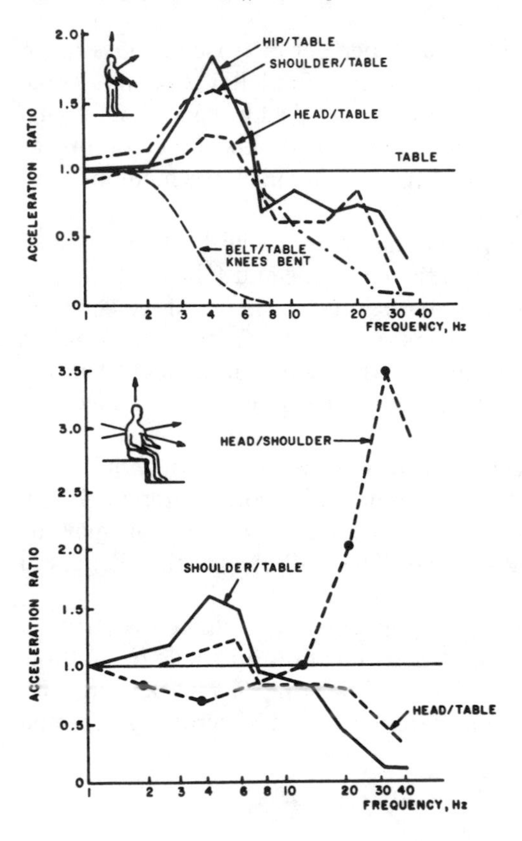

Figure 2.16
Top figure: Transmissibility of vertical vibration from table to various parts of the body of a standing human subject as a function of frequency. Bottom figure: Transmissibility of vertical vibration from table to various parts of a seated human subject as a function of frequency. (Both figures adapted from G. Rasmussen, "Human Body Vibration Exposure and its Measurement." Technical Review No. 1-1982, Bruel & Kjaer Technical Review.)

the mechanical effect of occupational vibrations on the lower back. Most of this work has been related to seating and motor vehicles.

Simons (1952) looked at the seated environment as a design problem. He suggested it was necessary to record the three dimensional motions, analyze the data with human tolerances in mind, and finally design a seat to protect the operator. Unfortunately, good vibration tolerance data has yet to be determined for all people. The most serious consequence of providing a seat in a vehicle is that man's most effective isolation mechanism, his legs, is lost.

Basic studies of vibration of the human have focused on the following mechanical parameters: resonant frequency, transmissibility, impedance, spinal muscle activity, and effects on materials in the spine.

The natural frequency is that frequency at which an object will freely vibrate after it has been struck mechanically, a bell being a good example. The frequency at which an object will freely vibrate is proportional to the square root of the stiffness divided by the mass in a single degree of freedom system (Steidel 1979). For a given mass, natural frequency depends on stiffness. A problem can be created when a structure having a particular natural frequency is moved by some periodic oscillating force at the structure's natural frequency. This condition, called resonance, has major mechanical consequences. In this situation, it takes very little additional energy to keep the structure vibrating at its natural frequency and can lead to the structure's mechanical failure (e.g., Tacoma-Narrows bridge collapse, an opera singer shattering a glass). Failure occurs because the structure oscillates at its maximum possible excursion, thus creating the greatest possible strains on its components. As a result, the structure is at its greatest susceptibility for fatigue failure.

In a vehicle, the primary source of vibration is the interaction of the vehicle and the ground surface; however, any component of the vehicle, engine, wheels, or drive shaft, for example, can be a source. In helicopters, the rotor and wind buffeting have vibration effects, and in semi-trailers, interaction between the trailer and the tractor may be an important source. Under some conditions, the natural frequency of the vehicle suspension is approached, resulting in a violent response of the vehicle. Over very smooth ground the vehicle component (tires, for example) may excite the chassis and again the vibration may be excessive.

The natural frequency of a single degree of freedom structure can be determined by two means: acceleration transmissibility and driving point impedance. Using the acceleration transmissibility method (transmissibility $= Acc_{out}/Acc_{in}$) one compares the output acceleration in the simple structure resulting from the input or driving acceleration. At resonance, the ratio of Acc_{out} exceeds Acc_{in}. In mechanical driving point impedance studies, the driving force is compared to the structure's resultant velocity. Resonance occurs when both the driving force and resultant velocity are in phase and the impedance/frequency curve reaches a maximum (Hixson 1976). Figure 2.15 illustrates the mechanical impedance of the body as a function of frequency and body posture (Rao and Ashley 1976).

Vibrational transmission can also be characterized by transfer functions, describing the relationship of input acceleration and measured output acceleration at a point in the body. In frequency ranges where attenuation is low, resonances occur causing increases in the transfer function magnitude. Figure 2.16 shows some data obtained when standing and sitting subjects are exposed to vertical vibrations. In a standing subject, the first resonance occurs at the hip, shoulder and head at about 5 Hz. Similarly, with subjects sitting, resonance occurs at the shoulders at 5 Hz and to some degree also at the head. Further, a significant resonance from shoulder to head occurs at about 30 Hz.

Mechanical studies to determine the resonant frequency of the seated human operator subjected to vertical vibration have been reported by Bastek *et al.* (1977), Coermann (1961), Coermann *et al.* (1960), Dieckmann (1958), Dupuis (1974), Edwards and Lange (1964), Ferraioli and Nightingale (1979), Goldman and von Gierke (1960), Guignard and Irving (1960), Loach (1958), Muller (1939), Panjabi *et al.* (1983), Pope *et al.* (1980), Radke (1956, 1964, 1973), Seidel *et al.* (1980), Stayner (1974), Stayner *et al.* (1975), Varterasian and Thompson (1978), Vogt *et al.* (1968), White *et al.* (1962) and Wilder *et al.* (1980, 1982, 1983). They generally found resonant frequencies to occur between 4 and 6 Hz because of the upper torso vibrating vertically with respect to the pelvis, and between 10 and 14 Hz representing a bending vibration of the upper torso with respect to the lumbar spine, producing a back slap against a seat back. Some of the above-mentioned studies also reported resonant frequencies of standing and supine subjects and resonant frequencies of seated subjects as affected by side to side or fore-aft vibrations (Hornick *et al.*, 1961).

The degree to which the operator moves in a vibration environment is demonstrated by the magnitude of the acceleration transmissibility at the frequency of interest. At the resonant frequency the acceleration transmissibility will be the greatest. Work reported by Bastek *et al.* (1977) Dieckmann (1958), Dupuis (1974), Ferraioli and Nightingale (1979), Goldman and von Gierke (1960), Griffin (1975), Guignard and Irving (1960), Hornick *et al.* (1961), Matthews (1964), Muller (1939), Panjabi *et al.* (1983), Pope *et al.* (1980), Radke (1956, 1964, 1973), Seidel *et al.* (1980), Simons (1952), Sjoflot and Suggs (1973), and Wilder *et al.* (1980, 1982, 1983) showed transmissibilities greater than 1.0 for the first resonant frequency of the seated operator. This indicates that the resultant motion exceeds the input motion and the biological structures are being stretched maximally. Using accelerometers implanted in the lumbar region, Panjabi *et al.* (1983) showed that the resonant frequency in the lumbar region of the vertically vibrated, seated operator was 4.5 Hz, indicating that maximum strain was occurring in the seated operator's lumbar region at resonances of 4.5 Hz.

Driving point mechanical impedance evaluation gives information concerning more than just frequencies of resonance and phase shift between driving force and resulting velocity. It can also show the mass, stiffness, and damping characteristics of the system being tested. Studies of impedance primarily of the seated human subjected to longitudinal vibration and secondarily of the human in other positions with other applied forces were reported by Bastek *et al.* (1977), Coermann (1961), Coermann *et al.* (1960), Dieckmann (1958), Edwards and Lange (1964), Goldman and von Gierke

(1960), Pope *et al.* (1980), Radke (1973), Seidel *et al.* (1980), Vogt *et al.* (1968) and Wilder *et al.* (1980, 1982, 1983). Changes in impedance of the subject may indicate a softening of the system and an increased susceptibility to damage as a result of material and muscular fatigue.

The muscular effort of driving is considerable. This is a result of the forces necessary to control the vehicle and resist its movements (Troup 1978). The components of vibration and shock may not be enough to cause acute injury, but lateral thrusts may add to the spinal stress. Thus there may be a symbiotic effect of vibration and posture.

SUMMARY

It appears from radiographic, disc pressure and myoelectric data that the load on the spine increases when sitting without a lumbar support, compared to when standing. Reduced low back stress levels can be expected, however, by the use of proper back supports. The most important factor in reducing low back stress is the backrest, and the most important parameter in backrest design is its inclination angle. By addition of a separate lumbar support, the stress on the back can be reduced further, particularly when sitting upright. Such support should be placed in the lumbar region to achieve a more normal lordotic curvature when sitting. In order to provide as much comfort as possible, it should be adjustable in both height and size. The evidence presented also shows that support of the arms and ensuring that the seat is adjustable to the proper height can reduce low-back stress further.

It should be clear from the preceding that moderate loads can produce high compression forces in the lumbar spinal discs. Both epidemiological and cadaver data indicate that compression forces in the ranges observed in industry can produce structural failures in the weight-bearing cartilage endplates. The resulting microfractures and scarring are believed by some researchers to be a major factor in accelerating the natural aging and degeneration of the spinal discs.

The lifting recommendations are based on a concern for minimizing compression forces incurred by the spinal column when lifting loads in the sagittal plane. It should be evident that much more research is needed to understand the pathobiomechanics of load lifting, especially with regard to asymmetric and highly dynamic motions. Simple biomechanical models of the type described are able, however, to assist in determining how the many anatomical, postural, and load-related variables combine to affect injury-producing stresses in common load-lifting situations. These models, though admittedly crude at this time, have provided much of the knowledge now used to develop the job design guidelines discussed later in this book.

Vibration affects the spine by causing vibrocreep and mechanical fatigue of spinal tissues. The postures in driving can cause muscular fatigue, and this is exacerbated by the vibrational environment. Many vehicles excite the natural frequencies of the human spinal system.

REFERENCES

Akerblom, B. 1948. *Standing and sitting posture. With special reference to the construction of chairs.* Nordiska Bokhandelin, Stockholm, doctoral dissertation.

______. 1969. Anatomische und Physiologische Grundlagen zur Gestaltung von Sitzen. In *Sitting Posture.* ed. E. Grandjean, pp. 6–17, London: Taylor.

Andersson, C. 1983. *A biomechanical model of the lumbosacral joint for lifting activities.* Doctoral dissertation, University of Michigan, Univ. Microfilms, Ann Arbor, Mi.

Andersson, G.B.J. 1974. *On myoelectric back muscle activity and lumbar disc pressure in sitting postures.* Doctoral dissertation. Gotab: Univ. of Gothenburg, Gothenburg, Sweden.

Andersson, G.B.J.; and Ortengren, R. 1974a. Myoelectric back muscle activity during sitting. *Scandinavian Journal of Rehabilitation Medicine*, Suppl. 3:73–90.

Andersson, G.B.J.; Jonsson, B.; and Ortengren, R. 1974b. Myoelectric activity in individual lumbar erector spinae muscles in sitting. A study with surface and wire electrodes. *Scand. J. Rehab. Med.*, Suppl. 3:91–108.

Andersson, G.B.J.; Ortengren, R.; Nachemson, A.; and Elfstrom, G. 1974c. Lumbar disc pressure and myoelectric back muscle activity during sitting. I. Studies on an experimental chair. *Scand. J. Rehab. Med.* 6 (3):104–14.

Andersson, G.B.J.; and Ortengren, R. 1974d. Lumbar disc pressure and myoelectric back muscle activity during sitting. II. Studies on an office chair. *Scan. J. Rehab. Med.* 6 (3):115–21.

Andersson, G.B.J.; and Ortengren, R. 1974e. Lumbar disc pressure and myoelectric back muscle activity during sitting. III. Studies on a wheelchair. *Scan. J. Rehab. Med.* 6 (3):122–27.

Andersson, G.B.J.; Ortengren, R.; Nachemson, A.; and Elfstrom, G. 1974f. Lumbar disc pressure and myoelectric back muscle activity during sitting. IV. Studies on a car driver's seat. *Scand. J. Rehab. Med.* 6 (3):128–33.

Andersson, Gunnar B.J.; Ortengren, Roland; Nachemson, Alf L.; Elfstrom, Gosta; and Broman, Holger 1975. The sitting posture: An electromyographic and discometric study. *Orthopedic Clinics of North America* 6 (1):105–20.

Andersson, G.B.J., Ortengren, R.; and Schultz, A. 1980. Analysis and measurement of the loads on the lumbar spine during work at a table. *Journal of Biomechanics* 13 (6):513–20.

Andersson, G.B.J.; Ortengren, R.; and Herberts, P. 1977. Quantitative electromyographic studies of back muscle activity related to posture and loading. *Orthop. Clin. North Am.* 8:85–96.

Andersson, G.B.J.; Ortengren, R; and Nachemson, A. 1976. Quantitative studies of back loads in lifting. *Spine* 1:178–85.

Andersson, G.B.J.; Murphy, R.W.; Ortengren, R.; and Nachemson, A. 1979. The influence of backrest inclination and lumbar support on the lumbar lordosis in sitting. *Spine* 4:52–58.

Armstrong, James R. 1965. *Lumbar disc lesions, pathogenesis and treatment of low back pain and sciatica.* 3rd ed. pp. 42–45. Baltimore, Md.: Williams and Wilkins.

Asmussen, E.; and Poulsen, E. 1968. On the role of the intra-abdominal pressure in relieving the back muscles while holding weights in a forward inclined position. *Commun.-Dan. Natl. Assoc. Infant. Paral.* 28:3–11.

Bartelink, D.L. 1957. The role of abdominal pressure in relieving the pressure on the lumbar intervertebral discs. *J. Bone Joint Surg.* 39B (4):718–25.

Bastek, R.; Buchholz, Ch.; Denisov, E.I.; Enderlein, G.; Kramer, H.; Malinskaja, N.N.; Meister, A.; Metz, A.; Mucke, R.; Rhein, A.; Rothe, R.; Seidel, H.; and Sroka, Ch. 1977. Comparison of the effects of sinusoidal and stochastic octave-band-wide vibrations–A multi-disciplinary study. Part I: Experimental arrangement and physical aspects, Part II: Physiological aspects, Part III: Psychological investigations, *Int'l. Arch. Occup. & Environmental Health*, 39:143–79.

Belytschko, T.B.; Andriacchi, T.P.; Schultz, A.B.; and Galante, J.O. 1973. Analog studies of forces in the human spine: computational techniques. *J. Biomech.* 6(4):361–72.

Brown, J.R. 1972. *Lifting as an industrial hazard.* Labour Safety Council of Ontario, Ontario Department of Labour, Toronto.

Burandt, U. 1969. Rontgenuntersuchung uber die Stellung von Becken und Wirbelsaule beim Sitzen auf vorgeneigten Flachen. *Ergonomics* 12:356–64.

Burandt, U.; and Grandjean, E. 1963. Sitting habits of office employees. *Ergonomics,* 6 (2):217–28.

Bush, Charles A. 1969. Study of pressures on skin under ischial tuberosities and thighs during sitting. *Archives of Phys. Med* 50 (4):207.

Carlsoo, S. 1961. The static muscle load in different work positions: An electromyographic study. *Ergonomics* 4:193.

Carlsoo, S. 1963. *Writing desk, chair and posture of work.* Dept. of Anatomy, Stockholm. (In Swedish)

Carlsoo, S. 1972. *How man moves.* London: Hanemann.

Chaffin, D.B. 1973. Localized muscle fatigue–Definition and measurement. *Journal of Occupational Medicine* 15:346.

Chaffin, D.B.; and Andersson, G.B.J. 1984. *Occupational Biomechanics,* New York: J. Wiley & Sons, Inc.

Chaffin, D.B., and Park, K.S. 1973. A longitudinal study of low-back pain as associated with occupational weight lifting factors. *Am. Ind. Hyg. Ass. J.* 34: 513–25.

Clauser, C.E.; McConville, J.T.; and Young, J.W. 1970. *Weight, volume, and center of mass of segments of the human body.* AMRL Tech Rep. 69–70. Ohio, Wright Patterson Air Force Base, Aerospace Medical Research Laboratory.

Coermann, Rolf R. 1961. The mechanical impedance of the human body in sitting and standing position at low frequencies. *Acoustic Systems Division Technical Report Rep #61–492,* Aerospace Medical Research Labs, Wright Patterson Air Force Base.

Coermann, R.R.; Ziegenrucker, G.H.; Wittwer, A.L.; and von Gierke, H.E. 1960. The passive dynamic mechanical properties of the human thorax-abdomen system and of the whole body system. *Aerospace Medicine* 31(6):443–55.

Davis, P.R.; Troup, J.D.G.; and Burnard, J.H. 1965. Movements of the thoracic and lumbar spine when lifting: A chrono-cyclophotographic study. *Anat.* (Lond) 99 (1):13–26.

Dempster, W.T. 1955. Space requirements of the seated operator. *WADC Tech. Rep. No. 55–159.* Wright Air Development Center.

Dempster, Wilfrid T. and Gaughran, George R.L. 1967. Properties of body segments based on size and weight. *Am. J. Anat.* 120 (1):33–54.

Dempster, Wilfrid T.; Sherr, Lawrence A.; and Priest, Judith G. 1964. Conversion scales estimating humeral and femoral lengths and lengths of functional segments in the limbs of American Caucasoid males. *Hum. Biol.* 36 (3): 246–62.

Dieckmann, D. 1958. A study of the influence of vibration on man. *Ergonomics* 1: 347–55.

Donish, E.R.: and Basmajian, J.V. 1972. Electromyography of deep back muscles in man. *Am. J. Anat.* 133:25.

Dupuis, H. 1974. Belastung durch mechanische Schwingungen und moegliche Gesundheitsschaedigungen im Bereich der Wirbelsaeule. *Fortschritte der Medizin* 92(14):618–20.

Edwards, R.G.; and Lange, K.O. 1964. A mechanical impedance investigation of human response to vibration. AMRL-TR-64-91, Final report. AD-609-006, Wright Patterson Air Force Base.

Engdahl, S. 1971. *School-chairs*. Swedish Furniture Research Institute. Report No. 24, Stockholm. (In Swedish)

Evans, F.G.; and Lissner, H.R. 1954. Strength of intervertebral disc. *Journal of Bone and Joint Surgery* 36-A: 185.

Ferraioli, A.; and Nightingale, J.M. 1979. A design of a vibrating platform and transmission of low frequency vertical vibrations of the seated man. *Technica Italiana* 44(3):213–19.

Floyd, W.F.; and Roberts, D.F. 1958. Anatomical and physiological principles in chair and table design. *Ergonomics* 2 (1):1–16.

Floyd, W.F.; and Silver P.H.S. 1955. The function of the erectores spinae muscles in certain movements and postures in man. *Journal of Physiology* (London), 129 (1):184–203.

Floyd, W.F.; and Ward, J.S. 1969. Anthropometric and physiological considerations in school, office and factory seating. In *Sitting Posture.* ed., E. Grandjean, pp. 18–25, London:Taylor and Francis.

Frymoyer, J.W.; Pope, M.H.; Costanza, M.C; Rosen, J.C.; Goggin, J.; and Wilder, D.G.; 1980. Epidemiologic studies of low back pain. *Spine* 5 (5):419–23.

Fung, Y.C. 1981. *Biomechanics: Mechanical properties of living tissues.* New York: Springer-Verlag.

Goldman, D.E.; and von Gierke, H.E. 1960. The effects of shock and vibration on man. Lecture and Review Series, No. 60-3, Naval Medical Research Institute, Bethesda, Maryland, 8 Jan., 1960. *American National Standards Institute ±S3-W-39.*

Griffin, M.J. 1975. Vertical vibration of seated subjects: Effects of posture, vibration level, and frequency. *Aviation, Space, and Environmental Medicine,* 46(3): 269–76.

Guignard, J.C.; and Irving, A. 1960. Effects of low frequency vibration on man. *Engineering* (London) 190:364–67.

Hall, M.A.W. 1972. Back pain and car-seat comfort. *Applied Ergonomics* 3 (2):82.

Herrin, G.D.; Chaffin D.B.; and Mach, R.S. 1974. *Criteria for research on the hazards of manual materials handling.* Workshop Proceedings on contract CDC-99–74–118, U.S. Dept. of Health and Human Services (NIOSH), Cincinnati, Oh.

Hertzberg, R.W., and Manson, J.A. 1980. *Fatigue of engineering plastics.* p. 64. New York: Academic Press, Inc.

Hixson, E.L. 1976. Mechanical impedance and mobility. *Shock and vibration handbook*, eds. C.M. Harris and C.E. Crede. New York: McGraw-Hill.

Hornick, R.J.; Boettcher, C.A.; and Simons, A.K. 1961. The effect of low frequency high amplitude, whole body, longitudinal and transverse vibration upon human performance. *Final report, contract No. DA-11-022-509-ORD-3300, Ordnance Project No: TEI-1000*, Bostrom Research Labs, Milwaukee, Wisconsin.

Jonsson, B. 1970. The functions of individual muscles in the lumbar part of the spinae muscle. *Electromyography* 10 (1):5–21.

Kazarian, L.E. 1972. Dynamic response characteristics of the human vertebral column. *Acta Orthop. Scand. Suppl.*146.

Keegan, J.J. 1953. Alterations of the lumbar curve related to posture and seating. *Journal of Bone and Joint Surgery* 35A (3):589–603.

Klausen, K. 1965. The form and function of the loaded human spine. *Acta Physiol. Scand.* 65:176.

Knutsson, Bertil; Lindh, Kuno; and Telhag, Hans. 1966. Sitting—An electromyographic and mechanical study. *Acta Orthop. Scand.* 37:415–28.

Kraus, H.; Robertson, G.H.; and Farfan, H.F. 1973. On the mechanics of weight lifting. *New England Bioengineering Conf. Proc. University of Vermont, April 1973*. New England Bioengineering Society.

Kroemer, K.H.E. 1963. Uber die Hohe von Schreibtischen. *Arbeitswissenschaft* 2 (4):132–40.

Kroemer, K.H.E. 1971. Seating in plant and office. *Amer. Industr. Hyg. Ass. J.* 32 (10):633–52.

Kroemer, K.H.E.; and Robinette, Joan C. 1969. Ergonomics in the design of office furniture. *Industr. Med. Surg.* 38 (4):115–25.

Krusen, Frank; Ellwood, P.M., Jr.; and Kottke, F.J. (eds.) 1965. *Handbook of physical medicine and rehabilitation.* Philadelphia: W.B. Saunders.

Lange, K.O.; and Edwards, R.G. 1970. Force input and Thoraco-abdominal strain resulting from sinusoidal motion imposed on the human body. *Aerospace Medicine* 41(5):538–43.

Laurig, W. 1969. Der Stehsitz als Physiologisch Gunstige Alternative zum Reinen Steharbeitsplatz. *Arbeitsmed Sozialmed Arbeitshyg.* 4:219.

Loach, J.C. 1958. Evaluation of riding of passenger coaches. *Railway Gazette*, 108 (5):130–32.

Lee, K 1982. *Biomechanical modeling of cart pushing and pulling*, doctoral dissertation, Univ. of Michigan, Ann Arbor, Mi.

Lundervold, Arne J.S. 1951a. Electromyographic investigations of position and manner of working in typewriting. *Acta Physiol. Scand.* 24 (Suppl.) 84:1–171.

Lundervold, A.J.S. 1951b. Electromyographic investigations during sedentary work, especially typewriting. *Brit. J. Phys. Med.* 14:32–36.

Mandal, A.C. 1975. Work-chair with tilting seat. *Lancet* 1 (7907):642–43.

Matthews, J. 1964. Ride comfort for tractor operators. II. Analysis of ride vibrations on pneumatic tyred tractors. *J. Agric. Eng'g Res.* 9(2):147–158.

McCormick, E.J. 1970. Man in motion. *Human factors engineering* (3rd edition), pp. 541–548. New York: McGraw-Hill.

Morris, J.M.; Benner, G.; and Lucas, D.B. 1962. An electromyographic study of the intrinsic muscles of the back in man. *J. Anat.* 96:509–20.

Morris, J.M.; Lucas, D.B. and Bresler, B. 1961. Role of the trunk in stability of the spine. *J. Bone Joint Surg.* 43A (3):327–51.

Muller, E.A. 1939. Die Wirkung sinusformiger Verticalschwingungen auf den sitzenden und stehenden Menschen. *Arb. Physiologie* 10:459.

Nachemson, Alf L; and Elfstrom, Gosta. 1970. Intravital dynamic pressure measurements in lumbar discs. A study of common movements, maneuvers and exercises. *Scand. J. Rehab. Med.* 2 (Suppl.1):1–40.

National Institute for Occupational Safety and Health. 1981. *A work practices guide for manual lifting,* Tech. Report No. 81–122, U. S. Dept. of Health and Human Services (NIOSH), Cincinnati, Oh.

Okushima, H. 1970. Study on hydrodynamic pressure of lumbar intervertebral disc. *Arch. Jap. Chir.* 39:45–57. (In Japanese).

Panjabi, M.M.; Andersson, G.B.J.; Jorneus, L; Hult, E.; and Mattson, L 1983. In vivo measurement of spinal column vibrations. Presented at and in *Proceedings of Am. Soc. Biomech.* Rochester, MN Sept. 28–30.

Park, K.; and Chaffin, D.B. 1974. A biomechanical evaluation of two methods of manual load lifting. *AIIE Trans.* 6(2):105–113.

Park, K.Y.S. 1973. A computerized simulation model of postures during manual materials handling. Doctoral dissertation, Univ. of Michigan, Ann Arbor, Michigan.

Perey, O. 1957. Fracture of the vertebral endplate in the lumbar spine: An experimental biomechanical investigation. *Acta Orthop. Scand. Suppl.* 25: 1–101.

Plagenhoef, S.C. 1966. Methods for obtaining kinetic data to analyze human motions. *Res. Quart. Amer. Assoc. Health Phys. Educ.* 37:103–112.

Pope, M.H.; Wilder, D.G.; and Frymoyer, J.W. 1980. Vibration as an aetiologic factor in low back pain. *Proc. of Inst. Mech. Engs. Conf. on Low Back Pain.* Inst. Mech. Engs. Paper #C121/80.

Radke, A.O. 1956. The application of human engineering data to vehicular seat design. Bostrom Research Labs, Rep. #117. Milwaukee, Wisconsin.

Radke, A.O. 1957. Vehicle vibration, man's environment. ASME 57A:54. Published in *Mech. Eng.* pp 38–41, July 1958.

Radke, A.O. 1964. The importance of seating in driver comfort and performance. SAE paper #838B.

Radke, A.O. 1973. International view of tractor seating. Soc. Automotive Engineers paper #730794. SAE, Inc., NY.

Rao, B.K.N.; and Ashley, C. 1976. Subjective effects of vibration. W. Tempest (Ed.), *Infrasound and low frequency vibration.* London: Academic Press.

Roaf, R. 1960. A study of the mechanics of spinal injuries. *J. Bone Joint Surg.* 42B(4):810–23.

Riddell, M.N.; Koo, G.P. and O'Toole, J.L 1966. Fatigue mechanisms of thermoplastics. Soc. Polymer Engineers, 22nd Annual Tech Conf. 12:2–6, *Polymer and Eng. Sci.* 6 (4):363–68.

Rosemeyer, B. 1971. Electromygraphische Untersuchungen der Rucken-und Schultermuskulatur im Stehen und Sitzen unter Berucksichtigung der Haltung des Autofahrers. *Arch. Orthop. Unfallchir.* 71:59–70.

Rosemeyer, B. 1972. Eine Methode zur Beckenfixierung im Arbeitssitz. *Z. Orthop.* 110:514.

Schanne, F.J., Jr. 1972. *A three-dimensional hand force capability model for a seated* person, Vols I, II. Doctoral dissertation, Univ. of Michigan, Ann Arbor, Mi.

Schoberth, H. 1962. *Sitzhaltung, Sitzschaden, Sitzmobel.* Berlin: Springer Verlag.

______. 1969. Die Wirbelsaule von Schulkindern–orthopadische Forderungen an Schulsitze. In E. Grandjean, ed., *Sitting Posture.* London: Taylor, pp. 98–111.

Schultz, A.B.; Andersson, G.B.J., Ortengren, R.; Bjork, R.; and Nordin, M. 1982. Analysis and quantitative myoelectric measurements of loads on the lumbar spine when holding weights in standing postures. *Spine* 7 (4):390–97.

Schultz, A.B.; Andersson, G.B.J.: Ortengren, R., Nachemson, A.; and Haderspeck, K. 1982a. Loads on the lumbar spine: validation of biomechanical analysis by measurements of intradiscal and myoelectric signals. *J. Bone Joint Surg.* 64A:713–20.

Schultz, A.B.; Andersson, G.B.J.; Ortengren, R.; Haderspeck, K.; and Nachemson, A. 1982b. Loads on the lumber spine. *J. Bone Joint Surg.* 64A:713–20.

Seidel, H.; Bastek, R.; Brauer, D.; Buchholz, Ch.; Meister, A.; Metz, A. M. and Rothe, R. 1980. On human response to prolonged repeated whole-body vibration. *Ergonomics* 23(3):191–211.

Simons, A.K. 1952. Tractor ride research. *S.A.E. Quarterly Transactions* 6:357–64.

Sjoflot, L.; and Suggs, C.W. 1973. Human reactions to whole-body transverse angular vibrations compared to linear vertical vibrations. *Ergonomics* 16:455–68.

Smith, Alan DeF.; Deery, Edwin M.; and Hagman, George L. 1944. Herniation of the nucleus pulposus–A study of one hundred cases treated by operation. *J. Bone Joint Surg.* 26 (4):821–28.

Sonoda, T. 1962. Studies on the compression, tension, and torsion strength of the human vertebral column. *J. Kyota Prefect Med. Univ.* 71:659.

Stayner, R.M. 1974. Vibration and the tractor driver. CIGR-COISTA Congress, Flerohof, Netherlands. (From Tractor Dept., National Institute Agricultural Engineering, Silsoe, Bedford, England).

Stayner, R.M.; Hilton, D.J.; and Moran P. 1975. Protecting the tractor driver from low frequency ride vibration. *Inst. Mechanical Engineers* paper #200/75, pp. 39–47.

Steidel, R.F. 1979. *An introduction to mechanical vibrations.* 2nd ed. Revised Printing. New York: John Wiley & Sons.

Tichauer, E.R. 1965. The biomechanics of the arm-back aggregate under industrial working conditions. *ASME Rep. No. 65-WA/HUE-1.*

Troup, J.D.G. 1978. Driver's back pain and its prevention. A review of the postural, vibratory, and muscular factors, together with the problem of transmitted road-shock. *Applied Ergonomics* 9(4):207–14.

Umezawa, F. 1971. The study of comfortable sitting postures. *J. Jap. Orthop. Assoc.*, 45:1015–22. (In Japanese)

Varterasian, J.H., and Thompson, R.R. 1978. The dynamic characteristics of automobile seats with human occupants. Society of Automotive Engineers paper

#770249, SAE, Inc.

Vogt, H.L.; Coermann, R.R.; and Fust, H.O. 1968. Mechanical impedance of the sitting human under sustained acceleration. *Aerospace Med.* 39(7):675–79.

von Gierke, H.E.; and Coermann, R.R. 1963. The Biodynamics of human response to vibration and impact. *Industr. Med. Surg.* 32(1):30–32.

Webb, P. (ed.) 1964. Bioastronautics data book. Scientific and Technical Information Division. NASA SP-3006, National Aeronautics and Space Administration, Washington, D.C., pp. 76–78.

Weisman, G.; Pope, M.H.; and Johnson, R.J. 1980. Cyclic loading in knee ligament injuries. *Am. J. Sports Med.* 8:24–30.

White, G.H.; Lange, K.O.; and Coermann, R.R. 1962. The effects of simulated buffeting on the internal pressure of man. *Human Factors* pp. 275–90.

Wilder, D.G.; Woodworth, B.B.; Frymoyer, J.W.; and Pope, M.H. 1980. Energy absorption in the human spine. Presented at 8th Northeast (New England) Bioengineering Conference, MIT, Cambridge. 27–28 March. In Proceedings, Igor Paul (ed.), pp. 442–45.

Wilder, D.G.; Frymoyer, J.W. and Pope, M.H. 1983. The effect of vibration on the spine of the seated individual. Pending in *Automedica.*

Wilder, D.G.; Frymoyer, J.W.; and Pope, M.H. 1983. The effect of vibration on the intervertebral motion segment. Presented at Northeast Bioengineering Conf., Dartmouth, Hanover, N.H. 15–16 March 1982. In Proceedings, E.W. Hansen (ed.), IEEE catalog #82CH1747-5, 1982, pp. 9–11.

Yamaguchi, Y.; Umezawa, F.; and Ishinada, Y. 1972. Sitting posture: an electromyographic study on healthy and notalgic people. *J. Jap. Orthop. Assoc.*, 46:277.

3 Clinical Classification

John W. Frymoyer and James Howe

OVERVIEW

The inability to give a precise diagnosis is the single greatest problem in the prevention, treatment, and prognosis of low back disease and disability evaluation of low back disease sufferers. In the general population, the likelihood of identifying a specific cause for low back disease is small, in the order of 5 to 10 percent (Frymoyer *et al.* 1980). Detailed analysis of patients with chronic low back disease indicates that a definite structural diagnosis can be reached in no more than 50 percent of all cases (Pope *et al.* 1979). The most common diagnoses include lumbosacral strain and sprain, postural LBP (low back pain), muscular insufficiency, "sprung back," sacroileitis, annular tears, bulging disc, degenerative spinal disease, or simply "low back syndrome" (Benn and Wood 1975). Because these diagnostic terms are ill defined, and a damaged anatomic structure cannot be identified, treating physicians often cannot agree upon the diagnosis in a particular patient. If the patient becomes well in three to six weeks, one can presume "the injury" has healed; if symptoms persist, the problem is identified as a disorder of greater seriousness, and further diagnostic studies may be indicated. The general attributes of LBP of undetermined etiology (termed here "idiopathic low back pain") are the absence of true sciatica, the increase of the pain with increased physical activity, and relief with rest (Bergquist-Ullman and Larsson 1977; Svensson 1981; Damkot *et al.* 1983). The onset of the pain may be either gradual or acute.

In the industrial setting, acute onset is often associated with some specific loading event. Acute back pain can also develop from a minor postural change, coughing, or be triggered by an unknown cause. Examination typically reveals loss of normal lumbar lordosis, varying degrees of back muscle tightness and spasm, and a restriction of spinal motion.

Neurologic signs and symptoms are absent, although elevation of the leg may produce accentuation of back complaints. Partial or complete relief of symptoms occurs when the patient lies down. Spinal radiographs will be normal other than for loss of lumbar lordosis, although other abnormalities may be identified which are unrelated to the low back symptoms. These radiographic findings will be discussed later in more detail.

The natural history of acute LBP episodes is shown in Figure 3.1. The majority of LBP patients experience complete relief of symptoms and are able to return to normal function. There is a 40 to 50 percent probability of recurring episodes, again usually associated with some change in spinal stress, or occurring during times of relative physical unfitness (Bergquist-Ullman and Larsson 1977; Troup 1978).

A smaller group of patients have persistent symptoms, lasting more than three months. In this group, diagnostic studies will pinpoint a precise pathologic causation in approximately 50 percent. The remaining patients pose the single greatest problem for those responsible for care, as well as for those determining disability. Not surprisingly, this patient population has often become labeled as suffering from "psychologic LBP," or "com-

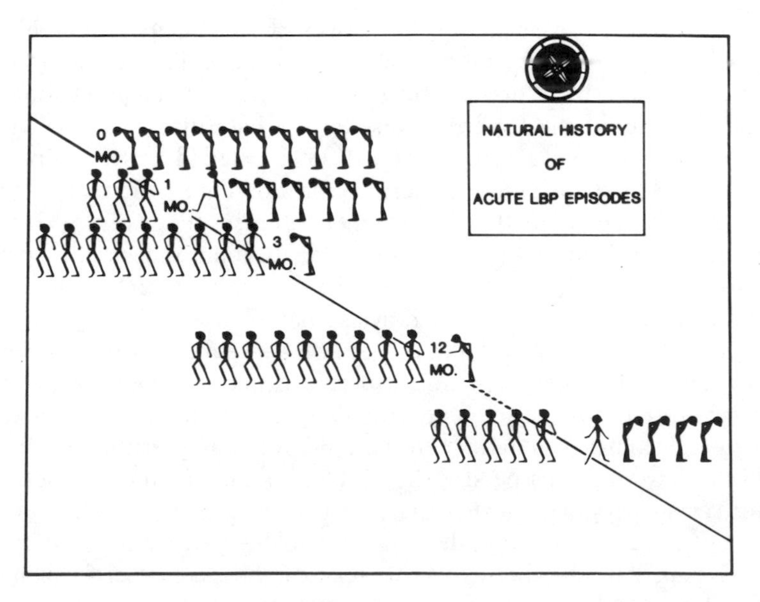

Figure 3.1
The natural history of low back pain. This illustration depicts relief of symptoms in low back pain patients. In 1 month, 35% experience relief of symptoms and are able to return to normal function. By 3 months, almost 90% (87%) are free from LBP. At one year, 96% have experienced relief of LBP, but unfortunately, 40 to 50% will experience recurring LBP episodes. Thus it is seen that most LBP is cured naturally in 3 months time, and that the prospects of LBP getting better after 3 months time are not promising.

pensation syndrome." Diagnostic evaluations often reveal evidence of "psychopathology" (Beals and Hickman 1972; White 1969; Southwick and White 1983). The most commonly used psychologic test in the United States is the Minnesota Multiphasic Personality Inventory (MMPI), a 389-item questionnaire. Figure 3.2 gives a typical example from the MMPI. Patients with chronic and disabling LBP often show abnormal findings in one or more scales of the MMPI such as hysteria, hypochondriasis, somatization and depression. These patients may also have a preoccupation with health, and anger at the health care delivery system (Svensson 1981; Frymoyer *et al.* 1983b). It should be emphasized, however, that it is unclear whether these psychologic abnormalities are the cause or the result of the continued LBP symptoms.

The most important objectives in patients with chronic LBP are to identify that the patients are not progressing according to the usual outcome of an acute LBP episode, to carry out sufficient diagnostic testing to be

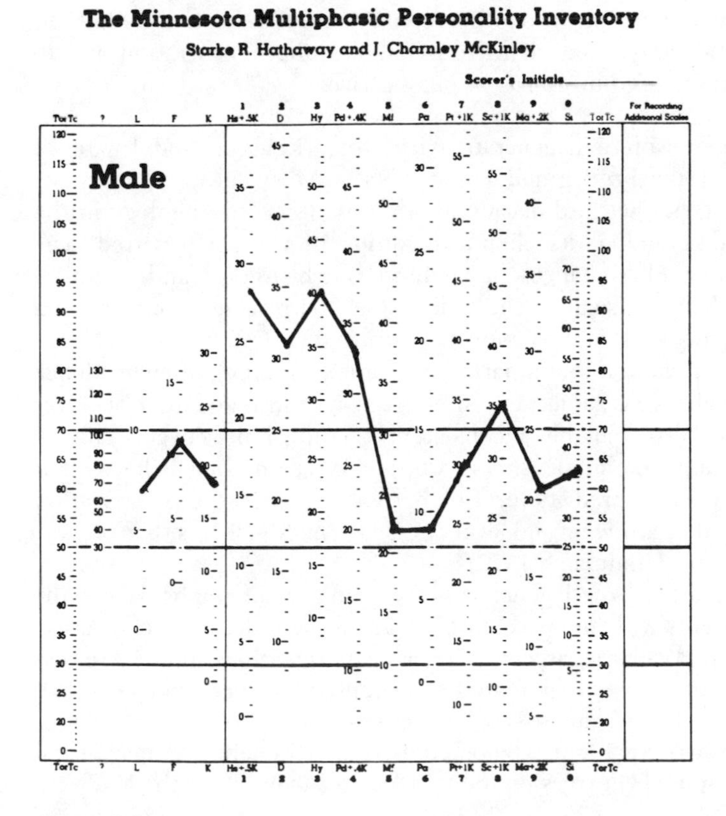

Figure 3.2
The typical low back pain patient profile on the MMPI.

certain there is not a remediable lesion, and to aggressively undertake rehabilitation. This concept is developed in Chapter 8.

LOW BACK PAIN OF KNOWN ETIOLOGY

One can conveniently distinguish the known causes of LBP in accordance with their underlying pathologic processes: degenerative, congenital, traumatic, inflammatory, neoplastic, and metabolic.

Degenerative Spinal Disease

Degenerative spinal disorders are the second only to idiopathic LBP as a diagnosis rendered for LBP. The term encompasses herniated nucleus pulposus, certain forms of spinal stenosis, segmental instability and spinal osteoarthritis. The diagnoses are usually based upon the clinical presentation, the spinal radiograph, and other confirmatory tests. It is critical to emphasize certain important characteristics of spine degeneration and to place this often misused diagnosis in perspective.

1. All human spines degenerate with time (Kellgren and Lawrence 1958; Sokoloff 1969; Friberg and Hirsch 1949). Most autopsy specimens have shown that, by the third decade alterations of the morphology in the intervertebral disc and of its chemical composition have occurred. The associated radiographic changes lag behind the histologic and chemical events. These changes usually include disc space narrowing at one or more levels and presence of spinal osteophytes (see Figure 3.3).

2. The interpretation that a particular spine has radiographic evidence of degenerative change is subject to rather extensive intra- and inter-observer errors. There is also a highly inconsistent relationship between clinical symptoms and radiographic signs. For example, it is commonly believed that a narrowed disc space predicts level of herniation. In fact, the correlation between disc space narrowing and level of disc herniation is less than 50 percent (Hakelius and Hindmarsh 1972).

3. The prevalence of degeneration in spinal radiographs is equally common in patients with and without LBP. From the industrial perspective, this point is of particular significance. Heavy laborers often show increased osteophyte formation when compared to sedentary workers, but low back complaints do not correlate to the presence or absence of osteophytes (Kellgren and Lawrence 1958). The osteophyte may benefit the mechanical function of the spine by increasing the surface area over which the loads are distributed.

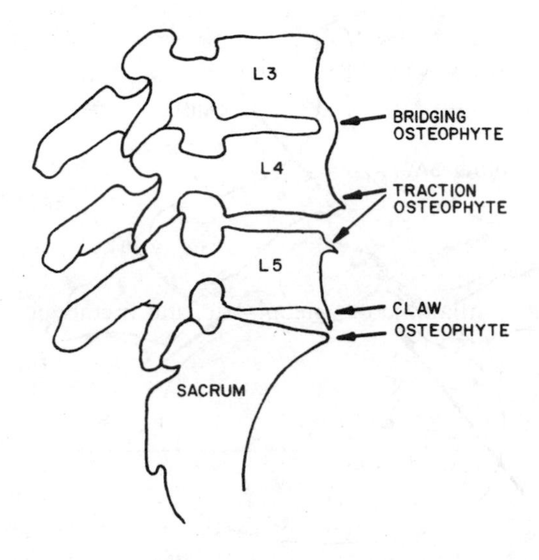

Figure 3.3
The spine viewed from the side. Three types of bony overgrowths are shown. At the L5/Sacrum level the *claw osteophyte* is directed towards the disc space; at the L4-5 level a traction osteophyte, while at L3-4 the osteophyte has bridged across the disc space. Bridging and claw osteophytes may often be asymptomatic, and are more common in those who have been involved in heavy work. Traction osteophytes usually imply the presence of abnormal motion at the level (segmental instability), and frequently are associated with LBP.

All of these points emphasize the importance of correlating radiographic changes with the clinical symptoms. Spinal degeneration is viewed by the lay public to be a progressive and unrelenting problem. The diagnosis may often lead to self-imposed and unnecessary restrictions in physical activity.

Herniated Nucleus Pulposus

The classic description of herniated nucleus pulposus (HNP) (Fig. 3.4), commonly called the slipped disc, was first published by Mixter and Barr (1934). The syndrome is characterized by sciatica, usually accompanied or preceded by LBP. Physical examination reveals the presence of one or more objective neurologic changes such as reflex asymmetry, sensory change in the distribution of a nerve root, or muscle weakness. Additionally, the clinical diagnosis requires the presence of a positive nerve root tension test of which the straight leg raising test is most commonly used (See Chapter 8 for a complete description of these tests). For the straight leg raising test to be considered positive, it should reproduce the sciatic pain below the level of

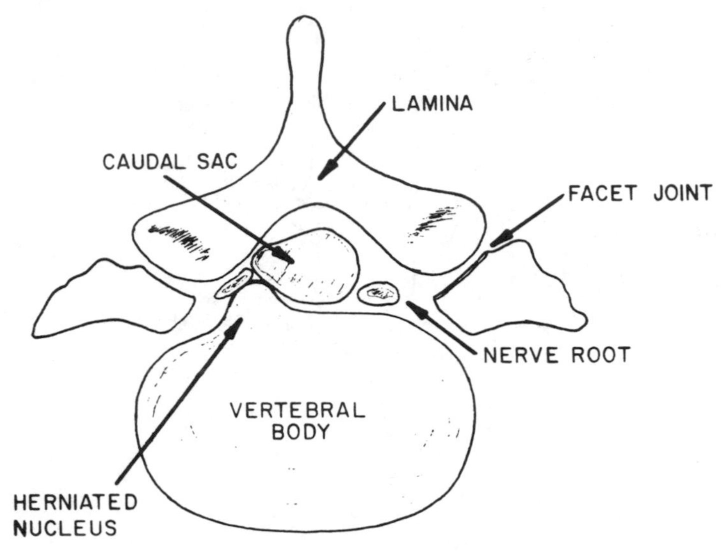

Figure 3.4
The view is identical with that seen in a CT scan of the spine. Note the presence of a herniated disc on the left, compressing the caudal sac. On the right a nerve root is shown, while on the left, the nerve root is not visible because of compression by the disc.

the knee. The degree of elevation necessary to produce sciatic symptoms roughly correlates with the degree of nerve root tension. Approximately 1 to 2 percent of patients who have lumbar disc herniations will have a massive extrusion of nuclear material sufficient to interfere with nerve control of bladder and bowel function (Spangfort 1972). Although not common, the cauda equina syndrome is a true surgical emergency since failure to decompress the lesion may result in permanent loss of bladder and bowel control.

It is important to emphasize that the majority of patients who meet the clinical criteria of HNP will recover from acute symptoms and have minimal residual functional or work capacity impairment. It is recognized that patients who have had a known disc herniation are more likely to have a recurrent herniation, and therefore reduction in excessive postural and lifting demands are advisable.

Segmental Instability

Segmental instability can be partially defined in both mechanical and clinical terms. Mechanically, instability indicates that application of a load to a structure (in this case the vertebral motion segment) will result in excessive displacement(s) of that structure in comparison to the normal. Clinically, there is still controversy over the proper criteria for the diagnosis of segmental instability. The available evidence suggests that segmental instability should be suspected when an individual has recurrent episodes of acute LBP, accompanied by postural disturbances such as rotoscoliosis (Kirkaldy-Willis and Farfan 1982; Macnab 1971). The clinical diagnosis of segmental instability is made if there are excessive displacements or angulations occurring with normal motion. The displacement and angulations are typically measured on lateral spinal radiographs taken when the patient is flexing or extending (Figure 3.5) (Knutsson 1944). Because patients with acute symptoms often have muscle spasm and guard their spinal motion, it may be some time before they can move sufficiently to demonstrate instability. Patients with segmental instability may have decreased capacity for repetitive twisting and manual handling tasks.

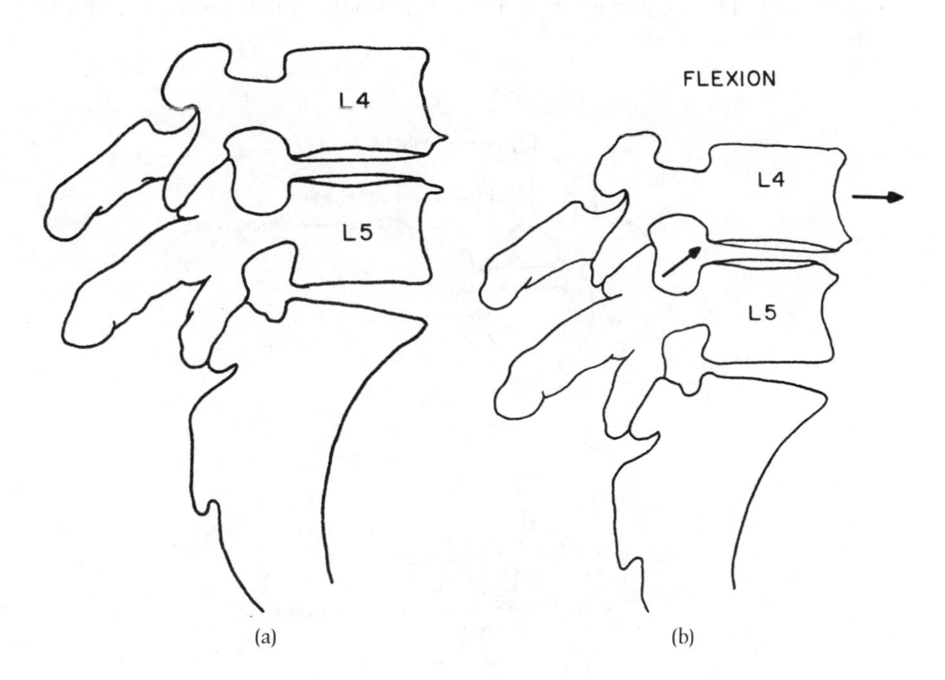

Figure 3.5
This demonstrates segmental instability at the L4-5 level. With the patient standing, note the presence of traction osteophytes (a). In forward flexion the disc space collapses anteriorly, and the L4 vertebra slides forward relative to the L5 vertebra (b).

Spinal Stenosis

Narrowing of the cross-sectional dimensions of the lumbar spinal canal may occur at a single level, or many levels of the lumbar spine (Fig. 3.6) (Verbeist 1954; Arnoldi *et al.* 1976; Paine 1976) and produce spinal stenosis. Typical symptoms of spinal stenosis include recurrent or continued LBP aggravated by body posture and physical exertion. As stenosis increases, extension of the spine becomes more painful, while relief may occur in flexion. Leg pain, classified as neurologic claudication, is the most typical symptom (see definitions in Introduction). Reflex motor and sensory changes are often confusing because of the involvement of several lumbar levels. Unlike disc herniations, nerve root signs and symptoms are frequently absent.

This narrowing of the canal may be congenital or degenerative (Arnoldi *et al.* 1976). Most commonly, spinal stenosis results from degenerative changes including posterior osteophytic projections into the spinal canal, hypertrophy of the articular facets, and buckling of the ligamentum flavum (Figure 3.7). These changes may involve a single spinal level or multiple levels.

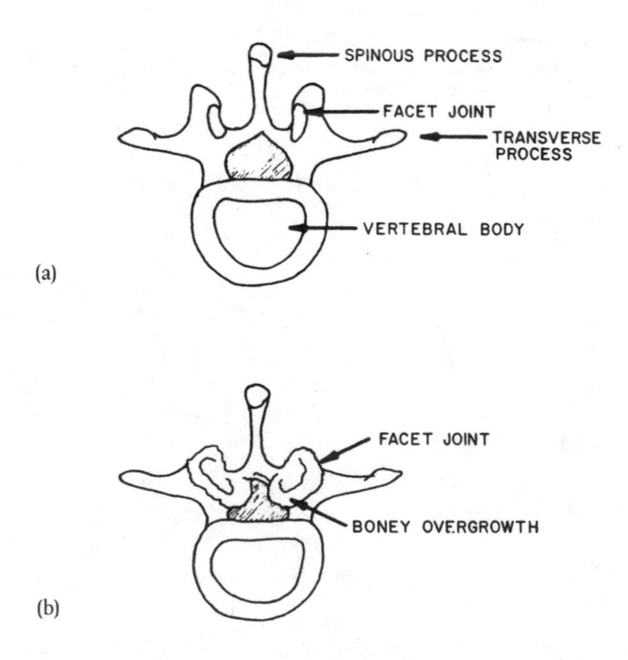

Figure 3.6
The spinal canal dimensions are shown by cross-hatching, and are bounded by the vertebral body in front, the pedicles to the sides, and the facet joints and lamina posteriorly (a). Notice that the osteophytic overgrowth of the facet joints affects the dimensions of the spinal canal reducing it particularly in the lateral recesses (b).

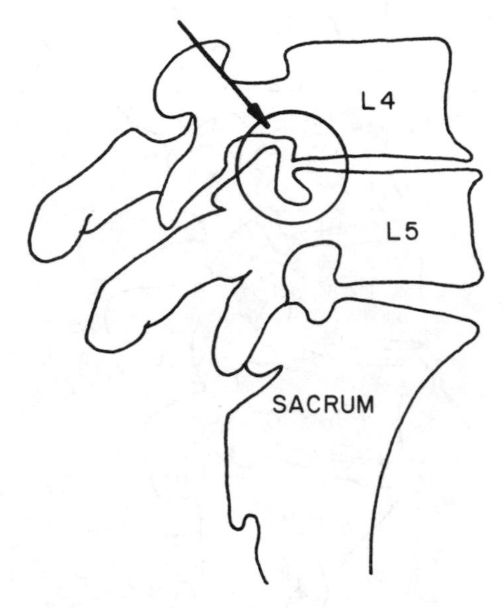

Figure 3.7
In this side profile of the lumbar spine, note there has been bony overgrowth not only anteriorly but posteriorly as emphasized by the circle. Because of the combination of vertebral spurs, the settling of the facets and the bony overgrowth adjacent to the facets, the available space for the exiting nerve root has been significantly compromised.

A second group of stenotic lesions are associated with focal degenerative disease. Degenerative spondylolisthesis (Fig. 3.8) usually occurs in females, typically during the fifth and sixth decades of life, is most common at L4-5, and frequently is associated with diabetes (Rosenberg 1975). In other patients, disc space collapse leads to backward displacement (retrospondylolisthesis) (Fig. 3.9). A third and less well defined group is termed rotational deformities and may produce stenosis. It is possible that these deformities may be preceded by segmental instability of the affected motion segment (Farfan 1969, Kirkaldy-Willis and Farfan 1982).

Ken (1983) has emphasized the interrelationship between spinal stenosis and disc herniations. When the spinal canal is of smaller dimensions, a relatively small disc herniation may produce significant nerve entrapment. Figure 3.10 summarizes the overall prevalence of the degenerative disorders.

Congenital Abnormalities

Abnormalities in embryonic growth of the spine may give rise to skeletal abnormalities, the majority of which do not cause low back pain.

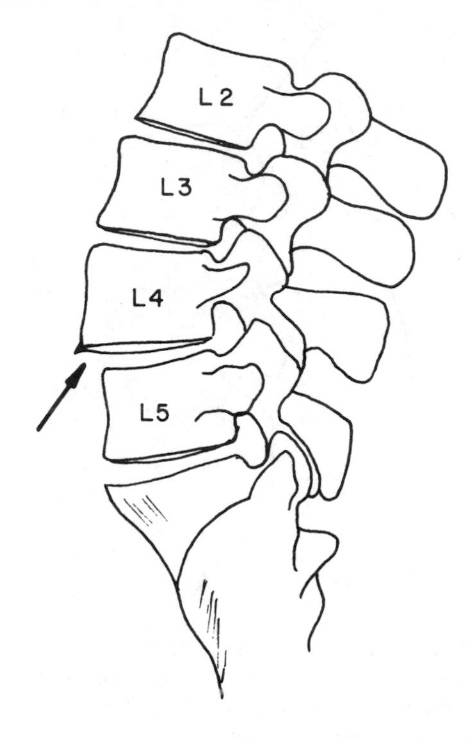

Figure 3.8
This is an example of degenerative spondylolisthesis, where the L4 vertebra has slid anteriorly relative to the L5 vertebra and sacrum. In this type of spondylolisthesis there are no defects in the posterior bony structures. Degeneration has occurred as evidenced by narrowing of the disc, and bony osteophytes anteriorly and posteriorly.

Because these abnormalities can be seen on spinal radiographs, they have been presumed to relate to the low back complaints, and sometimes are used as exclusion criteria in those industries that use radiographic screening for potential employees. Most of the common congenital abnormalities occur with equal prevalence in the LBP and non-LBP affected subgroups (see Table 3.1). This problem is further defined in Chapter 11, which discusses worker selection, and in Chapter 17 which describes the legal problems.

Spina Bifida Occulta

Defects in the formation of the neural arch with incomplete closure, and complete or partial absence of the spinous processes and lamina occur in 5 to 10 percent of all individuals (see Table 3.1) (Fig. 3.11). In a small

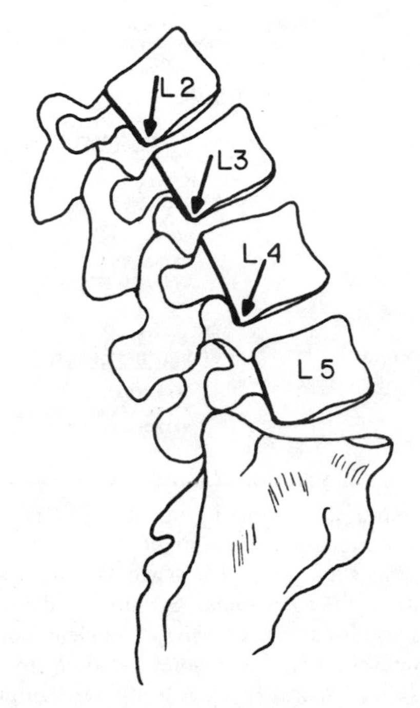

Figure 3.9
This is a lateral profile of a spine taken with the patient in extension. Note that at each of the levels shown, i.e., L2-3, L3-4, and L4-5 the vertebra above is slipping backward relative to the vertebra below. The posterior margins of the vertebral body have been accented to show this backward (retro) displacement.

subgroup (1/100,000) the bony defect is accompanied by neural elements herniating through the defect, a condition termed meningomylocele. The neurologic defects associated with meningomylocele are variable and to some degree dependent on the level and extent of the osseous defect. These patients usually have apparent defects during early childhood.

Segmentation Abnormalities

There are wide variations in the number and shape of the lumbar vertebrae. Lumbarization is a condition in which the first sacral segment has developed as a lumbar vertebra, resulting in six lumbar segments (Fig. 3.12). Conversely, sacralization means the fifth lumbar vertebrae has completely or partially been incorporated into the sacrum (Figs. 3.13 and 3.14). These lesions are equally prevalent in individuals with and without LBP (Frymoyer *et al.* 1984). In a small patient subgroup, a hemisacralized lumbar vertebra may result in the development of a false joint between the ilium and

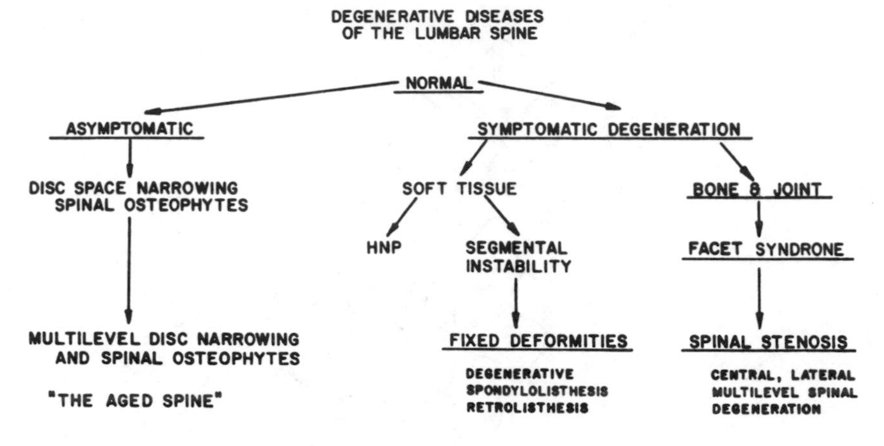

Figure 3.10
Summarized here are some of the possible manifestations of spinal degeneration, which can be asymptomatic, such as a single narrowed disc space and spinal osteophytes or in later stages multilevel disc space narrowing and spinal osteophytes which is frequently termed "the aged spine." Symptomatic degeneration may involve predominantly the soft tissues such as herniated nucleus pulposus, and segmental instability or predominantly bone and joint, leading to the facet syndrome. Compromising of the spinal canal may occur leading to symptoms either of LBP, or nerve root entrapment.

elongated transverse process of the fifth lumbar vertebra and may cause pain (Fig. 3.14). The other general relevance of segmentation abnormalities is that nerve root signs may be confusing when the patient has sciatica, and the precise localization of the affected level is more difficult to determine.

Spondylolysis, Spondylolisthesis

Spondylolisthesis is forward displacement of one vertebra relative to the next lower lumbar segment (Fig. 3.15). Careful analysis of clinical histories and radiographic appearance of patients with this deformity, indicate that spondylolisthesis may be subdivided into five types (Table 3.2) (Wiltse, Newman, and Macnab 1976). The most common form is isthmic spondylolisthesis, which involves an acquired or congenital defect in the neural arch, typically involving the pars interarticularis (Fig. 3.16). The defect may be unilateral or bilateral. If slippage has not occurred, the defect may not cause LBP. The presence of the defect varies widely in the population ranging from one percent in Negroes to 10 percent in Eskimos (Kettelkamp and Wright

TABLE 3.1
Lumbar Spine Radiographic Abnormalities in "Normal" Asymptomatic Populations

Year	*Author*	*Population*	*Number Subjects*	*Age Range*	*Osteo-arthrosis*	*Disc Space Narrow-ing*	*Hemi-Sacral-ization*	*Transitional Vertebrae*	*Spina Bifida*	*Facet Assymetry*	*Spondylo-listhesis*	*Scoliosis*
1968	Bremner*	Jamaican			65.0% (M)							
	Lawrence				56.0% (F)							
1954	Hult*	Swedish			77.0% (M)							
1969	Hirsch*	Swedish	9419		6.3%							
1968	Schein	U.S.	598 (M)		23.0%	2.5%		1.8%	1.7%		0.6%	
1954	Runge	U.S.	4654 (M)		19.2%	2.5%		6.7%	3.1%	25%	2.1%	6.1%
1969	Lawrence	English	1522		65.0% (M)							
					52.0% (F)							
1965	Fahrni	Indian			65.0%	10.0%						
1953	Splithoff*	U.S.	200		24.0%			10.0%	5.0%		2.5%	
1950	Allen	U.S.	3000		13.6%			6.5%	5.9%		3.0%	
1984	Frymoyer	U.S.										

*Includes patients with and without backache—no significant differences between the two.
Bremner and Lawrence in Jamaican Study: 30% had Grade III-IV changes.
Lawrence in English Study: 30% had Grade III-IV changes.
Source: The sources of this information are by author name and are cited in the bibliography.

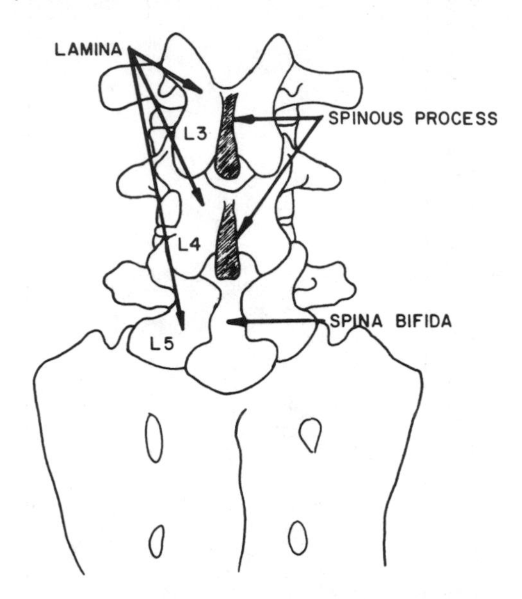

Figure 3.11
In this view of the spine, as seen from behind, the posterior vertebral structures (lamina and spinous processes) are intact at L3 and L4. Note that at L5 there is absence of a portion of the lamina and spinous process resulting in a spina bifida. This finding usually has no relationship to spinal symptoms.

1971). The most common level of involvement is at the fifth lumbar vertebra, and these lesions occur with greatest frequency in males. The lesion is viewed as a fatigue fracture caused by repetitive stress for the following reasons (Wiltse, Widell, and Jackson 1975):

1. Young individuals without the lesion can be seen to later develop the lesion.
2. In the early stages of the development of spondylolysis, increased radioisotope uptake sometimes can be seen, indicative of injury, and/or healing.
3. Healing of lesions can be shown in those patients identified early in the course of the problem and in whom appropriate bracing is prescribed.
4. The prevalence of the lesion is greater in certain stereotyped competitive physical and athletic activities, specifically American interior football linemen, female gymnasts, oarsmen, javelin throwers and backpackers.
5. The pattern of failure in the neural arch can be produced in the laboratory through chronic cyclic loading of vertebral specimens (Hutton, Stott and Cyron 1977).

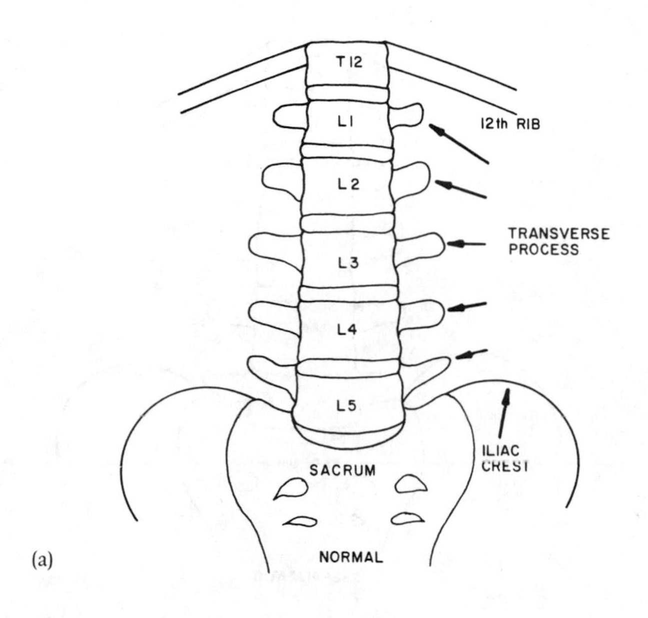

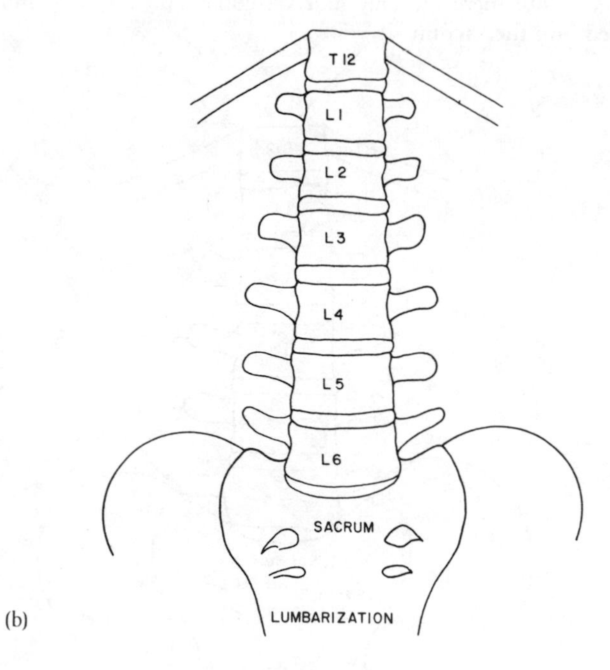

Figure 3.12
This view of the spine from behind shows the normal configuration (a). The five vertebrae are typically counted down using the 12th thoracic vertebra which has a rib as the demarcation between the thoracic and lumbar spines. In lumbarization (b), six vertebral bodies are identified.

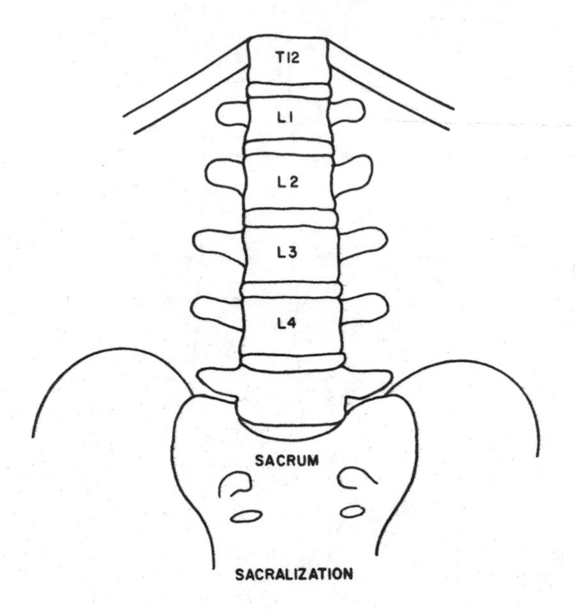

Figure 3.13
Sacralization. Note there are only four vertebrae, and the 5th lumbar vertebra is incorporated into the sacrum.

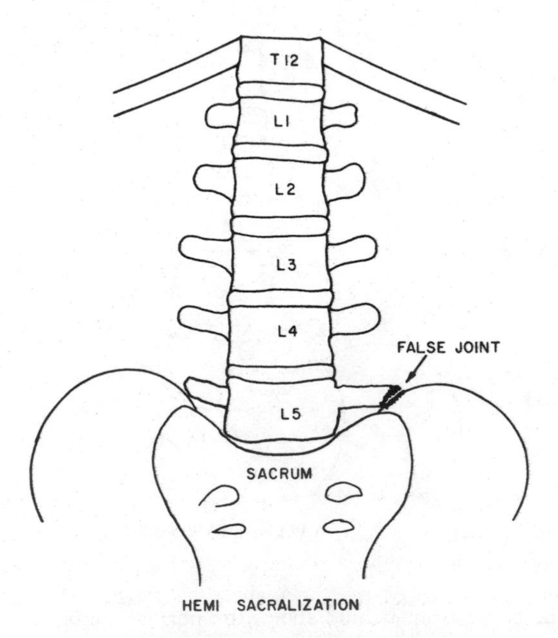

Figure 3.14
In this case of hemi-sacralization, the 5th lumbar vertebra, on the right shows the presence of a sacral-like structure which has resulted in a false joint.

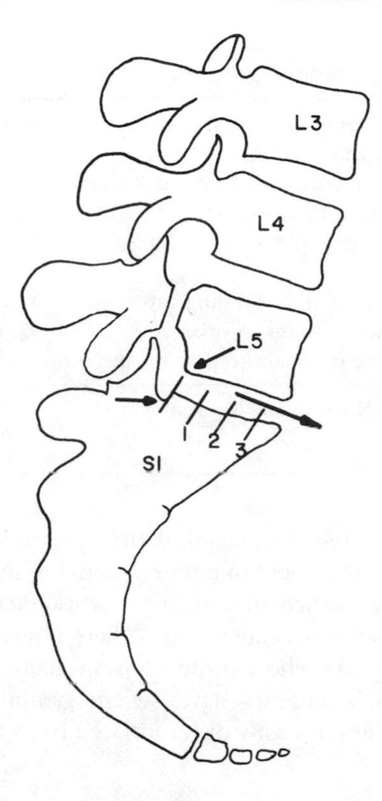

Figure 3.15
This lateral view of spondylolisthesis demonstrates that the L5 vertebra has slipped forward such that its posterior vertebral border margin has migrated approximately 25 percent anterior to the posterior border of the first sacral segment. Note the absence of continuity between the spinous process and the vertebral body with a defect in the isthmic portion of the lamina thus leading to the term isthmic spondylolisthesis. The degree of forward slip is usually based on the relationship of the L5 vertebra to the sacrum, and typically is graded first, second, third and fourth degree.

From the standpoint of industrial LBP, spondylolisthesis and spondylolysis are relevant in the following ways:

1. The group of patients with spondylolisthesis is in general more susceptible to recurrent LBP. However, screening radiography should be restricted to that group in which a palpable step deformity is revealed.

2. Spondylolysis is equally prevalent in subjects with and without LBP. To demonstrate its presence radiographically requires oblique films which produce an x-ray exposure at least twice that for routine AP and lateral radiographs (Scavone, Latshaw, and Rohrer 1981). We strongly advise against this form of evaluation except for highly selected cases.

TABLE 3.2
Spondylolisthesis can be subdivided into five types.

I. Dyplastic—In this type congenital abnormalities of the upper sacrum or the arch of L5 permit the olisthesis to occur.
II. Isthmic—The lesion is in the pars interarticularis. Three types can be recognized.
  a. Lytic-Fatigue fracture of the pars.
  b. Elongated but intact pars.
  c. Acute fracture
III. Degenerative—Due to long standing intersegmental instability.
IV. Traumatic—Due to fractures in other areas of the bony hook than the pars.
V. Pathological—There is generalized or localized bone disease.

Source: From Wiltse, Newman and Macnab 1976.

3. There is no evidence suggesting that any industrial exposures are associated with the development of either spondylolysis or spondylolisthesis, and therefore their presence in industrial workers who develop LBP is unrelated to known industrial causations. (There is a small group of patients with major spinal trauma who can develop an acute, traumatic spondylolisthesis. Because such patients have severe symptoms coincident with obvious injury, this lesion is easily differentiated from the other five types of spondylolisthesis.)

Spine Trauma

Acute spinal trauma indicates the application of a load exceeding the inherent strength of the tissues. The patterns of injury observed reflect the magnitude and velocity of the applied loads, and direction(s) of application (Holdsworth 1970; White and Panjabi 1978). Within the spinal column, bony structures rather than ligaments or discs are most susceptible to such acute mechanical overloads. In fact, it has proven impossible to produce disc herniations experimentally by the application of a single applied load, regardless of the direction of application, as discussed in Chapter 1. In all instances, a bony and/or cartilage injury results before any demonstrable injury occurs to the disc.

Vertebral Body Fracture—Fracture/Dislocation

These fractures in industry usually are the result of a fall or vehicular injury. Many classification systems have been devised which are based upon the direction of load application, the resultant stability of the involved spinal segments(s), and the associated neurologic defect. The most comprehensive scheme is that proposed by White and Panjabi (1978), to which the reader is

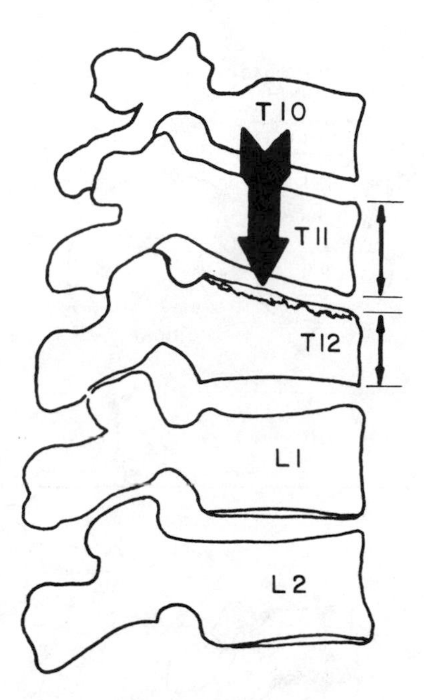

Figure 3.16
In this example of a spinal compression fracture, the arrow depicts the predominantly compressive downward forces which have combined with a flexion moment to produce anterior wedging of the body of the 12th thoracic vertebra. In this instance, the vertebral body height anteriorly has diminished 30 percent.

referred for a detailed analysis. For the purposes of this discussion, a somewhat simplified scheme of the characteristics of spinal fracture is outlined, based upon the direction of mechanical overload. The most common force application is an axially applied load, resulting usually from a fall in the seated position, or less commonly from a fall in which the person lands standing (the latter is often accompanied by hip and foot fractures). The result of axial compression loads is the typical compression fracture (Fig. 3.16). Under substantial compressive loads, bursting of the vertebral body may occur with bony fragments being driven posteriorly into the neural canal with subsequent neurologic injury.

Less common injury patterns involve a torsional force application, usually accompanying some element of axial compressive load. Because the spinal ligaments and facet joints are particularly vulnerable to this force application, fractures in association with ligamentous disruption may occur. At the extreme, dislocation of the affected motion segment may result. Such fractures have a high possibility of neurologic complications.

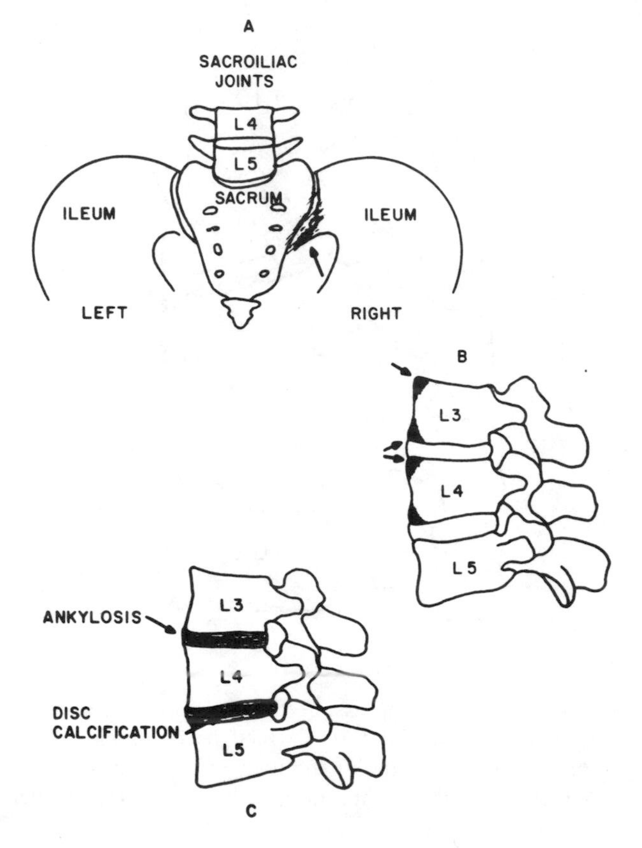

Figure 3.17
This is a view of the pelvis and lower lumbar spine viewed from in front. Note that the left sacroiliac joint has a smooth contour, while on the right erosive changes have occurred predominantly on the iliac side. The dark shading represents the sclerotic bone often seen. On the lateral view (b), the vertebral body has a squared appearance as depicted by the arrows, and often there is increased bony density represented by the shaded area. In c, these changes have progressed to complete ankylosis, with continuous bone visible along the anterior and posterior margins of L3, L4, and L5. Frequently calcification of the disc accompanies the late stage of ankylosing spondylitis.

A third and rare type of injury involves shearing forces. These injuries are most commonly associated with vehicular trauma, where the seat belts fix the lower spine, allowing shearing stresses to be applied.

Lastly, some industrial injuries involve a direct force applied to the back from a dorsal or dorsolateral position. Such injuries are most common in mining and quarrying operations. The most common injury is fracture of one or more of the transverse processes and occasionally of the spinous processes. In some instances, a transverse process fracture may be associated with injuries to the pelvis, chest wall, and kidneys.

Most components of the injury are radiographically visible, although the ligamentous injuries are often only surmised (White and Panjabi 1978). The treatment is usually specific and the prognosis is usually straightforward and depends upon the magnitude of the trauma, and the presence or absence of neurologic involvement. Fortunately, the majority of traumatic injuries are associated with the resumption of relatively normal function once healing has occurred. Chronic low back complaints are not usually excessive.

Inflammatory Lesions

Inflammatory lesions can be divided into two categories: those of infectious etiology and those of non-infectious etiology.

Infectious Etiologies

Acute or chronic bacterial infections of the spine are rare and may occur spontaneously or after spinal surgery. The most common bacteria is staphylococcus. Affected individuals often have an underlying disease process such as diabetes, are taking immunosuppressive drugs, or have undergone recent genitourinary surgery. One in 200 patients who have had disc excision develop an infection at the operated disc space (termed discitis) (Spangfort 1972; Pilgaard 1969; Frymoyer 1981). Characteristically, patients with spinal infections have severe LBP and muscle spasm. They are often very sensitive to pressure applied on the spinous processes. Unlike most back disorders, recumbency often does not relieve their symptoms. Laboratory abnormalities may include elevations of white blood count and sedimentation rate. Radiographs show erosion of the vertebral endplates and narrowing of the disc developing over a period of three to four weeks. The role of radioisotope scans is discussed in Chapter 8.

A second type of infection is caused by tuberculosis (Hoffman 1981). The problem is now most common in nonindustrialized countries and in miners and quarry workers in industrialized societies. Unlike other bacterial infections, tuberculosis may be more insidious at onset. Involvement of vertebrae and discs may result in collapse of vertebral bodies, and postural deformities, usually kyphosis. From the industrial perspective, bacterial, tuberculous and fungal infections of the spine should be kept in mind when a worker has incapacitating back pain, particularly when it is worse at rest and/or is accompanied by systemic symptoms such as fever, night sweats, and weight loss. The work potential of individuals who have been treated for these infections depends upon the degree of bony and neurologic involvement, the residual deformities, and treatment rendered (i.e., spine fusion). Typically, reductions in spinal loading will be required.

Non-Infectious Etiology

One of the most common, identifiable causes of LBP is spondyloarthropathy. The most prevalent type is ankylosing spondylitis which affects 1/100 males (Hollander 1966; Calin 1981a). These patients usually are 20 to 40 years old and characteristically complain of spinal stiffness and pain. A typical pattern of symptoms is pain, awakening the individual before his usual waking hour, marked spinal stiffness before arising from bed, and gradual mobilization and relief of symptoms over time (usually 1 to 2 hours). Physical examination often demonstrates marked restriction of motion in the spine as well as restriction of chest expansion (less than 2.5 cm). Radiographic studies (Fig. 3.17) eventually show erosive changes on the iliac side of the sacroiliac joints and a squared-off appearance of the vertebral body endplates. The sedimentation rate may be elevated. An underlying autoimmune disorder is thought to be responsible. A very high percentage of affected individuals have a specific serum antigen, the HLA-27B antigen. However, 5 to 12 percent of the male population has this antigen, and therefore a positive serology test without the signs and symptoms of ankylosing spondylitis is not diagnostic. The same clinical, radiographic and laboratory findings are also identified in patients with other "autoimmune" diseases, for example Reiter's Syndrome (a disease characterized by urethritis, uveitis, arthritis and sacroileitis) (Calin 1981b); inflammatory bowel disease such as ulcerative colitis (Good 1981) and psoriasis (Wright 1981).

Most patients with ankylosing spondylitis and its variants respond well to non-steroidal, anti-inflammatory drugs (Calin 1981a). The disease is often self-limited and has minimal implications for the individual's employability and physical tolerance. Physical therapy should be encouraged to maintain as large a range of spinal motion as possible. A small subgroup of patients with the spondyloarthropathies has a more unrelenting course accompanied by other joint involvement (typically hips), progressive stiffness and eventual ankylosis of their spine. The employability of this group is variable and depends upon job classification and the degree of impairment of spinal mobility. If hips are involved, this will also produce additional functional impairments.

Metabolic Disorders

The spine is subjected to a host of metabolic disorders which, with one exception, are fairly rare and are of minimal impact to industrial LBP. The exception is osteoporosis, a condition characterized by loss of total bone volume, usually occurring over time (Frost 1981). Hormonal causes (malfunctions of thyroid, adrenal glands and postmenopausal), exogenous causes (long-term administration of cortisone), and disuse (i.e., after prolonged recumbency for treatment of fracture) may contribute to the development of osteoporosis (Kleerekoper, Tolia, and Parfitt 1981). Most commonly, osteoporosis is identified in the sedentary, postmenopausal female. Fifty percent of women over the age of 45 have radiographic evidence of osteoporosis, and 30 percent are symptomatic (Iskrant and Smith 1969). Conversely, the male work force is usually unaffected.

Osteoporosis causes structural weakening of bone which renders the individual more susceptible to spinal compression fractures (Fig. 3.18). The diagnosis is most commonly considered in females in the fifth, sixth and seventh decades of life who incur a spinal fracture with minimal trauma. The spinal radiographs usually are diagnostic. When a worker has osteoporosis, reduction in excessive manual handling tasks is required.

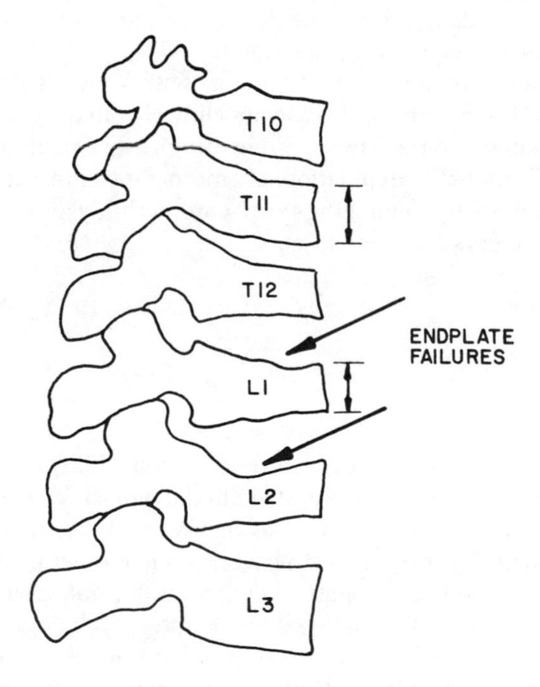

Figure 3.18
Spinal osteoporosis leads to weakening of the vertebral bodies. Note the loss of endplate contour depicted at both L1 and L2 due to compressive overloads, as well as diminishment of the anterior vertebral height of T 11 and 12, L1 and L2. Advanced changes of this type are associated with accentuation of the thoracic kyphosis (the so-called dowager's hump).

Spinal Neoplasms

Spinal neoplasms involving either osseous or neural structures can occur as the result of a metastatic spread from some other primary site or as a primary tumor within the spine. Primary osseous and neurologic tumors of the spine are relatively rare, whereas metastatic tumors are common. Bony metastases are frequent in female breast cancer, male prostatic cancer, and lung, kidney, and thyroid neoplasms in both sexes. Multiple myeloma is the most common primary malignant bone tumor. Regardless of the tumor type, onset of pain is usually insidious. Similar to the infectious processes, bony neoplasms are distinctive for the unrelenting nature of the pain and the lack of relief when recumbent. Occasionally, the first manifestation of a bone neoplasm may be a fracture that has occurred in structurally weakened bone.

The diagnosis of neoplasm should be considered in those patients with unrelenting back pain, particularly if there is a history of breast or lung tumor. The spinal radiograph may be distinctive, although in the early stages a bone scan may be required for diagnosis.

The continued employability of patients with known spinal neoplasms is a function of the underlying disease, its clinical course, and the degree of structural weakness of the spine. Some tumors (particularly breast) are responsive to hormonal manipulation, chemotherapy, and radiation therapy and may be compatible with long-term employability provided there are minimal requirements for manual labor.

SUMMARY

A major problem in many cases of LBP is the inability of the clinician to make a precise diagnosis. A definite structural diagnosis can be reached only in about 50 percent of patients. Consequently, many diagnostic categories are poorly defined. Most patients do have complete relief of symptoms in a reasonable time period. Those patients with pain of unknown etiology who have persistent symptoms usually demonstrate psychologic changes and represent the group with the greatest problems in terms of health care delivery, disability and cost. LBP of known etiology can be classified as congenital, degenerative, traumatic, inflammatory, neoplastic and metabolic. Degenerative disorders include herniated nucleus pulposus, spinal stenosis, segmental instability and osteoarthritis. Radiographic evidence of degeneration is equally prevalent in those with and without LBP. Histologically, by age 30, all spines show signs of degeneration. Congenital causes for LBP include spina bifida, segmentation abnormalities, spondylolysis, spondylolisthesis. The majority of minor congenital defects are not significant risk factors for LBP. However, some with spondylolisthesis are more susceptible to LBP. More precision in diagnosis will lead to improvements in prevention and treatment.

REFERENCES

Allen, M.L.; and Lindem, M.C. 1950. Significant roentgen findings in routine pre-employment examination of the lumbosacral spine: A preliminary report. *Am. J. Surg.* 80:762–66.

Arnoldi, C.C.; Brodsky, A.E.; Cauchoix, J.; Crock, H.V.; Dommisse, G.F.; Edgar, M.A.; Gargano, F.P.; Jacobson, R.E.; Kirkaldy-Willis, W.H.; Kurihara, A.; Langenskiold, A.; Macnab, I.; McIvor, G.W.D.; Newman, P.H.; Paine, K.W.E.; Russin, L.A.; Sheldon, J.; Tile, M.; Urist, M.R.; Wilson, W.E.; and Wiltse, L.L. 1976. Lumbar spinal stenosis and nerve root entrapment syndromes. Definition and classification. *Clin. Orthop.* 115:4–5.

Beals, R.K.; and Hickman, N.W. 1972. Industrial injuries of the back and extremities. Comprehensive evaluation–an aid in prognosis and management: A study of one hundred and eighty patients. *Journal of Bone and Joint Surgery* 54A(8): 1593–1611.

Benn, R.T.; and Wood, P.H.N. 1975. Pain in the back: An attempt to estimate the size of the problem. *Rheumatology and Rehabilitation* 14:121–28.

Bergquist-Ullman, M.; and Larsson, U. 1977. Acute low back pain in industry. *Acta* Orthop. Scand. (Suppl.) 170:1–117.

Bremner, J.M.; Lawrence, J.S.; and Miall, W. E. 1968. Degenerative joint disease in a Jamaican rural population. *Ann. Rheumat. Dis.* 27:326–32.

Calin, A.C. 1981a. Ankylosing spondylitis. In *Textbook of Rheumatology* eds. William N. Kelley; Edward D. Harris, Jr.; Shawn Ruddy; Clement B. Sledge. pp. 1017–32. Philadelphia: W.B. Saunders Co.

Calin, A. 1981. Reiter's syndrome. In *Textbook of Rheumatology* eds. W.N. Kelley, E.D. Harris, S. Ruddy, C.B. Sledge. pp. 1033–46. Philadelphia: W.B. Saunders Co.

Damkot, D.K.; Lord, J.; Pope, M.H.; Frymoyer, J.W. 1983. Relationship between work history, work environment, and low back pain in males. *Spine* (in press).

Fahrni, W.H.; and Trueman, G.E. 1965. Comparative radiological study of the spines of a primitive population with North Americans and Northern Europeans. *J. Bone Joint Surg.* 47B (3):552–55.

Farfan, H.F. 1969. Effects of torsion on the intervertebral joints. *Can. J. Surg.* 12: 336–41.

Friberg, S.; and Hirsch, C. 1949. Anatomical and clinical studies on lumbar disc degeneration. *Acta Orthop. Scand.* 19:222–42.

Frost, H.M. 1981. The evolution of pathophysiologic knowledge of osteoporoses. *Orthop. Clin. North Am.* 12 (3):475–83.

Frymoyer, J.W.; Pope, M.H.; Costanza, M.C.; Rosen J.C.; Goggin, J.E.; and Wilder, D.G. 1980. Epidemiologic studies of low-back pain. *Spine* 5(5):419–23.

Frymoyer, J.W. 1981. The role of spine fusion. Symposium–The role of spine function. Question 3. *Spine* 6 (3):284–90.

Frymoyer, J.W.; Pope, M.H.; Clements, J.H.; Wilder, D.G.; MacPherson, B.; and Ashikaga, T. 1983a. Risk factors in low-back pain. An epidemiological survey. *Journal of Bone and Joint Surgery* 65A(2):213–18.

Frymoyer, J.W.; Rosen, J.C.; Clements, J.H., and Pope, M.H. 1983b. Psychologic factors in low back pain patients. Submitted to *Clin. Orthop.*

Frymoyer, J.W.; Newberg, A.; Pope, M.H.; Wilder, D.G.; Clements, J.; and MacPherson, B. 1984. The relationship between spinal radiographs and LBP severity in males 18–55. *Journal Bone Joint Surgery*, in press.

Good, A.E. 1981. Enteropathic arthritis. In *Textbook of Rheumatology* eds. W.N. Kelley, E.D. Harris, S. Ruddy, C.B. Sledge. pp. 1063–75. Philadelphia: W.B. Saunders Co.

Hakelius, A.; and Hindmarsh, J. 1972. The significance of neurological signs and myelographic findings in the diagnosis of lumbar root compression. *Acta Orthopaedica Scandinavia* 43:239–46.

Hirsch, C.; Jonsson, R.; Lewin, T. 1969. Low-back symptoms in a Swedish female population. *Clin. Orthop.* 63:171–176.

Hoffman, G.S. 1981. Mycobacterial and fungal infections of bones and joints. In *Textbook of Rheumatology* eds. W.N. Kelley, E.D. Harris, S. Ruddy, C.B. Sledge. pp. 1573–85. Philadelphia: W.B. Saunders Co.

Holdsworth, F. 1970. Fractures, dislocations, and fracture-dislocations of the spine. *Journal of Bone and Joint Surgery* 52A(8):1534–51.

Hollander, J.L., ed. 1966. *Arthritis and allied conditions.* A textbook of rheumatology. 7th ed. Philadelphia: Lea & Febiger.

Hult, L. 1954. Cervical, dorsal and lumbar spinal syndromes. *Acta Orthop. Scand. Suppl.* 24:174–75.

Hutton, W.C.; Stott, J.R.R.; and Cyron, B.M. 1977. Is spondylolysis a fatigue fracture? *Spine* 2(3):202–209.

Iskrant, A.P.; and Smith, R.W., Jr. 1969. Osteoporosis in women 45 years and over related to subsequent fractures. Public Health Rep. 84(1)33–38.

Kellgren, J.H.; and Lawrence, J.S. 1958. Osteo-arthrosis and disk degeneration in an urban population. *Annals of the Rheumatic Diseases* 17:388–97.

Ken, Y.H. and Kirkaldy-Willis, W.H. 1983. The pathophysiology of degenerative disease of the lumbar spine. *Orthop. Clin. North. Am.* 14(3):491–504.

Kirkaldy-Willis, W.H.; and Farfan, H.F. 1982. Instability of the lumbar spine. *Clin. Orthop.* 165:110–123.

Kettelkamp, D.B.; and Wright, D.B. 1971. Spondylolysis in the Alaskan Eskimo. *J. Bone Joint Surg.* 53A:563.

Kirkaldy-Willis, W.H.; and Farfan, H.F. 1982. Instability of the lumbar spine. *Clin. Orthop.* 165:110–23.

Kleerekoper, M.; Tolia, K.; and Parfitt, A.M. 1981. Nutritional, endocrine, and demographic aspects of osteoporosis. *Orthop. Clin. North America* 12 (3): 547–58.

Knutsson, F. 1944. The instability associated with disk degeneration in the lumbar spine. *Acta Radiol.* 25:593–609.

Lawrence, J.S. 1969. Disc degeneration. Its frequency and relationship to symptoms. *Ann. Rheum. Dis.* 28:121–38.

Macnab, I. 1971. The traction spur. An indicator of segmental instability. *J. Bone Joint Surg.* 53A(4):663–70.

McCarty, D.J. (ed) 1979. *Arthritis and allied conditions. A textbook of rheumatology.* 9th ed. Philadelphia: Lea & Febiger.

Mixter, W.J.; and Barr, J.S. 1934. Rupture of the intervertebral disc with involvement of the spinal canal. *N.E.J.M.* 211 (5):210–15.

Mooney, V. 1983. The syndromes of low back disease. *Orthop. Clin. North Am.* 14 (3):505–16.

Nathan, H. 1962. Osteophytes of the vertebral column. An anatomical study of their development according to age, race, and sex with considerations as to their etiology and significance. *Journal of Bone and Joint Surgery* 44A(2):243–68.

Paine, K.W.E. 1976. Clinical features of lumbar spinal stenosis. *Clin. Orthop.* 115:77–82.

Pilgaard, S. 1969 Discitis (closed space infection) following removal of lumbar intervertebral disc. *Journal of Bone and Joint Surgery* 51A(4):713–16.

Pope, M.H.; Wilder, D.G.; Stokes, I.A.F.; and Frymoyer, J.W. 1979. Biomechanical testing as an aid to decision making in low-back pain patients. *Spine* 4: 135–40.

Pope, M.H.; Bevins, T.; Wilder, D.G.; and Frymoyer, J.W. 1983. The relationship between anthropometric, postural, muscular and mobility characteristics of males, ages 18–55. Submitted to *Spine.*

Rosenberg, N.J. 1975. Degenerative spondylolisthesis: Predisposing factors. *Journal Bone and Joint Surgery* 57A(4):467–74.

Runge, C.L 1954. Roentgenographic examination of the lumbosacral spine in routine pre-employment examination. *J. Bone Joint Surg.*, 36A:75–84.

Scavone, J.G.; Latshaw, R.F.; and Rohrer, G.V. 1981. Use of lumbar spine films. Statistical evaluation at a university teaching hospital. *JAMA* 246 (10):1105–1108.

Schein, A.J. 1968. Back and neck pain and associated nerve root irritation in the New York City Fire Department. *Clin. Orthop.* 59:119–24.

Sokoloff, L 1969. *The biology of degenerative joint disease.* Chicago: University of Chicago Press.

Southwick, S.M.; and White, A.A. 1983. Current concepts review. The use of psychological tests in the evaluation of low-back pain. *Journal of Bone and Joint Surgery* 65A:560–65.

Spangfort, E.V. 1972. The lumbar disc herniation. A computer aided analysis of 2504 operations. *Acta Orthop. Scand.* (Suppl.) 142:1–95.

Splithoff, C.A. 1953. Lumbosacral junction; roentgenographic comparison of patients with and without backaches. *J.A.M.A.* 152:1610–13.

Svensson, H.O. 1981. *Low back pain in relation to other diseases and cardiovascular risk factors.* Doctoral dissertation. Goteborg, Sweden.

Troup, J.D.G. 1978. Driver's back pain and its prevention. A review of the postural, vibratory and muscular factors, together with the problem of transmitted road-shock. *Appl Ergonom.* 9:207–14.

Verbiest, H. 1954. A radicular syndrome from developmental narrowing of the lumbar vertebral canal. *Journal of Bone and Joint Surgery* 36B(2):230–37.

White, A.A., III; and Panjabi, M.M., eds. 1978. *Clinical biomechanics of the spine*. Philadelphia: J.B. Lipincott Company.

White, A.W.M. 1969. Low back pain in men receiving Workmen's Compensation: A follow-up study. *Canad. Med. Assoc. J.* 101(2):61–67.

Wiltse, L.L.; Widell, E.H. Jr.; and Jackson, D.W. 1975. Fatigue fracture: The basic lesion in isthmic spondylolisthesis. *Journal of Bone and Joint Surgery* 57A(1): 17–22.

Wiltse, L.L.; Newman, P.H.; and Macnab, I. 1976. Classification of spondylolisis and spondylolisthesis. *Clin. Orthop.* 117:23–29.

Wright, V. 1981. Psoriatic arthritis. In *Textbook of Rheumatology* eds. W.N. Kelley, E.D. Harris, S. Ruddy, C.B. Sledge. pp. 1047–62. Philadelphia: W.B. Saunders Co.

PART II
The Problem

4 Epidemiology

Gunnar B.J. Andersson, Malcolm H. Pope, and John W. Frymoyer

Epidemiology is derived from the Greek: epi = on; demos = people; logos = study. It is often defined as the study of epidemics. This conjures up in one's imagination quarantine signs and the plague. In fact, in the view of many authorities, low back pain (LBP) has become epidemic in the twentieth century. For example, disabling low back episodes have increased 26 percent from 1974 to 1978, while the population of the United States increased only 7 percent.

Despite this concern, epidemiologic research in LBP is in its infancy, compared to the epidemiologic data available for other diseases such as cancer, infections, and cardiovascular malfunctions. Perhaps this is because LBP carries a negligible risk of mortality, and therefore has been of less public concern. Epidemiologic research in LBP has been, and still is, often hampered by methodologic problems in definition, classification and diagnosis, as discussed in Chapter 3. Methodologic problems also exist in quantifying physical exposures that might be of etiologic importance such as loads applied to the spine over time.

In general, data for epidemiologic studies can be obtained from official health registers or by retrospective, prospective, or cross-sectional surveys of general populations or specific industrial populations. Such data are useful in defining the magnitude of the problem, and also can be useful in planning health care facilities and other medical and social programs, including preventive programs. Such data also partially define the natural history of LBP episodes and can help in identifying those environmental or individual factors that may cause or contribute to LBP. These data also are useful in defining prevention strategies.

Because epidemiology is the source of much of our knowledge regarding LBP, we have divided this section into four chapters. The first defines the magnitude of the problem in society and industry, the second measures societal cost, the third discusses the role of the workplace, the fourth, the individual risk factors for LBP.

THE MAGNITUDE OF THE PROBLEM

The magnitude of any health problem is measured by prevalence and incidence. In a prevalence study, the presence of LBP and other important variables are determined at one point in time (point prevalence) or during one period of time (period prevalence) for each member of the population studied or for a representative sample. Incidence can be defined as the number of people who develop LBP over a specified time period such as their lifetimes (lifetime incidence, which is synonymous with lifetime prevalence) or a single year (annual incidence). In short, prevalence means all cases of LBP, whereas incidence means all new cases of LBP.

Table 4.1 presents the prevalence and incidence of LBP as determined by different studies. The prevalence rates vary from a low of 12.0 percent to a high of 35.0 percent. Some authors report a higher prevalence in females, but others could not demonstrate a difference. The lifetime incidence rates are higher and range from 48.8 percent to 69.9 percent. Frymoyer *et al.* (1983) determined the lifetime incidence of LBP in males 18 to 55 years old and subdivided the symptoms by their intensity. They found that 69.9 percent of this population had experienced some form of back pain. Of those who had experienced LBP, 46.3 percent had moderate LBP, while 23.5 percent had LBP which they rated as severe.

These data presented in Table 4.1 have been attained from retrospective, cross-sectional, and some prospective studies performed in the United States, Great Britain, Scandinavia, and the Netherlands. These studies allow estimates of the frequency of occurrence of LBP and its impact on society.

NATIONAL STUDIES

We will separately consider information obtained from different countries, because the differing socioeconomic factors of these populations may influence the results.

United States

There are 2.5 million low back injured and 1.2 million low back disabled adults in the United States. LBP is the diagnosis in 10 percent of all chronic health conditions (Kelsey, Pastides and Bigbee 1978; Kelsey and White 1980). There are other epidemiologic data demonstrating the importance of

TABLE 4.1
Prevalence and Life-time Incidence of LBP as Determined by Different Studies.

Study	Life-time Incidence (%)	*Prevalence (%)*		*Study Group*			Comment
		Point	*Period*	*N*	*Age*	*Sex*	
Biering-Sorensen (1982)	62.6	12.0	–	449	30–60	M	
	61.4	15.2	–	479	30–60	F	
Hirsch *et al.* (1969)	48.8	–	–	692	15–72	F	
Hult (1954)	60.0	–	–	1193	25–59	M	Industrial population
Frymoyer *et al.* (1983)	69.9	–	–	1221	28–55	M	
Nagi *et al.* (1973)	–	18.0	–	1135	18–64	MF	
Svensson & Andersson	61	–	31	716	40–47	M	1 month period
(1983; 1982)	67	–	35	1640	38–64	F	1 month period
Valkenburg & Haanen	51.4	22.2	–	3091	20–	M	
(1982)	57.8	30.2	–	3493	20–	F	
Magora & Taustein (1969)	–	12.9	–	3316		MF	8 workgroup
Gyntelberg (1974)	–	–	25		40–59	M	1 year period

Source: The sources of these data are the individual studies, cited in the Bibliography by author and year.

LBP. Impairments of the back are the most frequent cause of activity limitation in persons under age 64. In subjects aged 25 to 44, a decrease in work capacity was caused by LBP. An average of 28.6 days/100 workers was lost each year and there was an average of nine days of confinement to bed. Each year 113,000 laminectomies and "disc excisions" as well as 34,000 other lumbar spine operations are performed.

United Kingdom

Benn and Wood (1975) found that the number of sickness absence episodes/1000 persons was 11 for women and 22.6 for men. Wood (1976) later attempted to calculate the impact of back problems on medical and social services (Table 4.2). Other surveys have shown that 25 percent of all laboring men are affected by low back disorders each year (Haber 1971). Annually, one man out of 25 workers changes his job because of a back condition. In 1979, 79,000 persons were chronically disabled (Harris, 1971). Estimating the British population at 50 million, this means that chronic low back disability was less than in the United States (.158 percent vs. .4 percent). On any one day, .05 percent of the work force has been chronically disabled for more than six months by a back problem (Wood and Badley 1980) which is closer to the United States' figures.

Sweden

In a ten-year period from 1961 to 1971, 12.5 percent of all annual sickness absence days were related to low back disorders. This means that one percent of all work days were lost annually under the heading of low back conditions (Helander 1973). The average sickness absence period was 36 days which is quite similar to the 28.6 days for the United States and the

TABLE 4.2
The Back Patient's Need for Medical and Social Services (expressed as rates per 1,000 persons both sexes at risk per year.)*

	# Subjects per 1,000 per year	
Handicapped/pension	2.0	
General Practitioner	20.0	(58 visits)
Referrals	9.0	
Admissions	1.0	
Operations	0.1	
Spinal Braces/Corsets	7.0	

*Based on Benn and Wood, 1975.

32.6 days for Great Britain. Forty percent of the low back affected workers were disabled for less than one week, while 9.9 percent were disabled for greater than six months. No other disease category was responsible for a greater number of days lost from work. The number of sickness absence periods in 1970 was 10/100 men and 6/100 women (Andersson and Svensson 1979). LBP accounts for many cases of early retirement and disability pensions. In any given year, 25 percent of all new pension cases resulted from a chronic low back condition. This amounts to 12,500 new retirements per annum.

CROSS-SECTIONAL STUDIES

A cross-sectional epidemiologic study is one in which a population is studied at a single point in time, attempting to evaluate all members of that population. In the past few years, at least seven cross-sectional studies have been performed. In addition there have been eight different work groups that have been studied by a cross-sectional epidemiologic design. In one additional study, subjects sick-listed for LBP were compared to a matched control group. Again, because of possible influences resulting from socio-economic conditions, these studies have been arranged by country.

United States

Frymoyer *et al.* (1981, 1983) performed a retrospective and cross-sectional analysis in the United States of 1,221 males 18 to 55 years of age who had enrolled in a family practice facility from 1975–1978. Almost 70 percent had had LBP (Table 4.1). When the data from that study were extrapolated to the 50 million working American males in the age group 18 to 55, it was calculated that 17 million workdays are lost annually. Patients with severe LBP had significantly more leg complaints, sought more medical care and treatment for LBP, and had lost more time from work for this reason when compared to subjects with no or moderate LBP. Sciatica-like symptoms had been present in 28.9 percent of the males with moderate LBP and 54.5 percent of the males with severe LBP. Objective reports of numbness were present in 14.0 percent of the males with moderate LBP and 37.4 percent of those with severe LBP, while weakness was reported by 17.9 percent of those with moderate LBP and 44.0 percent with severe LBP. The utilization of health care services is given in Table 4.3. It is clear that a very high percentage of individuals with severe LBP required care from health care practitioners and that a variety of medical treatments was also required.

TABLE 4.3
Type of Health-Care Services and Treatment Utilized by Men with Low Back Pain

	Percent Moderate (n = 565)	*Percent Severe (n = 288)*
Health care practitioner		
Family physician	30.5	66.7
Orthopaedic surgeon	8.8	32.3
Neurosurgeon	2.7	9.5
Osteopath	7.0	23.8
Chiropractor	12.7	27.5
Physiotherapist	3.8	16.1
Other	5.0	12.1
Treatment		
Bed rest	35.1	72.8
Muscle relaxant	17.1	52.6
Prescription pain medication	21.1	58.0
Physiotherapy	9.5	23.9
Back support	11.3	37.4
Other non-surgical treatment	12.6	27.4
Lumbar spine surgery	2.0	10.5

Source: Frymoyer *et al.*, 1983.

Kelsey (1975) sampled 20–64-year-olds residing in the New Haven (Connecticut) area who had lumbar x-rays taken over a two-year period for suspected herniated nucleus pulposus. She divided the sample into those with surgically confirmed herniated discs and those who had probable or possible herniated discs based on clinical signs and symptoms. She was able to define a variety of risk factors that related to the diagnosis of herniated lumbar disc as discussed further in Chapter 7.

In 1973, Nagi, Riley and Newby determined the prevalence rates of persistent back pain of persons between 18 and 64 years, residing in Columbus, Ohio. A random sample of 1,135 subjects was studied, of whom 203 (18 percent) reported "often being bothered with pain in the back." Of those with back problems, 62 percent had had a spine radiograph; 26 percent had worn a back support; and four percent had had back operations. The four percent who had back operations corresponds to the observation of Frymoyer *et al.* (1980) that three percent of their population had had back operations. Based on these statistics and other calculations, it is estimated that 113,000 disc operations and 34,000 other lumbar spinal surgical procedures are performed annually.

Scandinavia

Hirsch, Jonsson and Lewin (1969) interviewed 692 women (15 to 72 years of age) from three census districts in Goteborg, Sweden, selected at random to represent the adult Swedish female population. The lifetime incidence of LBP was found in 48.8 percent of all women and increased with increasing age up to 55 years, after which no further increase was noted.

Svensson and Andersson (Svensson 1981, 1982; Svensson and Andersson 1982, 1983; Svensson *et al.* 1983; Andersson *et al.* 1983) studied a randomized sample of 940 40 to 47 year old men in Goteborg, Sweden. Seven hundred and sixteen men were personally interviewed, and information about the remaining 234 was obtained from the Swedish National Health Insurance Office. Over their entire working lifetime 96 percent of the men had been off work at some time for some disease or injury, 74 percent because of disease or injury to the locomotor system. Thirty-three percent of sickness absence episodes were spine-related, but these constituted 47 percent of all sickness absence days. Total disability existed in 3.6 percent of the participants, and 4 percent had been off work more than three months because of LBP in the three years preceding the study. Forty percent had sciatica. Forty percent had consulted a physician, 3.5 percent had been admitted to a hospital, and 0.8 percent had been operated on because of their LBP.

The same study design was later used to survey 1,640 women in Goteborg, Sweden (Svensson and Andersson 1982). Nineteen percent had been off work because of LBP in the preceding three-year period, 3.5 percent for three months or longer.

Horal (1969) and Westrin (1970, 1973) studied a random sample of subjects who in 1964 had been sick listed for LBP by physicians in Goteborg, Sweden. They were compared to a control group, matched with respect to sex, age and sickness benefit, but not previously sick listed for LBP. Of the total group, 212 pairs of probands and controls were studied by Horal, and shortly thereafter 214 by Westrin (78 percent of the base material). Table 4.4 shows that 95 percent of the probands had had LBP in the preceding 3 to 4 years and that 52 percent had ongoing pain at the time of the interview. In the control group, the corresponding figures were 49 percent and 27 percent respectively. This means that once LBP is experienced, it has a greater probability of recurrence.

Biering-Sorensen (1982) sampled 82 percent of all 30 to 60-year-old inhabitants in Glostrup, Denmark. There were 449 men and 479 women. An extensive questionnaire of low back problems was part of the study as well as objective measurements of spine function. Twelve months after the examination 99 percent of the study population completed a follow-up questionnaire on LBP occurring in the intervening period. The lifetime prevalence, one-year incidence of LBP appears in Figure 4.1 along with the one-year period and point prevalence data. In general, increasing age was

TABLE 4.4
LBP as Subjects Sick listed for LBP (Probands) and in (Controls)

	Probands percent	*Controls percent*
LBP previous 3–4 yrs.	95	49
Thereof:		
Duration more than 1 wk.	83	21
Medication	73	6
Physiotherapy	47	3
Brace	18	1
LBP at examination	53	27

Source: Westrin 1973.

associated with increasing episodes of LBP. Work absence at some time was reported by 22.5 percent of those who had LBP, 10 percent had needed some job adjustment, and 63 percent had changed their jobs because of this symptom. Of those who had LBP at some time, 60 percent had consulted a physician, 25 percent a specialist, and 15 percent a chiropractor (Biering-Sorensen 1983b). About 30 percent had had radiographs taken of the lumbar spine, 4.5 percent had been admitted to a hospital and one percent had been operated on because of LBP.

Israel

Magora and Taustein (1969) performed geoethnic, psychosocial, economic and occupational investigations on 3,316 individuals taken at random. Present and past LBP was determined. Four hundred and twenty-nine (12.9 percent) were found to suffer from LBP at the time of the survey (point prevalence), and 92 percent (394) of those had their pain on and off from six months to 11 years or more before the investigation. The majority of the subjects with LBP did not take sick-leave (57.8 percent) and, of those who did, 29.4 percent had absence periods from 1 to 10 days.

The Netherlands

Valkenburg and Haanen (1982) reported on a study of 3,091 men and 3,493 women 20 years of age and older performed in 1975 through 1978 in the Dutch city of Zoetermeer. A questionnaire, physical examination, and radiographs were obtained. The prevalence of LBP in men and women increased slightly with age up to 65 years, and thereafter decreased (Table 4.5). Disc prolapse, defined by clinical signs and symptoms, was found in 1.9

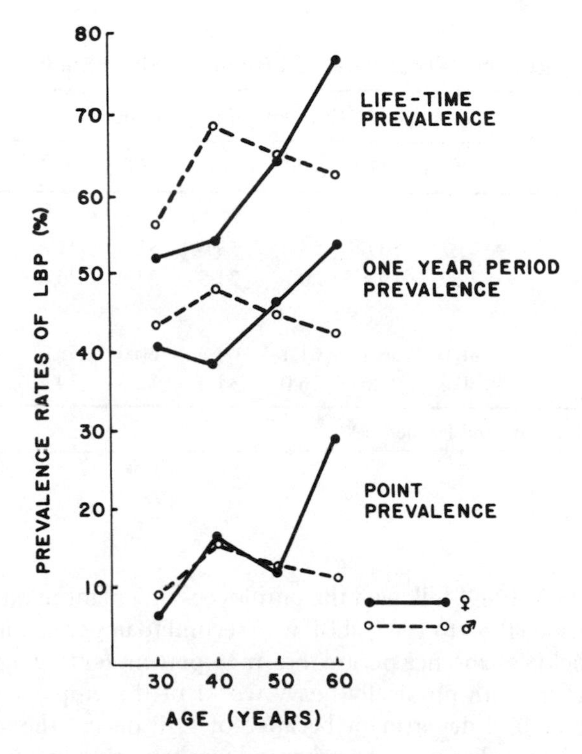

Figure 4.1
The lifetime prevalence and incidence of low back pain with one year period and point prevalence data (Biering-Sorenson *et al.*, 1982).

percent of men and 2.2 percent of women. Considerable disability was attributed to LBP. Eighty-five percent had recurrences; 30 percent had LBP for more than three months, and 30 percent had been bedridden by their symptoms. Nearly half of the men and one-third of the women reported that they had been unfit to work because of LBP at some time, and eight percent of the men as well as four percent of the women had changed their jobs because of LBP. Twenty-eight percent of the men and 42 percent of the women had consulted a physician for LBP.

THE INDUSTRIAL PROBLEM

There are numerous reports focussing particularly on the prevalence of low back pain in the workplace. Much of these data are difficult to evaluate because the work environment is highly selective (for review see Anderson 1980, Andersson 1981, Snook 1982). Again we will present these data by country.

TABLE 4.5
Age and Sex Specific Prevalence Rates of LBP in the EPOZ Study

	Frequency in Age Groups (0/0%)							
Prevalence	*20*	*25*	*35*	*45*	*55*	*65*	*75*	*All*
Men								
Life-time	51.7	50.6	53.8	53.0	53.8	41.8	32.6	51.3
Point	22.2	19.5	20.7	23.5	23.0	26.6	17.0	15.2
Women								
Life-Time	46.0	56.1	61.1	64.9	60.0	52.7	46.4	57.8
Point	30.2	23.6	26.0	31.4	32.6	34.4	33.4	28.4

Source: Valkenburg and Haanen, 1982.

United States

Rowe (1963, 1969) followed the employees at a plant in New York over a ten-year period (1956 to 1965). LBP was second to upper respiratory illness in terms of sickness absence period. From 35 percent (sitting workers) to 47 percent (workers with physically heavy work) of the employees had made visits to the medical department because of LBP during the study period. The yearly sickness absence per worker (all workers included) was estimated to be four hours. Recurrences were frequent, occurring in about 85 percent.

Snook (1982) has compiled data from the Liberty Mutual Insurance Company indicating annual rates of LBP for individual workers ranging from less than one percent to over 15 percent. Kelsey and White (1980) estimates that about two percent of all employees in the United States have a compensable back injury each year.

United Kingdom

In a British survey of 2,685 male postal workers, Anderson (1971) found that 23 percent had LBP. Seven percent of the total population were referred to the hospital for their complaint, and two percent were admitted because of LBP. The annual absenteeism from work for LBP is 70 weeks/100 men employed (Anderson 1980) which is a very high figure.

Afacan (1982) surveyed the medical records of twelve collieries of the National Coal Board. The study group comprised 12,125 men, 9,414 of whom were underground workers. Over one year (February 1976 through January 1977) 14.8 percent of the total work force were absent from work because of back injuries. The number of new sickness absences was 19.1 for every 100 men employed.

Scandinavia

The prevalence of back pain in different Scandinavian industries has been reported by several investigators. Hult (1954) found that 60 percent of a sample of Swedish males with different jobs (1,193 persons) had had back pain at some time. Four percent had been off work more than six months because of these symptoms and 11 percent had disability which extended between three weeks and six months. In the Swedish building industry Ostlund (1975) found that 22 percent had lost work time because of LBP during the preceding year, and 33.5 percent had received sickness allowance because of back pain at some time. Corresponding figures from the forestry industry showed an incidence of 37.5 percent and a prevalence of 18 percent (Tufvesson 1973). A Danish study of LBP among males aged 40 to 59 report that 25 percent of all had LBP in the previous year, eight percent severe enough to warrant bed rest or absence from work (Gyntelberg 1974).

Other Countries

As reported previously in this chapter, Magora and Taustein (1969) found prevalence rates from 6.4 to 21.6 per 100 males and females in their survey of 3,316 Israeli workers in eight different jobs, while Ikata (1965) in a study of 1,110 Japanese workers in ten jobs reported on sciatica in between 5.2 percent and 22.4 percent.

SUMMARY

In spite of the different socioeconomic conditions in different countries, it is surprising how many similarities there are. There are several problems in the validity and reliability of LBP data. The main difficulty in using the national statistics is that they are based on reported information and may not truly reflect the problem. Further, classification systems vary, so that data retrieval becomes difficult, and erroneous diagnoses can be given. Additionally, there is a tendency to over-report severe LBP in comparison to milder forms. Social and personal factors enter the tendency to report sick. Svensson (1981) found that men who had been sick listed for LBP on average had 30 percent more sickness absence episodes, and 70 percent more sickness absence days than men who had not been sick listed because of LBP.

The reproducibility of autoamnestic information concerning LBP has recently been estimated by Biering-Sorensen (1983). In 6- and 12-month follow-up studies he analyzed a general population of 30-, 40-, 50- and 60-year-old men and women and a population of 20 to 68-year-old male hospital porters. At an interval of about six months, the question of ever having had LBP was answered Yes/No in a ratio of approximately 2:1 and 84 percent answered consistently on the two occasions. Affirmative or negative answers concerning previous lumbar spinal x-ray examination were contradicted at one-year's interval by 11 percent. After six months, two-fifths of the subjects reproduced their statement of age at onset of LBP within one year. The cumulative incidence curves of LBP estimated by age at onset varied systematically between the 30-year-olds and 60-year-olds, suggesting forgetful behavior. The annual risk of first time experience of LBP was seemingly about four-fold higher during the follow-up year than during the seven preceding years.

Westrin (1973) and Horal (1969) studied the same group of subjects but on different occasions. Most interviews took place on the same day. The concordance between the two interviewers regarding lifetime incidence was about 87 percent. Svensson and Andersson (1982) compared their interview data to insurance data. Twenty-seven percent of men who had stated that they had never had LBP in their lives had in fact been sick listed for LBP. This means that any data on frequency of LBP must be handled with caution.

REFERENCES

Afacan, A.S. 1982. Sickness absence due to back lesions in coal miners. *J. Soc. Occup. Med.* 32 (1):26–31.

Anderson, J.A. 1971. Rheumatism in industry: A review. *Br. J. Ind. Med.* 28:103–21.

Anderson, J.A.D. 1980. Back pain and occupation. In Jayson MIV, ed., *The Lumbar Spine and Back Pain.* 2nd ed. Pitman, London Medical, pp. 57–82.

Andersson G.B.J. 1981. Epidemiologic aspects on low back pain in industry. *Spine* 6 (1):53–60.

Andersson, G.; and Svensson, H.O. 1979. Prevalence of low-back pain. *Social Planerings—och Rational iseringsinstitut Rapport* 22:11–23 (In Swedish).

Andersson, G.B.J.; Svensson, H.O.; and Oden, A. 1983. The intensity of work recovery in low back pain. *Spine*. In press.

Benn, R.T.; and Wood, P.H.N. 1975. Pain in the back: An attempt to estimate the size of the problem. *Rheumatol. Rehabil.* 14 (3):121–28.

Biering-Sorensen, F. 1982. Low back trouble in a general population of 30-, 40-, 50-, and 60-year old men and women. Study design, representativeness and basic results. *Dan. Med. Bull.* 29 (6):289–99.

Biering-Sorensen, F. 1983a. A prospective study of low back pain in a general population. I. Occurrence, recurrence and aetiology. *Scand. J. Rehabil. Med.* 15 (2):71–79.

Biering-Sorensen, F. 1983b. A prospective study of low back pain in a general population. III. Medical Service-Work Consequence. *Scand. J. Rehabil. Med.* 15 (2):89–6.

Biering-Sorensen, F.; and Hilden, J. 1983. Reproducibility of the history of low back trouble. *Spine.* In press.

Frymoyer, J.W.; Pope, M.H.; Costanza, M.C.; Rosen J.C.; Goggin, J.E.; and Wilder, D.G. 1980. Epidemiologic studies of low-back pain. *Spine* 5 (5):419–23.

Frymoyer J.W.; Pope, M.H.; Clements, J.H.; Wilder, D.G.; McPherson, B.; and Ashikaga, T. 1983. Risk factors in low back pain. *J. Bone Joint Surg.* 65A (2): 213–18.

Gyntelberg, F. 1974. One year incidence of low back pain among male residents of Copenhagen aged 40–59. *Dan. Med. Bull.* 21 (1):30–36.

Haber, L.D. 1971. Disabling effects of chronic disease and impairment. *J. Chronic Dis.* 24 (7/8):469–87.

Harris, A.I. 1971. *Handicapped and impaired in Great Britain. Part 1. Social Survey Division*. London, Office of Population of Censuses and Surveys, Her Majesty's Stationary Office.

Helander, E. 1973. Back pain and work disability. *Socialmed. Tidskr.* 50:398–404 (In Swedish).

Hirsch, C.; Jonsson, B.; and Lewin, T. 1969. Low-back symptoms in a Swedish female population. *Clin. Orthop.* 63:171–76.

Horal, J. 1969. The clinical appearance of low back pain disorders in the city of Gothenburg, Sweden. Comparisons of incapacitated probands and matched controls. *Acta Orthop. Scand.* (Suppl.) 118:1–109.

Hult, L. 1954. Cervical, dorsal, and lumbar spinal syndromes. *Acta Orthop. Scand.* (Suppl) 17:1–102.

Ikata, T. 1965. Statistical and dynamic studies of lesions due to overloading on the spine. *Shikoku Acta. Med.* 40:262–86.

Kelsey, Jennifer L. 1975. An epidemiological study of acute herniated lumbar intervertebral discs. *Rheumatol. Rehabil.* 14 (3):144–59.

Kelsey, Jennifer L.; Pastides, Harris; and Bigbee, Gerald E., Jr. 1978. *Musculo-skeletal disorders: Their frequency of occurrence and their impact on the population of the United States*. New York: Prodist.

Kelsey, J.L.; and White, A.A., III. 1980. Epidemiology and impact on low back pain. *Spine* 5 (2):133–42.

Magora, A.; and Taustein, I. 1969. An investigation of the problem of sick-leave in the patient suffering from low back pain. *Industr. Med. Surg.* 38 (11):398–408.

Nagi, S.Z.; Riley, L.E.; and Newby, L.G. 1973. A social epidemiology of back pain in a general population. *J. Chron. Dis.* 26 (12):769–79.

Ostlund, E.W. 1975. *Personal communication.*

Rowe, M.L. 1963. Preliminary statistical study of low back pain. *J. Occup. Med.* 5 (7):336–41.

______. 1965. Disc surgery and chronic low back pain. *J. Occup. Med.* 7 (5): 196–202.

______. 1969. Low back pain in industry. A position paper. *J. Occup. Med.* 11 (4): 161–69.

Snook, S.H. 1982. Low back pain in industry. In *Symposium on idiopathic low back pain*, eds., A.A. White and S.L. Gordon, pp. 23–28, St. Louis: Mosby.

Svensson, Hans-Olof; and Andersson, Gunnar B.J. 1982. Low back pain in forty to forty-seven year old men. I. Frequency of occurrence and impact on medical services. *Scand. J. Rehabil. Med.* 14 (2):47–53.

Svensson, Hans-Olof.; and Andersson, G.B.J. 1983. Low back pain in forty to forty-seven year old men. Work history and work environment factors. *Spine*, 8 (3):272–76.

Taylor, P.J. 1968. Personal factors associated with sickness absence. A study of 194 men with contrasting sickness absence experience in a refinery population. *Brit. J. Indus. Med.* 25:106–18.

Tufvesson, B. 1973. Stockholm, Swedish Work Environment Fund. (Unpublished data).

Valkenburg, H.A.; and Haanen, H.C.M. 1982. The epidemiology of low back pain. In *Symposium on idiopathic low back pain*, eds., A.A. White and S.L. Gordon, pp. 9–22, St. Louis: Mosby.

Westrin, C.G. 1970. *Sicklisting because of low back pain. A nosologic and medical insurance investigation.* Goteborg. (doctoral dissertation, in Swedish).

———. 1973. Low back sicklisting. A nosological and medical insurance investigation. *Scand. J. Soc. Med.* (Suppl) 7:1–116.

Wood, P.H.N. 1976. Epidemiology of back pain. The lumbar spine and back pain. In *The lumbar spine and back pain*, ed., M. Jayson, pp. 13–17, London: Pitman.

Wood, P.H.N.; and Badley, E.M. 1980. Epidemiology of back pain. In *The lumbar spine and back pain*, ed., M. Jayson 2nd ed., pp. 29–55, London: Pitman.

5 Cost

Stover H. Snook and Roger C. Jensen

INTRODUCTION

It is difficult to determine accurately the cost of work-related low back pain (LBP) because of the many variables that are involved. For example, payments for LBP may come from many sources, such as workers' compensation insurance (including federal, state and private insurers), group and individual health insurance plans, and social security benefits. The amount of workers' compensation payments, if any, vary according to the state law, the personnel policies of the employer, and the inflationary cycles of the economy. The form of payment may be a lump sum, an annuity, or investments in rehabilitation or retraining. The requests for payments are influenced by labor negotiations, plant closings, and the state of the economy. Finally, there are the indirect costs associated with production losses and personal suffering.

Most of the cost information that does exist for LBP comes from the various systems of workers' compensation. Consequently, this chapter will concentrate on compensation costs. It should be recognized, however, that these costs are only part of the total cost. All workers are not covered by compensation insurance, and all back injuries do not qualify for compensation payments.

COMPENSATION COSTS

Total Compensation Costs in the United States

In a survey of nine states that publish detailed workers' compensation statistics, back injuries averaged 21 percent of all compensable work injuries, ranging from 15 percent in Kansas to 24 percent in California

(Antonakes 1981). This figure is similar to the 19.3 percent reported by Klein, Jensen and Sanderson (in press 1983), and the 22 percent reported by Leavitt, Johnston and Beyer more than 10 years earlier (1971). However, the cost of compensable back injuries in the same nine states averaged 33 percent of compensation and medical costs for all occupational injuries, ranging from 26 percent in New York to 42 percent in Kansas. This figure is supported by the National Council on Compensation Insurance (Antonakes 1981), which has also estimated the cost of back injuries to be 33 percent of all compensation costs. In 1978, the Social Security Administration reported that the total cost of occupational injuries was $8 billion in the United States. One estimate of the total compensation costs for back pain in the country may be obtained by taking 33 percent of $8 billion—or $2.7 billion.

Another estimate of total compensation costs can be derived from the experience of the Liberty Mutual Insurance Company. Liberty Mutual has been the largest underwriter of workers' compensation insurance in the United States for almost 50 years. The company reported that in 1980 they paid $217,441,000 for compensable back pain; almost one million dollars each working day (Antonakes 1981). Liberty Mutual represents about nine percent of the insured workers' compensation market. The entire industry represents about 52 percent of the market, with the balance consisting of 35 percent for state and federal funds and 13 percent for self-insurers. Assuming that other insurance carriers have had the same experience as Liberty Mutual, the total workers' compensation costs for back pain in the United States can be estimated at $4.6 billion.

Cost per Case of Back Pain

Table 5.1 summarizes some of the recent estimates that have been made for the cost per case of compensable back pain. The effects of inflation are quite evident in Table 5.1. The estimate by Leavitt, Johnston and Beyer (1971) is based upon 100 disability cases of back pain occurring in 1967. Snook's estimate is based upon 191 cases occurring in 1976 (Snook 1980).

TABLE 5.1
Cost per Case of Compensable Back Pain

Year	*Source*	*Mean Cost*	*Median Cost*
1967	Leavitt, Johnston & Beyer (1971)	$ 2911	$ 404
1976	Snook (1980)	4500	566
1978	Antonakes (1981)	6600	–
1979	Klein, Jensen & Sanderson (1983)	5081	–
1979	National Council on Compensation Insurance (1983)	5500	–

Antonakes (1981) based his estimate on 1978 data from the Social Security Administration and the National Safety Council. The estimate attributed to Klein, Jensen and Sanderson (1983) is for all forms of back injuries based on reported workers' compensation claims closed during 1979 in five states. The estimate from the National Council on Compensation Insurance (1983) is based upon first, second and third reportings of Massachusetts cases beginning in 1979. It seems reasonable to conclude that the current (1983) mean cost of compensable back pain is about $6,000 per case.

Two of the references in Table 5.1 show median cost per case in addition to the mean cost. The large discrepancy between the mean and the median indicates that back pain costs are not normally distributed. A few high cost cases account for most of the cost. Snook (1980) found that 25 percent of the cases accounted for 90 percent of the cost.

Distribution of Back Pain Costs

Leavitt, Johnston and Beyer (1971) and Snook (1980), in Table 5.1, show median cost per case in addition to the mean cost. The median is the dollar value which 50 percent of the back pain cases were equal to or less than. Fifty percent of the back pain cases are equal to or less than the median cost. The large discrepancy between the mean and the median indicates that back pain costs are not normally distributed. A few high cost cases account for most of the expense. Snook (1980) found that 25 percent of the cases accounted for 90 percent of the cost. Leavitt, Johnston and Beyer (1972) report that 25 percent of their cases accounted for 87 percent of the cost. The high cost cases were characterized by greater degrees of hospitalization, surgery, litigation, and psychological impairment.

Leavitt, Johnston and Beyer (1971) reported that medical costs accounted for only one-third of the total costs; and disability payments accounted for the remaining two-thirds (see Table 5.2). It is interesting to note that hospital costs comprised one-third of the medical dollar in spite of the fact that only 30 percent of the injured were hospitalized. Physician fees also comprised one-third of the medical dollar. Two-thirds of the disability dollars was for permanent disability accounted, with one-third for temporary disability. Antonakes (1981) presents data to show that, as the duration of disability for back pain increases, the cost rises at an accelerating rate. Most of the added expense is for permanent partial and permanent total disability. Costs increase dramatically when surgery is performed, which in most states is accompanied by payment for permanent partial disability. For example, permanent partial payments as a result of laminectomies generally range from 15 to 25 percent of permanent total disability.

The results of an assessment of more recent workers' compensation costs are summarized in Tables 5.3 and 5.4 (Klein *et al.*, in press). The tables are based on cost data for cases closed in 1979 in five states: Arkansas, Delaware, North Carolina, New York and Virginia. The payments for

TABLE 5.2
Percentage of Costs by Type of Treatment and Compensation

Back Pain Costs	%	%
Medical Costs		33
Physicians' Fees	11	
Hospital Costs (not Including Drugs or Physical Therapy)	11	
Diagnostic Tests	4	
Physical Therapy	3	
Drugs	2	
Appliances	2	
Disability		67
Temporary	22	
Permanent	45	
Total Costs		100

Source: Adapted from Leavitt, Johnston & Beyer 1971.

medical care for injuries involving the back are shown in Table 5.3 according to the nature of the injury. It should be noted that these figures do not include the cost of medical care provided by the employer's in-house medical personnel.

Table 5.4 lists the payments to partially indemnify workers who lost wages because of a work-related back injury. Those injuries reported to be dislocated backs obviously involved the most prolonged period of disability, having an average wage indemnification cost of close to $20,000 per case.

TABLE 5.3
Mean Medical Costs for Workers' Compensation Claims Closed During 1979 in Five States for Back Injuries

Nature of Injury	*Number of Reported Cases with Cost Data*	*Mean Cost per Case*
Inflamed Joint	200	$ 4,689
Dislocation	762	3,533
Fracture	344	1,888
Strain/Sprain	11,740	470
Laceration	24	425
Contusion	578	303
TOTAL	13,648	$ 731

Source: Klein, Jensen and Sanderson 1983.

TABLE 5.4
Mean Indemnity Compensation Costs for Workers' Compensation Claims Closed During 1979 in Five States for Back Injuries

Nature of Injury	*Number of Reported Cases with Cost Data*	*Mean Cost per Case*
Dislocation	2,906	19,536
Inflamed Joint	235	7,120
Fracture	848	6,710
Nerve Involvement	132	5,045
Strain/Sprain	33,794	3,063
Laceration	40	2,712
Contusion	1,101	1,439
Burn/Scald	23	891
TOTAL	39,079	4,351

Source: From Klein, Jensen & Sanderson 1983.

OTHER COSTS

According to the National Safety Council (1974), workers' compensation costs for medical care and wage indemnification only represent a portion of the total costs of occupational injuries. Some of the less obvious costs associated with occupational LBP include:

1. Medical treatment and rehabilitation provided at a plant dispensary.
2. Wages paid to other workers during the time their work was interrupted because of the injury of their co-worker.
3. Wages paid to the injured worker between the time of injury and the time when workers' compensation payments began (the time is usually called the "waiting period").
4. Wages paid to the supervisor for time spent assisting the injured, investigating the incident, preparing a report, and training a replacement employee.
5. Cost for paying full wages to a replacement worker during the learning period when work output is less than it would have been if the experienced worker had not been injured.
6. Wages paid to clerks and others to prepare and process compensation application forms.

The above costs are often difficult to quantify. One British report estimates that back pain costs the community more than $300 million a year in lost productivity: the equivalent of the output of a British town of 120,000 people (Department of Health and Social Security 1979). The community also loses in lost tax revenue and community service. For example, litigation of compensation claims contributes to the overloading of the legal system.

In a study of 1,221 males Frymoyer *et al.* (1983) were able to extrapolate that the total annual cost for work loss in this group was 11 billion dollars. 1.7 percent (the chronic group) accounted for $8 billion (75 percent) of the total. The most pertinent finding in Leavitt's study was that high cost cases were associated with "lag times" (Leavitt *et al.* 1971, Part 2). Time was lost by delays from first examination to first referral, date of injury to last day worked, delays in hospitalization, and delays in the performance of surgery. If surgery were mentioned but never performed, the patients tended to fall into a high cost group. The high cost group also had an average of 5.2 referrals compared to an average of 1.6 referrals for the other lower cost group. It is important to emphasize that the actual costs of any physician's consultation is negligible compared to the costs attendant to delays with continued work loss. As the duration of disability increased, there was a progressively lesser chance for successful restoration of function. McGill (1968) found that patients whose symptoms had continued and were disabled for greater than six months had a 50 percent chance of successful rehabilitation; at one year this figure was reduced to 20 percent, and at two years the chances for successful rehabilitation were virtually nil.

SUMMARY

Payments to those suffering from LBP come from many sources. This complicates our determination of the total cost. This chapter has concentrated on compensation costs, which are only one small part of the total societal cost of LBP. Back injuries average 21 percent of all compensable work injuries but average 33 percent of the cost. The total compensation costs for LBP are estimated to be $4.6 billion and the cost per case $6,000. However, 25 percent of the cases account for 90 percent of the cost. As the duration of disability increases, the total costs accelerate. Medical costs account for 33 percent of the total cost and disability payments the remainder.

REFERENCES

Antonakes, J.A. 1981. Claims costs of back pain. *Best's Review*. September.

Great Britain Department of Health and Social Security. 1979. *Working group on back pain*. London: Her Majesty's Stationery Offices.

Klein, B.P.; Jensen, R.C.; and Sanderson, L.M. 1983. *Assessment of worker's compensation claims for back strains/sprains.* In press.

Leavitt, S.S.; Johnston, T.L.; and Beyer, R.D. 1971. The process of recovery: Patterns in industrial back injury. Part I: Costs and other quantitative measures of effort. *Industrial Medicine and Surgery* 40 (8):7–14.

Leavitt, S.S.; Johnston, T.L.; and Beyer, R.D. 1971. The process of recovery: Patterns in industrial back injury. Part 2. Predicting outcomes from early case data. *Industrial Medicine* 40 (9):7–15.

Leavitt, S.S.; Johnston, T.L.; and Beyer, R.D. 1972. The process of recovery: Patterns in industrial back injury. Part 4: Mapping the health care process. *Industrial Medicine and Surgery*, 41 (2):5–9.

McGill, C.M. 1968. Industrial back problems. A control program. *J. Occ. Med.* 10 (4):174–78.

National Council on Compensation Insurance. 1983. Detailed claim information, lower back injuries, State of Massachusetts, breakdown by class code. Unpublished data. *National Council on Compensation Insurance,* New York.

National Safety Council. 1974. *Accident prevention manual for industrial operations.* 7th Edition, pp. 162–169, National Safety Council: Chicago.

Snook, S.H. 1980. Unpublished data. Liberty Mutual Insurance Company, Hopkinton, MA.

PART III
Etiology

6 The Workplace

Malcolm H. Pope, Gunnar B. J. Andersson, and Don B. Chaffin

INTRODUCTION

We have learned in Chapters 4 and 5 of the magnitude of low back pain (LBP) in terms of suffering, work loss, and cost. In this chapter we will consider the role of the workplace in the causality of LBP. This is extremely important because the greatest potential for LBP prevention exists in the workplace. This chapter is divided into a number of subsections which consider the general relationship of LBP to physical and psychological factors in the workplace, as well as to the work environment in general. Emphasis is placed on physical work factors and specifically on work posture, lifting, pulling, pushing, and cyclic loading.

PHYSICAL WORK FACTORS

A few words on present injury models are useful in the discussion of the importance of specific physical work factors. A musculoskeletal injury can be triggered by a direct trauma, a single overexertion, or repetitive loading. The strengths of the various tissues are influenced by such factors as age, fatigue, and concomitant diseases and thus the loading level at which an injury occurs can vary greatly. In the case of direct trauma, several different structures can be hurt at the same time. For example, a blow to the back can fracture vertebral bodies, and at the same time cause muscular, ligamentous and neurologic damage. In single overexertion injuries, such as from a single heavy lift, the injury usually occurs at one site only. For example, a fissure can occur in an intervertebral disc or a muscle can rupture. Repetitive loading can cause fatigue failure. Again, failure will usually occur at one location. It is important to remember here that tissue will heal if given sufficient time. For example, a fatigue failure crack may not propagate to fracture if rest periods allow healing to occur. Temporal factors, as well as healing properties, will therefore be critical to such failure. It should also be remembered that a fatigue failure injury may not become obvious until complete failure occurs and that the final event may be trivial and occur outside the workplace.

The third injury mechanism is continued static loading of tissues. Interference with the circulation of blood within a muscle, and thereby also with oxygen supply and removal of waste products, occur already at contraction levels as low as 10 percent of maximum contraction. Strong contractions rapidly become fatiguing and painful and secondary changes can occur in the muscle. Tendons are also similarly influenced by sustained tension and may necrotize, in part setting the stage for a total failure at much lower load levels than otherwise.

Heavy exertions during work are common in the United States and elsewhere. The National Institutes for Occupational Safety and Health (NIOSH, 1981) have stated that approximately one-third of the U.S. workforce is currently required to exert significant strength as part of their jobs. In the same report the following statistics were presented:

1. Overexertion was claimed as the cause of LBP by over 60 percent of LBP patients.
2. Overexertion injuries of all types in the United States occur to about 500,000 workers per year (which is in about 1 of 20 workers each year).
3. Overexertion injuries account for about one-fourth of all reported occupational injuries in the United States, with some industries reporting that over half of the total reported injuries are caused by overexertion.
4. Approximately two-thirds of overexertion injury claims involved lifting loads, and about 20 percent involved pushing or pulling loads.
5. Less than one-third of the patients with LBP from overexertion and with significant time loss from work eventually returned to their previous job.

In a British survey of 2,685 men, Anderson (1971) found that 30 percent had backache, 76 percent of whom (23 percent of all men) had pain in the lumbar region. Twenty-two percent of the low-back pain sufferers were referred to hospital, and 6 percent were admitted for treatment (i.e., 7 percent and 2 percent of the total population respectively) (Anderson 1971). The annual rate of absenteeism from work in Great Britain is calculated to be 70 weeks per 100 men employees. Magora and Taustein (1969) found prevalence rates of 6.4 and 21.6 per 100 men and women in a survey of 3,316 Israeli workers in eight different jobs, while Ikata (1965), in a study of 1,110 Japanese workers in ten jobs, reported sciatica in 5.2 to 22.4 percent.

Despite these impressive statistics, it is often difficult to relate the workplace to the complaint of LBP in a specific worker. Rowe (1973) found only 20 percent of industrial LBP sufferers had recognizable trauma at the onset of symptoms. Brown (1973, 1975) and Magora (1974) have indicated that specific lifting or bending episodes account for approximately one-third of the work-related cases. Brown (1975) found that another third of the cases are attributable to some other specific occupational event, and one-third have no remembered incident or work demand. Magora (1974) showed that the most expensive injuries for industry are those for which there is a traumatic event associated with the onset. These events are generally identified as suddenly-applied loads, slips, and falls. In a recent study of 8,000 people, Frymoyer *et al.* (1980, 1983) showed that the workplace environment was related to LBP, although other factors such as personal life style, psychological stress and driving were also implicated. In a separate case control study of this population, Damkot *et al.* (1982) identified differences between those with LBP and those without in specific lifting postures, requirements for pushing and pulling objects, and repetitive load bearing requirements.

Brown (1973), Chaffin (1974) Hult (1954), Magora (1973, 1974), Rowe (1973), Snook (1978a) and Andersson (1981) have all discussed the relationship between LBP and physically demanding work. Magora (1974) showed that LBP sufferers are found in sedentary as well as physically demanding occupations.

LIFTING

From an epidemiological perspective, the NIOSH *Guide* (1981) cites studies revealing that musculoskeletal injury rates (i.e., number of injuries per man-hours on job) and severity rates (i.e., number of hours lost because of injury per man-hours on job) increases significantly when:

1. Heavy objects are lifted.
2. The object is bulky.
3. The object is lifted from the floor.
4. Objects are frequently lifted.

Snook *et al.* (1978a, 1978b) and Chaffin *et al.* (1973, 1974) have separately identified means of determining the lifting capacity of a worker and developed guidelines. These are discussed in Chapter 10 to 14 (Snook *et al.* 1978a, 1978b; Chaffin *et al.* 1973, 1974). Magora (1974) and Andersson, Ortengren and Schultz (1980) found that when a worker exceeds that capacity, symptomatic LBP is more likely to occur. These data also suggest the specific method of lifting may be relevant to low back complaints.

Bergquist-Ullman and Larson (1977) found a strong relationship between LBP caused by lifting and the duration of sickness absence.

Chaffin and Park (1973) found that over a one-year period LBP was three times greater in workers who had strengths less than those required by their jobs. This was validated in a later study (Chaffin *et al.* 1977). Snook, Campanelli and Hart (1978b) concluded that the proper design of lifting tasks could reduce up to one-third the incidence of LBP; however, simply training workers in good lifting technique was ineffective. The "good" lifting technique is usually defined as an erect back and squat lift posture. Damkot *et al.* (1982) found significant differences between pain groups (severe, moderate or no LBP) in regard to the way they lifted. The severe LBP group tended to lift with much more bending of the legs whereas those without LBP tended to use various lifting techniques, depending upon the specific work situation. Lifting instructions had been given to 70 percent of the no pain group, 82.6 percent of the moderate pain group, and 92.6 percent of the high pain group. Perhaps the clinical lifting instructions were faulty. Both theoretical and laboratory research suggest this may be true. In a theoretical analysis Roozbazar (1980) found that the bent knee lifting method produced less mechanical stress on the spine, but only if intra-abdominal pressure (IAP) was included. In many postures IAP does not have a major role in spinal support (Gilbertson, Krag, and Pope 1983). Figure 6.1 demonstrates that the bent leg lift may increase the moment and the disc load. Extensive investigations, particularly by the group in Gothenburg, (Nachemson and Elfstrom 1970; Andersson 1974) have demonstrated the relationship between various lifting postures and increased or decreased intradiscal pressures. The moment (the product of the weight and the distance from the spinal axis) rather than the lifting method has been found to be most important in affecting these pressures.

Long-term physiologic changes accompanied heavy lifting. Hult (1954) showed that long-term heavy lifting was related to osteophyte formation. One proposed mechanism for osteophytic formation relates to the annular bulging that occurs in lifting or bending. Fibers of the annulus attaching at the disc margins are placed under tension, stimulating new bone formation at that site (Macnab 1971). These osteophytes actually increase the cross-sectional area of vertebrae and may reduce the disc stresses. In that light, the osteophyte may be a physiologic adaptation to the requirement for lifting (Macnab 1971).

Several investigations indicate an increase in sickness absence because of LBP as well as an increase in low back symptoms in jobs generally considered to be physically heavy. Some of those papers are summarized by Andersson (1981) and in Chapters 2 and 12 of this volume. A few will be discussed here. In the previously mentioned study by Hult (1954) the frequency of LBP was 64.4 percent in subjects with physically heavy work, 52.7 percent in other types of work. Severe back pain was present in 6.8

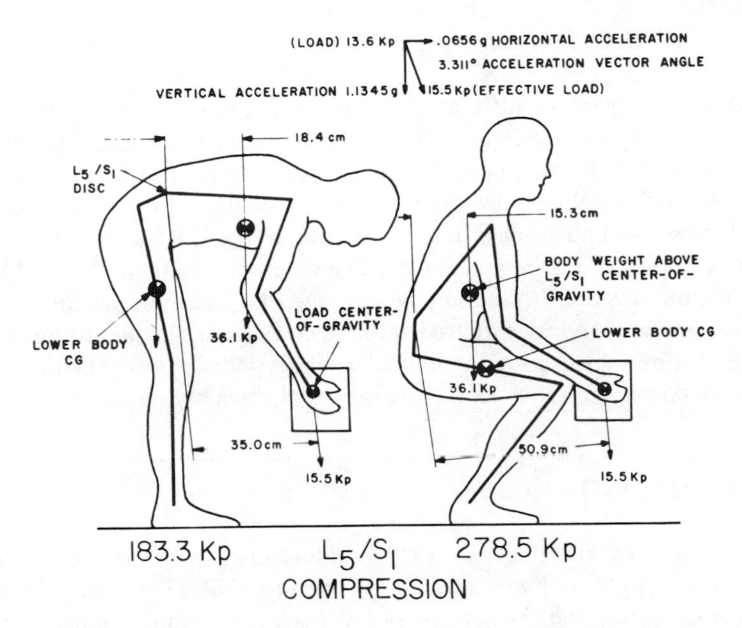

Figure 6.1
Mathematical model showing that the bent knee lifting method may produce a greater moment and disc load than lifting with the back.

percent of those with light physical work and 10.6 percent with heavy physical work. Differences were more pronounced when work absence was considered; 43.5 percent (heavy) compared to 25.5 percent (light) had been off work because of back pain. Lawrence (1955) studied 362 workers in four different jobs. LBP was found in 41 percent of those working in physically heavy jobs, 38.68 percent in miners and 29 percent in subjects with light physical work. In his study of sciatica, Ikata (1965) reported a point of prevalence of 22.4 percent in heavy jobs, 5.2 percent in light, while Magora (1970) found LBP in 21.6 percent of subjects with heavy industrial work and 10.4 percent in bank employees.

Snook (1982) found that a worker was three times more susceptible to compensable low back injury if exposed to excessive manual handling tasks. Unskilled laborers had the highest prevalence rate for disc prolapse and lumbago in the Dutch study by Valkenburg and Haanen (1982). Svensson and Andersson (1983) found heavy physical work to be strongly associated with the occurrence of LBP, and the highest prevalence of LBP in their cross-sectional study was in men with professions involving physically heavy work. In addition to the above mentioned studies, there are other investigations in which physically heavy work is reported not to be associated with back pain.

PULLING AND PUSHING

There is relatively little information regarding pushing and pulling activities and their role in work-related LBP. In one recent study (Damkot *et al.* 1982), a measure of pushing exposure was derived by multiplying the weight of pushed objects by the number of pushing efforts required each day. There were significant differences between the LBP group and controls in the frequency of pushing activities. The controls averaged 326 weightday units, and the severe LBP respondents averaged 1,612 weightday units. There was also a tendency for increased severity of LBP for those with increased pulling requirements. In a theoretical analysis White and Panjabi (1978) have shown that high disc loads accompany these activities.

CYCLIC LOADING

Many jobs expose the worker to small but repetitive loadings. There are a variety of epidemiologic studies confirming the relationship between vehicular exposure and the severity of low back complaints. Typically, these stresses involve vibration. There are many ways vibration may be applied to the spine. This occurs most commonly in vehicles. Kelsey (1975) found that truck drivers were four times more likely than others to have had a disc herniation. In commuters traveling more than 20 miles/day, she found two to four times as many low back complaints and twice the incidence of herniated nucleus pulposus when compared with a less frequently driving population. Also, those who drive tractors or trucks had two to four times as many low back complaints as those who do not. Pope, Wilder and Frymoyer (1980) analyzed vibration and found that many vehicles vibrate at a fundamental frequency similar to the body's natural frequency. The vibration levels of many vehicles are shown in Figure 6.2. The ISO standards for vibration are reproduced in Figure 6.3. It can be seen that many vehicles exceed those standards. Studies of the effects of vibration on seated individuals demonstrate a mechanism by which LBP may occur. If one measures the transmission of vibration through the body, certain frequencies will be accompanied by enhanced transmission and greater energy absorption. These particular frequencies are a function of the material properties of the spine and its supportive structures. This phenomenon of enhanced transmission and energy absorption is termed "resonance" and represents a frequency at which there are potential destructive forces.

Figure 6.4 demonstrates measurements of transmission through the spinal column of individuals subjected to vibration in the seated posture. Note the enhanced transmission that occurs at 4 to 6 Hz and, 9 to 11 Hz. Direct measurements of many vehicles reveal the dominant frequency of those vehicles to be at the 4 to 6 Hz range.

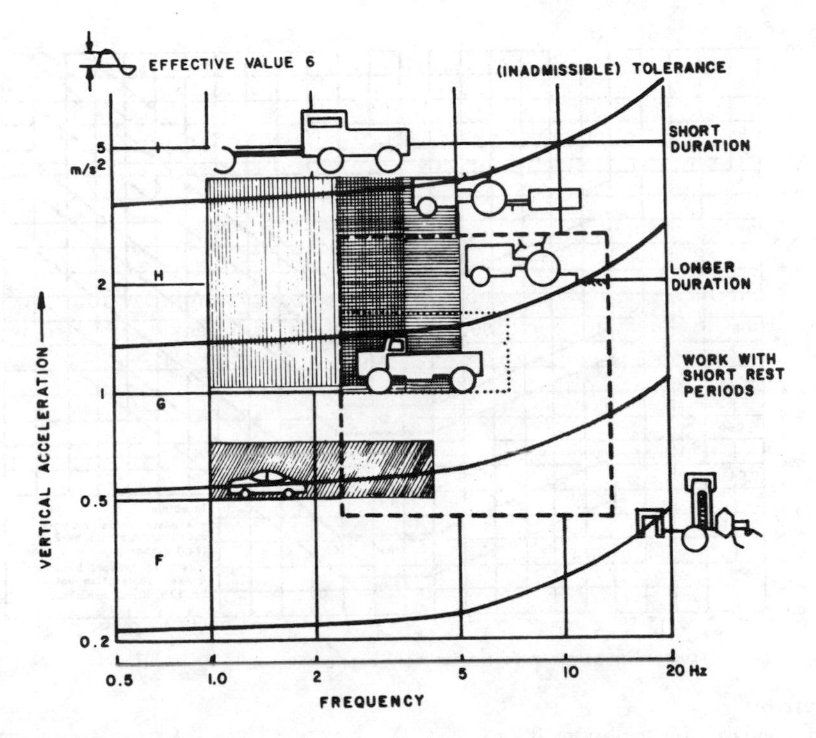

Figure 6.2
Vibration levels of different vehicles. (VDI–Richtlinie 2057; Beurteilung der Einwirkung Mechanisher Schwingongen auf den Menscher. okt., 1963)

The implication of these observations is: Vehicle operators are being exposed to vibration at the resonating frequencies of their spines. It remains to be firmly established that this relationship can be responsible for spinal degeneration; however, it is a convenient explanations for the greater prevalence of degenerative lesions in industrialized societies.

There is also evidence that vibration affects other spinal structures. Exposure to vibration leads also to fatigue of the paraspinal muscles as measured by frequency shifts of the EMG. These observations suggest that the muscles become more vulnerable and less capable of load bearing after vibration. Damkot *et al.* (1982) stated that it is probable that exposure to vehicular vibration, lifting, pulling and pushing, and vehicle driving combine as important risk factors for LBP. Thus truck drivers may have a reduced capacity for material handling (Mital *et al.* 1971). Pope, Wilder and Frymoyer (1980) showed that cyclic loading can result in ligament fatigue and disc herniations (see Chapter 3). Workers using hand-held vibrating machinery (particularly chainsaws) are observed to have accelerated osteoporosis, and carpal and phalangeal degenerative lesions.

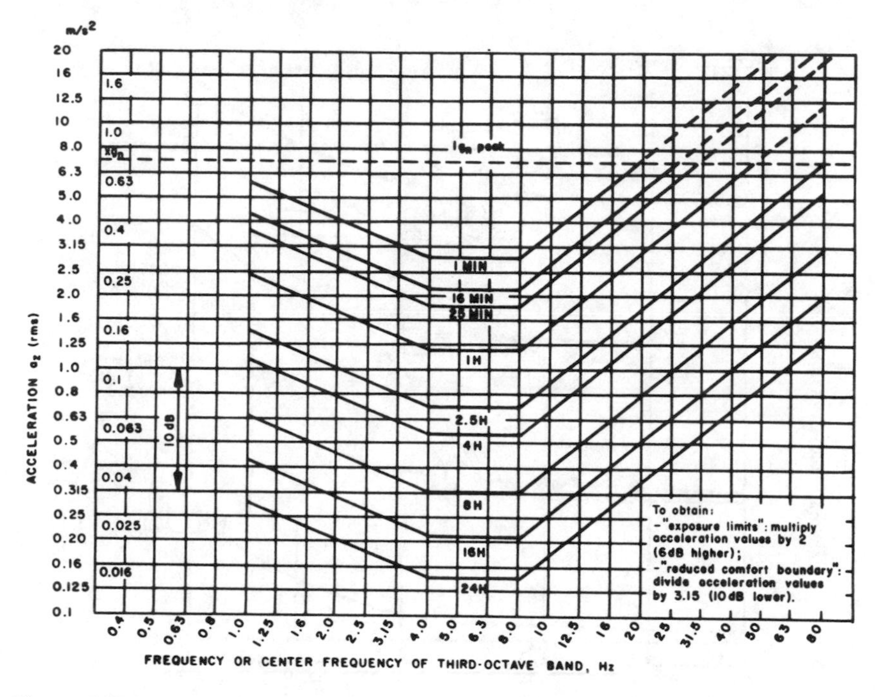

Figure 6.3
ISO standards for vibration. (ISO ref. no. 2631-1978(E), Second Edition 1978-01-15.)

POSTURE

Posture is defined as the overall positioning of the body and may be influenced by sitting, reaching, or the way in which we stand. There are several studies indicating an increased risk of low back pain in subjects who perform work in a predominantly sitting posture (Hult 1954 a, b; Lawrence 1955; Kroemer and Robinette 1969; Magora 1972). These studies also show an increase in back symptomology in subjects with pre-existing LBP when required to sit for prolonged periods. Kelsey (1975) found that men who spend more than half their workday in a car have a threefold increased risk of disc herniation. Whether this is caused by the sitting postures or vibration is difficult to establish in these studies. Other studies did not find seating to be a risk factor for LBP (Westrin 1973; Bergquist-Ullman and Larsson 1977; Svensson and Andersson 1983). This may be true because sitting work is often physically lighter. Regardless, it is important to change one's work posture. Postural fatigue and sickness absence decrease when such changes are required (Griffing 1960; Kroemer and Robinette 1969; Magora 1972).

Nachemson and Elfstrom (1970) and Andersson (1974) have extensively evaluated seating postures by electromyographic and intradiscal pressure methods. Their work may explain why both sedentary and other

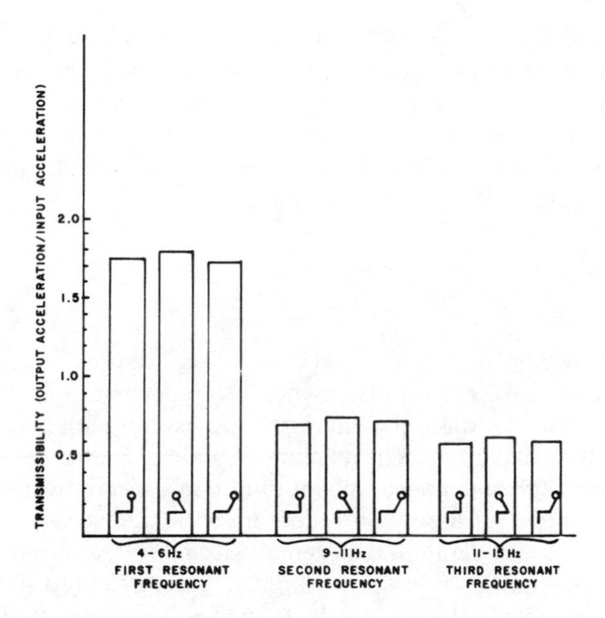

Figure 6.4
Measurements of transmission through the spinal column of seated individuals. Transmissibility is a measure of the enhancement of the vibrational input. In this case, the vertical acceleration of the head (output acceleration) is divided by the vertical acceleration of the seat (input acceleration) to obtain the transmissibility in three seated postures.

workers get LBP. The disc pressure is similar in the worker bending forward at 20 and the worker sitting unsupported. In any posture, lifting raises the disc pressure and muscle forces. Their work has identified optimal seating postures which will be discussed in Chapter 9.

Other postural factors are reported, by Corlett and Bishop (1976), to be related to muscular fatigue and LBP in the workplace. Damkot *et al.* (1982) found significant differences in the incidence of LBP as a function of stretching versus reaching demands. Forty percent of the group without LBP and 36 percent of the moderate LBP group were required to stretch and reach, whereas 59 percent of the severe LBP group had these job requirements. Furthermore, those with severe LBP were more likely to reach with their arms fully extended than were the other two groups.

PSYCHOLOGICAL WORK FACTORS

Of different psychological work factors, monotony at work and work dissatisfaction have been found to increase the risk of LBP. Monotony was primarily related to LBP in the study by Svensson and Andersson (1983), and

Bergquist-Ullman and Larsson (1979) found that workers with monotonous jobs requiring less concentration had a longer sickness absence following LBP than the others. Diminished work satisfaction has been found by several to be related to an increased risk of LBP (Westrin 1970, Magora 1973). According to Taylor (1968), sickness absence is increased in subjects with work dissatisfaction, regardless of diagnosis.

SUMMARY

Injuries leading to LBP can occur by direct trauma, overexertion or repetitive trauma. Overexertion is claimed by 60 percent of LBP patients as the cause of injury. Of these patients with overexertion injuries, 66 percent implicated lifting and 20 percent pushing or pulling. It is, however, difficult to relate the workplace to the complaint of LBP in a specific worker, and LBP is found quite often in those with sedentary occupations.

The incidence, severity and potential disability are all related to the demands on the individual in the workplace. Among the factors implicated are the requirements for lifting (particularly when compared to the worker's lifting capacity), pushing and pulling, posture, and cyclic loading. Injury and severity rates increase when the load-moment increases and when the loads are lifted frequently. Decreasing the moment is more important than adopting a specific lifting style. Drivers of heavy vehicles have two to four times the average incidence of serious LBP. This is probably due to the cyclic loading environment. Many vehicles excite the patients' natural frequency, leading to muscle fatigue and tissue strain. Identifying these risk factors is the first step in developing the prevention strategies discussed in Chapter 14.

REFERENCES

Anderson, J.A.D. 1971. Rheumatism in industry. A review. *Br. J. Ind. Med.* 28(2):103–121.

Andersson, G.B.J. 1974. *On myoelectric back muscle activity and lumbar disc pressure in sitting postures.* Thesis, Univ. Goteborg.

Andersson, G.B.J.; Ortengren, R.; and Schultz, A. 1980. Analysis and measurement of the loads on the lumbar spine during work at a table. *J. Biomech.* 13(6):513–520.

Andersson, G.B.J. 1981. Epidemiologic aspects of low back pain in industry. *Spine.* 6(1):53–60.

Bergquist-Ullman, Marianne; and Larson, Ulf. 1977. Acute low back pain in industry. A controlled prospective study with special reference to therapy and confounding factors. *Acta Orthop Scand* (Suppl) 170:1–117.

Brown, John R. 1973. Lifting as an industrial hazard. *J. Am. Ind. Hyg. Assoc.* 34(7):292–97.

Brown, John R. 1975. Factors contributing to the development of low-back pain in industrial workers. *J. Am. Ind. Hyg. Assoc.* 36(1):26–31.

Chaffin, Don B.; and Park, K.Y.S. 1973. A longitudinal study of low- back pain as associated with occupational weight lifting factors. *J. Am. Ind. Hyg. Assoc.* 34(12):513–25.

Chaffin, Don B. 1974. Human strength capability and low back pain. *J. Occup. Med.* 16(4):248–54.

Chaffin, Don B.; Herrin, Gary D.; Keyserling, W.M.; and Garg, Arun. 1977. A method for evaluating the biomechanical stresses resulting from manual materials handling jobs. *J. Am. Ind. Hyg. Assoc.* 38(12):662–75.

Corlett EN, Bishop RP. 1976. A technique for assessing postural discomfort. *Ergonomics* 19(2):175–8.

Damkot, D.K.; Pope, M.H.; Lord, J.; and Frymoyer, J.W. 1982. The relationship between work history, work environment and low back pain in males. In *Proc. Int. Soc. Study Lumbar Spine.* Submitted to *Spine*, 1982.

Frymoyer, J.W.; Pope, M.H.; Costanza, M.C.; Rosen, J.C.; Goggin, J.E.; and Wilder, D.G. 1980. Epidemiologic studies of low-back pain. *Spine* 5(5):419-23.

Frymoyer, J.W.; Pope, M.H.; Clements, J.H.; Wilder, D.G.; MacPherson, B.; and Ashikaga, T. 1983. Risk factors in low-back pain: An epidemiologic study. *J. Bone Joint Surg.* 65A(2):213–18.

Garg, A.; and Ayoub, M.M. 1980. What criteria exist for determining how much load can be lifted safely? *Human Factors* 22 (4):475–86.

Gilbertson, L.G.; Krag, M.H.; and Pope, M.H. 1983. Investigation of the effect of intra-abdominal pressure on the load bearing of the spine. *Trans. Orthop. Research Soc.* 8:177.

Griffing, J.P. 1960. The occupational back. In *Mod. Occup. Med.* Philadelphia: Lea & Febinger. pp. 219-27.

Hult. L. 1954. Cervical, dorsal and lumbar spinal syndromes. *Acta Orthop. Scand.* (Suppl) 17:1–102.

Ikata, T. 1965. Statistical and dynamic studies of lesions due to overloading on the spine. *Shikoku Acta Med.* 40:262–86.

Kelsey, J.L. 1975. An epidemiological study of acute herniated lumbar intervertebral discs. *Rheum. Rehab.* 14(3):144–59.

Krag, M.; Wilder, D.G.; and Pope, M.H. 1983. Internal strain and nuclear movements of the intervertebral disc. Submitted to the International Society for the Study of the Lumbar Spine, 1983.

Kroemer, K.H.; and Robinette, J.C. 1969. Ergonomics in the design of office furniture. *Industr. Med. Surg.* 38:115–25.

Lawrence, J.S. 1955. Rheumatism in coal miners: Occupational factors. *Brit. J. Industr. Med.* 12:149–61.

Macnab, I. 1971. The traction spur. An indicator of segmental instability. *J. Bone Joint Surg.* 53A(4):663–70.

Magora, A. 1973. Investigation of the relation between low back pain and occupation: IV. Physical requirements: bending, rotation, reaching and sudden maximal effort. *Scand. J. Rehab. Med.* 5(4):186–90.

Magora, Alexander. 1974. Investigation of the relation between low back pain and occupation: VI. Medical history and symptoms. *Scand. J. Rehab. Med.* 6:81–88.

Magora, A.; and Taustein, J. 1969. An investigation of the problem of sick leave in the patient suffering from low back pain. *Industr. Med. Surg.* 38:398–408.

Mital, M.A.; Ayoub, M.M.; Asfour, S.S.; and Bethea, N.J. 1971. Relationship between lifting capacity and injury in occupations requiring lifting. Proceedings of the Fifteenth Annual Meeting of the Human Factors Society. pp. 469–73.

Nachemson, A.L; and Elfstrom, E. 1970. Intravital dynamic pressure measurements in lumbar discs. A study of common movements, maneuvers and exercises. *Scand J Rehab Med.* 2 (Suppl 1):1–40.

Nachemson, A. 1971. Low back pain: Its etiology and treatment. *Clin. Med.* 78: 18–24.

Pope, M.H.; Wilder, D.G.; and Frymoyer, J.W. 1980. Vibration as an aetiologic factor in low back pain. Proc Conf Engineering Aspects of the Spine. Meeting Brit Orth Assoc and Inst Mech Eng., London, U.K.

Roozbazar A: Biomechanics of lifting. In *Biomechanics IV, 1974.* Univ Park Press, pp. 37–45, 1980.

Rowe, M.L. 1973. Preliminary statistical study of low back pain. *J. Occup. Med.* 5 (7):336–41.

Snook, Stover H. 1978a. The design of manual handling tasks. *Ergonomics* 21(12):963–85.

Snook, S.H.; Campanelli, R.A.; and Hart, J.W. 1978b. A study of three preventive approaches to low back injury. *J. Occup. Med.* 20 (7):478–81.

Snook, S.H. 1982. Low back pain in industry. In *Symposium on idiopathic low back pain.* A.A. White and S.L. Gordon (eds.). St. Louis: Mosby. pp. 23–28.

Svensson, Hans O. 1982. Low back pain in 40–47 year old men. II. Socioeconomic factors and previous sickness absence. *Scand. J. Rehab. Med.* 14(2):55–60.

Taylor, P.J. 1968. Sickness absence resistance. *Trans. Soc. Occup. Med.* 18:96–100.

Valkenburg, H.A.; and Haanen, H.C.M. 1982. The epidemiology of low back pain. In *Symposium on low back pain.* A.A. White and S.L. Gordon (eds.). St. Louis: Mosby. pp. 9–22.

Westrin, C.G. 1970. Low back sick-listing: A nosological and medical insurance investigation. *Acta Soc. Med. Scand.* 2–3:127–34.

Westrin, C.G. 1973. Low back sick-listing: A nosological and medical insurance investigation. *Scand. J. Soc. Med.* Suppl. 7:1–116.

White, A.A.; and Panjabi, M.M. 1978. *Clinical biomechanics of the spine.* Philadelphia: J.B. Lippincott, Co.

7 The Patients

Gunnar B.J. Andersson and Malcolm H. Pope

It is clearly important to understand which individual characteristics increase the risk for low back pain (LBP) as well as influence the disability from a LBP episode. Epidemiologic surveys provide the greatest insight. It is the purpose of this chapter to summarize some of our current knowledge on individual factors and LBP. It will become obvious that our knowledge is far from complete. In many surveys, there is poor definition of LBP, and the individual characteristics are poorly quantified. Further, the statistical analyses are often univariate so that any association found can be secondary. Factors that will be considered include age, sex, body characteristics (anthropometry), posture, spinal mobility, muscle strength and physical fitness, and psychosocial characteristics.

AGE

Both LBP in general and disc herniations are influenced by age. Although data on LBP in general lacks precision, there are quite accurate data on the age at which disc surgery is performed.

The first attack of LBP usually occurs early in life (Bergquist-Ullman and Larsson 1977; Svensson and Andersson 1983; Biering-Sorensen 1982). Hirsch, Jonsson and Lewin (1969), for example, found that 18 percent of a cross-sectional group of 15 to 24 year-old women had back pain. A comparable finding was identified in males by Hult (1954) and Horal (1969). The maximum period of symptoms appears to be in the age range of 35 to 55 years, a point previously discussed in Chapter 4. Biering-Sorensen found the lifetime incidence, one-year incidence and point prevalence to increase consistently with increasing age in females. In males, the highest rates were found at age 40 and did not show the increase thereafter which was identified in females. These differences may be caused by increasing osteoporosis in women as they get older, or men may place less physical demands on their backs after the age of 40. Hirsch, Jonsson and Lewin (1969), on the other hand, found a gradual increase in the number of females with low back disorders until age 55. Thereafter, the incidence did not increase. Similar findings are reported from the Dutch study of Valkenburg and Haanen (1982).

Most patients who have operations for disc herniations are between 35 and 45. Spangfort (1972) found the mean age at operation in Sweden to be 40.8 years (41.0 years for women, 40.7 for men). These data are shown in Figure 7.1. Spangfort's data are highly consistent with all previous reports for disc surgery from all parts of the world. Interestingly, the lumbar level involved by herniations was also age-related. As seen in Figure 7.2, the incidence of L5–S1 herniations is low in juveniles, increases to about age 30 and then decreases constantly with age, whereas the reverse pattern was found for L4–5 herniations. The incidence of lumbar herniations at the L2–3 and L3–4 levels increased with age (Figure 7.3). This pattern is consistent with the evaluation of disc degeneration, which starts at first at the L5–S1 level and then moves upwards to higher lumbar interspaces. Spangfort's data are quite similar to that obtained in other countries (Hrubec and Nashold 1975). The operation data are also consistent with data by Kelsey and Ostfeld (1975) and Valkenburg and Haanen (1982) on subjects with clinically diagnosed herniated lumbar discs.

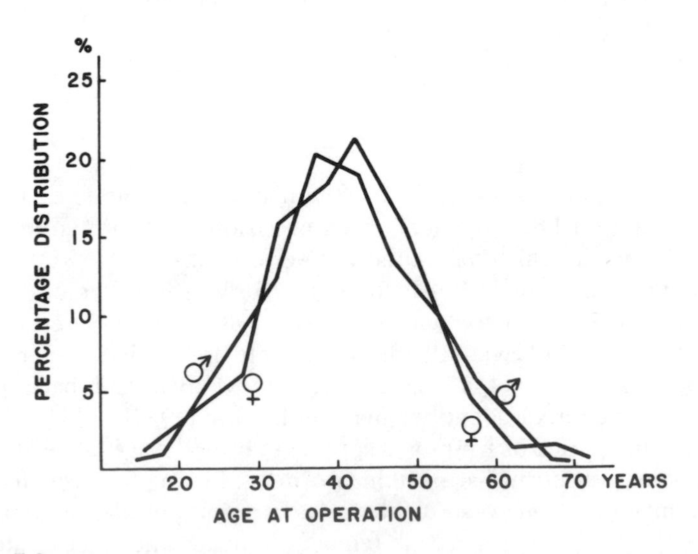

Figure 7.1
The distribution in age groups by sex in the total material (After Spangfort, 1972)

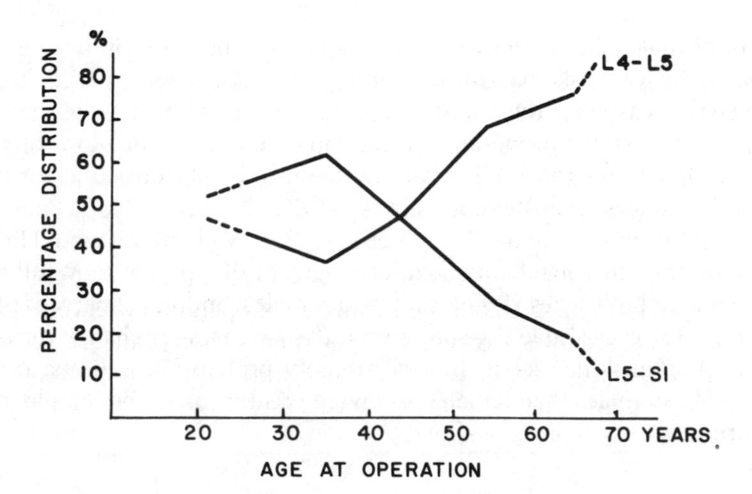

Figure 7.2
The level of herniation in patients with complete herniations by age at operation (percentage distribution) (After Spangfort, 1972)

GENDER

LBP seems as frequent in females as in males (Horal 1969; Valkenburg and Haanen 1982; Svensson and Andersson 1982). When the work situation is taken into account, however, the pattern changes. Magora (1970) found that 35 percent of women in physically heavy jobs had LBP compared to 19.1

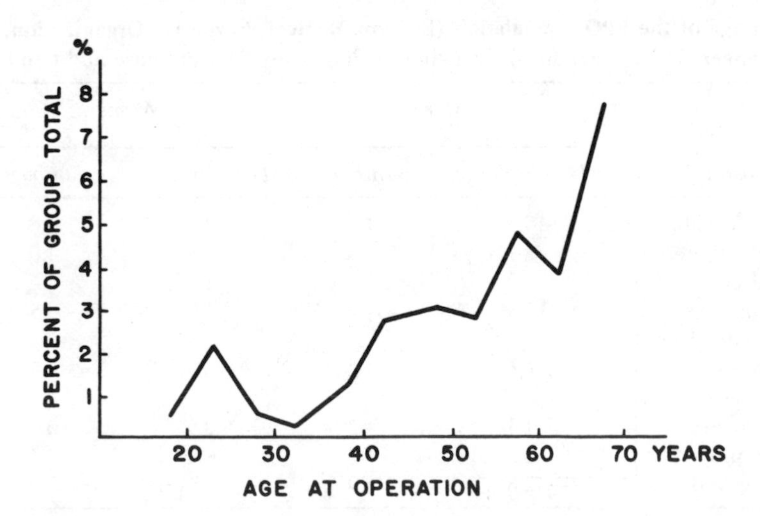

Figure 7.3
The incidence of high lumbar herniations (above L4-5 level) by age at operation. (After Spangfort 1972)

percent of males. In Sweden, there are higher absence rates in women with physically heavy work than in men doing the same jobs. Strength factors could be the reason for these differences. For example, mismatches between the requirements for lifting and lifting capacity may occur more often in females. Operations for disc herniations seem to be performed about twice as often in males as in females (Brown 1973; Lawrence 1977; Spangfort 1972; Kelsey and Ostfeld 1975; Weber 1983). Valkenburg and Haanen (1982), on the other hand, found clinical signs of disc prolapse overall to be equivalent for both sexes (1.9 percent of the males, and in 2.2 percent of the females). However, Table 7.1 shows how the percentages change for those employed as unskilled labor. In this category prolapse and lumbago were both higher in males than females and were greater than other employment categories.

ANTHROPOMETRY

Anthropometric data are conflicting, but in general indicate that there is no strong correlation between height, weight, body build and LBP (Hult 1954; Hirsch, Jonsson and Lewin 1969; Horal 1969; Rowe 1965; 1969; Chaffin and Park 1973; Svensson and Andersson 1983). However, tall people have been found to have a higher than average risk of back pain in some

TABLE 7.1
Percentage of the EPOZ population (Epidemiological Preventive Organization, Zoetermeer, The Netherlands) with clinical diagnoses Disc Prolapse and Lumbago

	Men		*Women*	
Occupation	*Disc Prolapse*	*Lumbago*	*Disc Prolapse*	*Lumbago*
Unskilled Labor	5.9	5.9	2.7	4.0
Skilled Labor	2.1	3.3	2.3	4.0
Low level employees	1.5	4.0	1.0	5.5
Intermediate level employees	0.3	3.2	0.9	4.7
Higher level employees	1.3	4.2	2.5	0.0
Self-employed	2.2	3.7	0.0	4.4
Unemployed[1]	2.6	4.9	2.6	5.6

[1]House-wives and pensioners.
Source: Valkenburg and Haanen 1982

studies (Gyntelberg 1974; Kelsey 1975; Lawrence 1966, Tauber 1970, Biering-Sorensen 1983). Back pain has been found more frequently in the very obese (Ikata 1965). If height and weight are combined using a technique called the Davenport Index, these differences no longer exist (Pope *et al.* in press) (see Table 7.1).

Westrin (1973) studied several other anthropometric factors such as length of tibia and femur and width of femoral condyles and malleoli. None was related to risk of LBP. Biering-Sorensen (1983) measured the femoral epicondylar width, the leg length, the length of the upper part of the body, and the so-called Rohrers Index (weight in grams × 100 (height in meters^3). None of these measurements had prognostic value either for the first-time experience or for the recurrence of LBP.

POSTURE

Postural deformities such as scoliosis, kyphosis, hypo- or hyperlordosis, and leg length discrepancy do not seem to predispose to LBP (Hult 1954, Horal 1969; Rowe 1969; Hodgson *et al.* 1974; Magora 1975; Sorensen 1964). Scoliosis frequently has been suggested as a cause of LBP, but no hard evidence of a true association with LBP has been accumulated (Nilsonne and Lundgren 1968; Nachemson 1968; Collis and Ponseti 1969). Some evidence indicates that back pain may be more prevalent when the degree of scoliosis is greater than 80 degrees, particularly when the curve involves the lumbar spine (Kostuik, Israel, and Hall 1973; Bradford, Moe, and Winter 1975). Kostuik and Bentivoglio (1982) reviewed 5,000 intravenous pyelograms in adults over the age of 20. An increase in the amount of pain was found with increase in the magnitude of scoliotic curvatures, but was not related to age.

The degree of lordosis or kyphosis was unrelated to LBP in studies by Hult (1954), Horal (1959) and Rowe (1969). Magora (1975) found flattening of the spine (i.e., decreased lordosis) in patients with LBP. Because loss of the lumbar lordosis was frequently accompanied by muscle spasm, he concluded that the loss of lordosis was related to pain.

The possible association between LBP and leg length discrepancies has been the subject of several studies and much controversy. Leg length discrepancy is quite common and appears to be related to an increase of LBP in a study by Giles and Taylor (1981). Other studies have shown no correlation (Hult 1954; Rowe 1965; Horal 1969; Magora 1975; Bjonness 1975; Biering-Sorensen 1983). One of the major problems in determining the value of the leg length data is the accuracy of different measurement methods (including radiography) which are poor.

SPINE MOBILITY

Spine motions are reduced in most subjects with back pain (Magora 1975; Pope *et al.* 1980). Although all motions are restricted, flexion is usually most severely affected. Bergquist-Ullman and Larsson (1977) did not find decreased lumbar flexion to have any influence on the duration of a LBP episode, nor did it influence the risk of recurrence. Biering-Sorensen (1983) found that reduced flexibility of the back was more pronounced in subjects who experienced recurrence of LBP in the year following examination.

MUSCLE STRENGTH

Measurements of trunk strength have been used to determine if there are differences between subjects with and without LBP. Clinical observations led Rowe (1963) and Bergquist-Ullman and Larsson (1977) to conclude that abdominal and spinal extensor muscle strength was decreased in patients with LBP. Many other investigators (Addison and Schultz 1980; Alston *et al.* 1966; Berkson *et al.* 1979; Hasue *et al.* 1980; McNeill *et al.* 1980; Nachemson and Lindh 1969; Nordgren *et al.* 1980; Nummi *et al.* 1978; Pedersen *et al.* 1975; Onishi and Nomura 1973) have established that patients with LBP have lower mean trunk strength than healthy subjects. Some investigators have found the extensors to be comparatively more influenced (weaker) than the flexors (Schultz 1982), while others have identified LBP sufferers to have relatively greater extensor strength (Biering-Sorensen 1983; Pope 1983). None of the above studies clarifies whether a muscle weakness or imbalance is primary or secondary to LBP.

PHYSICAL FITNESS

Is an individual less likely to have LBP if he is fit? Cady *et al.* (1979) concluded from a study of Los Angeles firefighters that physical fitness and conditioning have significant preventive effects on back injuries. A similar finding has been reported by Imrie (1983) in his study of Toronto ambulance medics. Chaffin *et al.* (1978) and Keyserling, Herrin and Chaffin (1980) have used pre-employment strength testing procedures and found that the risk of a back injury increases three-fold when the job requirement exceeds the strength capability on an isometric simulation of the job.

Svensson *et al.* (1983) found that subjects with LBP had higher physical activity at work, while they had lower physical activity during their leisure time. This may be because LBP prevented or made physical activities more difficult. Bergquist-Ullman and Larsson (1977) found no difference in the rates of recovery from acute LBP episodes with improved physical fitness.

Another measure of physical fitness is participation in other activities such as sports. Frymoyer *et al.* (1983) found no differences between occurrences of LBP and a variety of sporting activities including tennis, football, baseball, downhill skiing, snowmobiling and basketball. Cross-country skiing and jogging were associated with complaints of moderate, non-disabling LBP. The reasons for this association have been analyzed (Frymoyer *et al.* 1982). Some sports have been associated with specific structural diagnoses. Thus, Kelsey (1975) found insufficient physical exercise, as well as participation in some sports (baseball, golf and bowling) were marginally associated with the risk for the development of a prolapsed lumbar disc. Bowlers, gymnasts, American interior football linemen, javelin throwers, and backpackers are all at increased risk for spondylolisthesis, as previously discussed in Chapters 1 and 3.

RADIOGRAPHIC FACTORS

For the purposes of this section, radiographic factors will be divided into degenerative changes and selected abnormalities.

Degenerative Changes

The relationship between the occurrence of *disc degeneration* and low back pain is controversial. It is obvious from many different studies that disc degeneration *per se* is not symptomatic, and is part of a general age process (Figure 7.4). Back pain, however, appears to be more frequent in subjects with severe degenerative changes, involving several discs (Bistrom 1954; Caplan, Friedman and Connelly 1966; Hult 1954; Lawrence 1966; 1969; Magora and Schwartz 1976; Rowe 1963, 1965, 1969; Torgerson and Dotter 1976; Wiikeri *et al.* 1978). In moderate or light degeneration the situation is less clear, and most literature reports that the correlation is negative (Hirsch *et al.* 1969; Horal 1969; Hult 1954; Hussar and Guller 1956; Magora and Schwartz 1976; Splithoff 1953). There is evidence that disc degeneration is more frequent in individuals with heavy manual work (Hult 1954; Kellgren and Lawrence 1958; Lawrence 1969), although the nature of the stress inducing the degenerative changes is not clear (Caplan, Freedman and Connelly 1966; Kelsey 1978). Obesity does not appear to be related to degenerative disc disease, (Lawrence, 1961), nor does generalized osteo-arthritis (Kellgren and Lawrence 1958).

A large number of studies have been carried out to the purpose of establishing if a relationship exists between *skeletal defects*, congenital or acquired, and LBP. Further information on this topic appears in Chapters 3 and 11. Thus, only a short summary of some aspects will be given here, as we interpret them from the current back literature. The prevalence of each

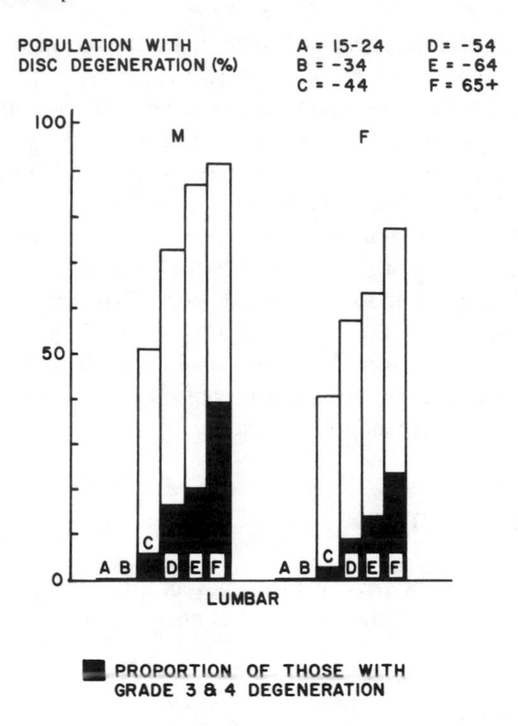

Figure 7.4
Prevalence of radiographic disc degeneration in Great Britain. (After Lawrence, 1969)

defect is so small that studies are difficult to perform. Ninety-eight percent of the defects are located in the L4, L5 and upper sacral segments (Ad Hoc Committee on Low Back X-Rays 1964; Moreton 1969). As with disc degeneration it has been clearly established that different defects do not necessarily give rise to pain (Horal 1969; Hult 1954; LaRocca and Macnab 1969; Redfield 1971; Rowe 1963; Splithoff 1953). Some are associated with an increased risk, however.

Low back pain appears to be more common in subjects with spondylolisthesis (Fischer, Friedman and Van DeMark 1958; Hult 1954; Horal 1969; Kettelkamp and Wright 1971; Wiltse 1971; Magora and Schwartz 1980), but there are several reports indicating the opposite (LaRocca and MacNab 1969; Rowe 1963; Splithoff 1953). Scheuermann's disease and severe lumbar scoliosis are conditions where an increased risk has often been claimed, but where the association has not been clearly established (Sorensen 1964).

Sacralization or the presence of a lumbosacral transitional vertebra are other abnormalities in association with which an increased prevalence of back pain has sometimes been established (Paillas, Winninger and Louis 1969; Tilley 1970), but where other studies have failed to confirm a positive correlation. Spina bifida occulta appears not to be more frequent in low back sufferers than in healthy controls.

Advanced *osteoporosis* with fractures of a macro- or a micro-type is known to be painful. Studies of young or moderately old populations have not indicated that osteoporosis is more common in subjects with low back symptoms than in subjects without (Bistrom 1954; Horal 1969; LaRocca and MacNab 1969).

Psychological Factors and Psychiatric Problems

An important consideration in any disease is the mental health of the patient as well as their psychologic adaptation to the disease process. This is certainly true in low back disease. In a comparison of back pain patients and controls, Westrin (1970, 1973) found severe mental health problems were no more common in the LBP group than in the controls. Psychological factors, on the other hand, were found to differ significantly. A poorer intellectual capacity, a lesser ability to establish emotional contacts, and a less philosophic attitude were characteristics of the back pain sufferer. Several studies indicate that psychological tests can be used as predictors of treatment and rehabilitation outcome. The most widely used test is the Minnesota Multiphasic Personality Inventory (MMPI). When the MMPI scores are elevated in hysteria, hypochondriasis, somatization, and depression, there is less likelihood of good recovery from chemonucleolysis (Wiltse and Rocchio 1975) or after operative treatment for lumbar disc disease (Wilfling, Klonoff and Kokan 1973). Workmen's compensation recipients with these same psychological symptoms have a longer period of unemployment and less potential for re-employment (Beals and Hickman 1972). After a back injury, psychological distress is also associated with a greater restriction in the range of spinal motion (Pope *et al.* 1980).

What are the characteristics of these individuals with "poor adjustment"? Much of the research into psychological factors in LBP has divided the population into the "organic" pain patient versus the "functional" pain patient (Carr, Brownsberger and Rutherford 1966; Cox, Chapman and Black 1978). The organic patient is usually defined as one with anatomical findings that could reasonably cause the symptoms reported; he copes with his pain well and has an appropriate degree of disability commensurate with his symptoms. In contrast, the functional LBP patient exaggerates his pain reports, often abuses alcohol or medications, and frequently does not respond appropriately to treatment. Physical findings are often exaggerated, a point discussed in detail in Chapter 8. Obviously, there are problems with this dichotomy. Patients do not neatly fall into "organic" or "inorganic" categories. The present diagnostic tools do not permit us to diagnose accurately 50 percent of the population. The non-organic or so-called functional patient is assumed to have created the pain out of a psychologic disturbance (Phillips 1964). Probably the lack of response to treatment causes a transformation into the chronic LBP patient, often after multiple unsuccessful surgeries (Mooney, Cairns and Robertson 1975) characterized by dependency and depression (Sternbach *et al.* 1973). Since this patient population is so difficult to manage, attempts have been made to discriminate between the good and poor surgical candidate (Blumetti and Modesti 1976; Gentry *et al.* 1977; Jamison *et al.* 1976; Kelsey and Hardy 1975; Pheasant *et al.* 1979; Waring, Weisz and Bailey 1976; Wiltse and Rocchio 1975). Similarly, attempts have been made to predict the appropriately-disabled from the inappropriately-disabled (Magora 1970; Shaffer, Nussbaum and Little 1972), the good medical management risk (McCreary, Turner and Dawson 1979) and the good risk for acupuncture (Hossenlopp, Leiber and Mo 1976). Methods used have included the MMPI, the Cornell Medical Index (Wiltse and Rocchio 1975) and other questionnaires (Forrest and Wolkind 1974; Pilowsky and Spence 1975; Woodforde and Merskey 1972).

Some suggest that psychologic profiles may not accurately predict who is at risk for LBP (Freeman, Calyson and Louks 1976; Rosen, Frymoyer and Clements 1980). Obviously, psychological factors are associated with LBP and with its prognosis. Several studies have demonstrated that the MMPI of patients with poor treatment outcomes changes during the course of their illness (Beals and Hickman 1972; Gentry *et al.* 1977; Jamison *et al.* 1976; McCreary, Turner and Dawson 1979; Wiltse and Rocchio 1975). However, which comes first, the back disorder or psychological problems? Other theories suggest that the individual prone to back pain for psychological reasons is a passive individual who is afraid of interpersonal conflict. Therefore, the social and occupational withdrawal of the back patient is motivated by a desire to escape emotional stress in these situations. Recent Swedish studies have shown that those with the least desirable working environment are less likely to return to work.

Social Factors

A high incidence of social problems have been found in back patients. Their general social situation and economy is on the average less good and there is a greater proportion of back patients suffering from drug and alcohol abuse (Magora 1973; Westrin 1970). Divorces and family problems are also more frequent. The educational level has been found to be lower in some of the studies. Whether some social factors are the cause or the result of LBP is difficult to assess.

Other Factors

Smoking was identified as significantly associated with LBP episodes in both Vermont (Frymoyer *et al.* 1983) and Sweden (Svensson 1983). In a study of Swedish industrial workers, Svensson (1983) speculated that coughing led to increased intradiscal pressure and thus to increased spine loadings and LBP. This conclusion is supported by a Danish study (Biering-Sorenson 1983) that identified coughing but not smoking, as important in the etiology of low back complaints. However, the Vermont study (Frymoyer *et al.* 1983) indicated that coughing alone was insufficient to account for the observed differences in LBP complaints. It might be speculated that smokers have emotional, recreational, or occupational differences from non-smokers, although multivariate analysis did not confirm that speculation.

SUMMARY

Many individual factors are associated with the incidence and prevalence of low back complaints. For convenience, these factors are summarized completely in Table 7.2. LBP is first experienced early in life, and the incidence increases with age. There is no increase after age 40 in males, and in females there is no increase after age 55. The highest incidence of operation for disc herniations is at age 40 for both sexes. LBP is equally common in both sexes. However, 35 percent of women in physically heavy jobs had LBP compared to 19 percent of males. There is little evidence that anthropometric factors, such as height, weight, body build or limb length, or body indices have any relationship to LBP. Postural deformities, other than severe scoliosis, do not predispose to LBP. Decreased mobility, particularly flexion, and decreased strength is found in those with LBP. It is not known whether these attributes are primary or secondary to LBP. The role of improved physical fitness in preventing LBP is not completely clear, although there is some evidence in the favor of improved fitness. The risk of a back injury markedly increases if the strength requirements of the job exceed the worker's strength. Disc degeneration is greater in those involved in heavy work. Osteoporosis and spondylolisthesis are both associated with increased LBP. Severe mental health problems are no more common in LBP patients but changes in the MMPI are observed. Those with elevated MMPI scores have a poorer prognosis. Social problems are greater in LBP patients. LBP patients have increased exposure to smoking but the etiology is unclear.

TABLE 7.2
Risk Facators

Risk Factors	*Associated with LBP*	*Unassociated*
Constitutional	Age Physical fitness Abdominal muscle strength Flexor/Extensor balance Muscular insufficiency	Sex Weight Height Davenport Index
Postural-Structural	Severe scoliosis Some congenital abnormalities Narrowed spinal canal	Lordosis
Radiographic	Only specific structural abnormalities, i.e., spondylolysis, fractures, multi-level degenerative disc disease, spondyloarthropathies	Disc space narrowing Schmorl's nodes Spina bifida Osteophytes
Environmental	Smoking	
Occupational	Heavy lifting Vibration (vehicles & non-vehicles) Requirements for heavy, physical lifting activity, prolonged sitting and other body postures	
Recreational	Golfing, tennis, football, gymnastics, jogging, cross-country skiing	Snowmobiling Downhill skiing Ice Hockey Baseball, other sports
Psychosocial	Anxiety Depression Hypochondriasis Somatization Work dissatisfaction	Psychoses & most neuroses
Other	Multiple births in the female Some familia clustering	

Source: Frymoyer, J.W. 1984. Helping your patients avoid low back pain. *J. Musculoskeletal Med.* 1:65–74.

REFERENCES

Ad hoc committee on low back x-rays. 1964. Low-back x-rays. Criteria for their use in placement examinations in industry. *Journal of Occupational Medicine* 6 (9):373–80.

Addison, R., and Schultz A. 1980. Trunk strengths in patients seeking hospitalization for chronic low back disorders. *Spine* 5(6):539–44.

Alston, W.; Carlson, K.E.; Feldman, D.J.; Grimm, Z.; and Gerontinos, E. 1966. A quantitative study of muscle factors in the chronic low back syndrome. *J. Amer. Geriatr. Soc.* 14 (10):1041–47.

Beals, R.K.; and Hickman, N.W. 1972. Industrial injuries of the back and extremities. *J. Bone Joint Surg* 54A (8):1593–1611.

Bergquist-Ullman, M.; and Larsson, U. 1977. Acute low back pain in industry. A controlled prospective study with special reference to therapy and confounding factors. *Acta Orthop. Scand. Suppl* 170:1–117.

Berkson, M.; Schultz, A.; Nachemson, A.; and Andersson, G.B.J. 1977. Voluntary strengths of male adults with acute low back syndromes. *Clin. Orthop* 129:84–95.

Biering-Sorenson, F. 1982. Low back trouble in a general population of 30-, 40-, 50-, and 60- year old men and women. Study design, representativeness and basic results. *Dan. Med. Bull* 29 (6):289–99.

Biering-Sorensen, F. 1983. The prognostic value of the low back history and physical measurements. Unpublished doctoral dissertation, University of Copenhagen.

Biering-Sorensen, F. 1984. Physical measurements as risk indicators for low back trouble over a one year period. *Spine*, to be published.

Bistrom, O. 1954. Congenital anomalies of the lumbar spine of persons with painless backs. *Ann. Chir. Gynaecol. Fenn* 43:102–115.

Bjonness, T. 1975. Low back pain in persons with congenital club foot. *Scand. J. Rehab. Med* 7 (4):163–65.

Blumetti, A.E.; and Modesti, L.M. 1976. Psychological predictors of success or failure of surgical intervention for intractable back pain. In *Advances in pain research and therapy*, Vol. 1, eds. J.J. Bonica and D.G. Albe-Fessard. pp. 323–25, New York: Raven Press.

Bradford, D.S.; Moe, J.H.; and Winter, R.B. 1975. Scoliosis and kyphosis. Operative management of idiopathic scoliosis. *The Spine*. 2nd ed., vol. 1, pp. 347–48 ed., R.H. Rothman and F.A. Simeone, Philadelphia: W.B. Saunders Company.

Brown, J.R. 1973. Lifting as an industrial hazard. *Amer. Industr. Hyg. Assoc. J* 34 (2):292–97.

Cady, L.D.; Bischoff, D.P.; O'Connel, E.R.; Thomas, P.C.; and Allan, J.H. 1979. Strength and fitness and subsequent back injuries in fire-fighters. *Journal of Occupational Medicine* 21 (4):269–72.

Caplan, P.S.; Freedman, L.M.J.; and Connelly, T.P. 1966. Degenerative joint disease of the lumbar spine in coal miners—a clinical and x-ray study. *Arthritis. Rheum* 9 (5):693–702.

Carr, J.E.; Brownsberger, C.N.; and Rutherford, R.D. 1966. Characteristics of symptom-matched psychogenic and "real" pain patients on the MMPI. *Proc. 74th Ann. Convention of the Amer. Psychol. Assoc.* 1:215–16.

Chaffin, D.B.; and Park, K.S. 1973. A longitudinal study of low back pain as associated with occupational weight lifting factors. *A.I.H.A.J.* 34:513–25.

Chaffin, D.B.; Herrin G.D.; and Keyserling, W.M. 1978. Preemployment strength testing. An updated position. *Journal of Occupational Medicine* 20 (6):403–408.

Collis, D.K.; and Ponseti, I.V. 1969. Long term follow-up of patients with idiopathic scoliosis not treated surgically. *J. Bone Joint Surg* 51A (3):425–45.

Cox, G.B.; Chapman, C.R.; and Black, R.F. 1978. The MMPI and chronic pain: The diagnosis of psychogenic pain. *J. Behav. Med* 1 (4):437–43.

Fisher, F.J.; Friedman, M.M.; and Demark, R.E. Van. 1958. Roentgenographic abnormalities in soldiers with low back pain: A comparative study. *Amer. J. Roentgen.* 79 (4):673–76.

Forrest, A.J.; and Wolkind, S.N. 1974. Masked depression in man with low back pain. *Rheumatol. and Rehabil.* 13 (3):148–53.

Freeman, C.; Calyson, D.; and Louks, J. 1976. The use of the MMPI with chronic low back pain patients with a mixed diagnosis. *J. Clin. Psychol.* 32:532–36.

Frymoyer, J.W.; Pope, M.H.; and Kristiansen, T. 1982. Skiing and spinal trauma. *Clin. Sports Med.* 1 (2):304–18.

Frymoyer, J.W.; Pope, M.H.; Clements, J.H.; Wilder D.G.; McPherson, B.; and Ashikaga, T. 1983. Risk factors in low back pain. An epidemiological survey. *J. Bone Joint Surg.* 65-A (2):213–18.

Gentry, W.D.; Newman, M.C.; Goldner, J.L.; and von Baeyer, C. 1977. Relation between graduated spinal block technique and MMPI for diagnosis and prognosis of chronic low back pain. *Spine* 2 (3):210–13.

Giles, L.G.F., and Taylor, J.R. 1981. Low-back pain associated with leg length inequality. *Spine* 6 (5):510–21.

Gyntelberg, F. 1974. One year incidence of low back pain among male residents of Copenhagen aged 40-59. *Dan. Med. Bull* 21 (1):30–36.

Hasue, M.; Fujiwara, M.; and Kikuchi, S. 1980. A new method of quantitative measurement of abdominal and back muscle strength. *Spine* 5 (2):143–48.

Hirsch, C.; Jonsson, B.; and Lewin, T. 1969. Low back symptoms in a Swedish female population. *Clin. Orthop* 63:171–76.

Hodgson, S.; Shannon, H.S.; and Troup, J.D.G. 1974. The prevention of spinal disorders in dock workers. *Report to National Dock Labour Board* London, U.K.

Horal, J. 1969. The clinical appearance of low back disorders in the city of Gothenburg, Sweden. *Acta Orthop. Scand. Suppl* , 118:1–109.

Horal, J. 1969. The clinical appearance of low back disorders in the city of Gothenburg, Sweden. Comparisons of incapacitated probands with matched controls. *Acta Orthop. Scand. Suppl* , 118:1–109.

Hossenlopp, C.M.; Leiber, L.; and Mo, B. 1976. Psychological factors in the effectiveness of acupuncture for chronic pain. In *Advances in pain research and therapy*, Vol. 1, eds. J.J. Bonica and D.G. Albe-Fessard. pp. 803–809, New York: Raven Press.

Hrubec, Z.; and Nashold, B.S., Jr. 1975. Epidemiology of lumbar disc lesions in the military in World War II. *Am. J. Epidem* 102 (5):366–76.

Hult, L. 1954. Cervical, dorsal, and lumbar spinal syndromes. *Acta. Orthop. Scand.* Suppl. 17:1–102.

Hussar, A.E.; and Guller, E.J. 1956. Correlation of pain and the roentgenographic findings of spondylosis of the cervical and lumbar spine. *Am. J. Med. Sci.* 232:518–27.

Ikata, T. 1965. Statistical and dynamic studies of lesions due to overloading on the spine. *Shikoku Acta Med.* 40:262–86.

Imrie, D. 1983. Personal communication.

Jamison, K.; Ferrer-Biechner, M.T.; Brechner, V.L.; and McCreary, C.P. 1976. Correlation of personality profile with pain syndrome. In *Advances in pain research and therapy.* Vol. 1, pp. 317–21, eds., J.J. Bonica and D.G. Albe-Fessard, New York: Raven Press.

Kellgren, J.H.; and Lawrence J.S. 1958. Osteoarthrosis and disk degeneration in an urban population. *Ann. Rheum. Dis.* 17 (4):388–97.

Kelsey, J.L. 1975. An epidemiological study of the relationship between occupations and acute herniated lumbar intervertebral discs. *Int. J. Epidemiol* 4 (3):197–205.

Kelsey, J.L. 1978. Epidemiology of radiculopathies. *Adv. Neurol.* 19:385–98.

Kelsey, J.L.; and Hardy, R.J. 1975. Driving of motor vehicles as a risk factor for acute herniated lumbar intervertebral disc. *Am. J. Epidemiol.* 102 (1):63–73.

Kelsey, J.L., and Ostfeld, A.M. 1975. Demographic characteristics of persons with acute herniated lumbar intervertebral disc. *J. Chron. Dis.* 28 (1):37–50.

Kettelkamp, D.B.; and Wright, D.G. 1971. Spondylolysis in the Alaskan Eskimo. *J. Bone Joint Surg* 53A (3):563–66.

Keyserling, W.M.; Herrin, G.D.; and Chaffin, D.B. 1980. Isometric strength testing as a means of controlling medical incidents on strenous jobs. *J. Occup. Med.* 22 (5):332–36.

Kostuik, J.P.; Israel, J.; and Hall, J.E. 1973. Scoliosis surgery in adults. *Clin. Orthop.* 93:225–34.

Kostuik, J.P.; and Bentivoglio, J. 1982. *The incidence of low back pain in adult scoliosis.* Manuscript submitted for publication.

LaRocca, H.; and Macnab, I. 1969. Value of pre-employment radiographic assessment of the lumbar spine. *Canad. Med. Ass. J.* 101 (7):383–88.

Lawrence, J.S. 1961. Rheumatism in cotton operatives. *Br. J. Indust. Med.* 18 (4):270–76.

______. 1955. Rheumatism in coal miners, Part III. Occupational factors. *Br. J. Indust. Med.* 12:249–61.

Lawrence, J.S.; Molyreux, M.K.; and Dingwall-Fordyce, I. 1966. Rheumatism in foundary workers. *Br. J. Industr. Med.* 23 (1):42–52.

Lawrence, J.S. 1969. Disc degeneration. Its frequency and relationship to symptoms. *Ann. Rheum. Dis.* 28:121–38.

Magora, A. 1970. Investigation of the relation between low back pain and occupation. 2. Work history. *Industr. Med. Surg.* 39 (12):504–10.

______. 1973. Investigation of the relation between low back pain and occupation. 5. Psychological aspects. *Scand. J. Rehabil. Med.* 5 (4):191–96.

______. 1975. Investigation of the relation between low back pain and occupation. 7. Neurologic and orthopedic conditions. *Scand. J. Rehabil. Med.* 7 (4):146–51.

Magora, A.; and Schwartz, A. 1976. Relation between the low back pain syndrome and x-ray findings. 1. Degenerative osteoarthritis. *Scand. J. Rehabil. Med.* 8 (3-4):115–25.

Magora, A.; and Schwartz, A. 1978. Relation between low back pain syndrome and x-ray findings. 3. Transitional vertebra (mainly sacralization). *Scand. J. Rehabil. Med.* 10 (3):135–45.

Magora, A.; and Schartz, A. 1980. Relation between the low back pain syndrome and x-ray findings. 4. Lysis and olisthesis. *Scand. J. Rehabil. Med.* 12 (2):47–52.

McCreary, C.; Turner, J.; and Dawson, E. 1979. The MMPI as a predictor of response to conservative treatment for low back pain. *J. Clin. Psychol.* 35 (2):278–84.

McNeill, T.; Warwick, D.; Andersson, G.; and Schultz, A. 1980. Trunk strengths in attempted flexion, extension, and lateral bending in healthy subjects and patients with low back disorders. *Spine* 5 (6):529–38.

Mooney, V.; Cairns, D.; and Robertson, J. 1975. The psychological evaluation and treatment of the chronic back pain patient. A new approach (part 2). *Orthopaedic Nurses' Assoc. J* 2 (8):187–89.

Moreton, R.D. 1969. So-called normal backs. *Industr. Med. Surg.* 38 (7):216–19.

Nachemson, A.L. 1968. Back problems in childhood and adolescence. *Lakartidningen* 65:2831–43. (In Swedish)

Nachemson, A.H.; and Lindh, M. 1969. Measurement of abdominal and back muscle strength with and without low back pain. *Scand. J. Rehabil. Med.* 1 (2):60–65.

Nilsonne, U.; and Lundgren, K.D. 1968. Long-term prognosis in idiopathic scoliosis. *Acta Orthop. Scand.* 39 (4):456–65.

Nordgren, B.; Schek, R.; and Linrotn, K. 1980. Evaluation and prediction of back pain during military field service. *Scand. J. Rehabil. Med.* 12 (1):1–8.

Nummi, J.; Jarvinen, T.; Stambej, U.; Wickstrom, G. 1978. Diminished dynamic performance capacity of back and abdominal muscles in concrete reinforcement workers. *Scand. J. Work Environ. Health.* 4 Suppl. 1:39–46.

Onishi, N., and Nomara, H. 1973. Low back pain in relation to physical work capacity and local tenderness. *J. Human Ergol.* 2:119–32.

Paillas, J.E.; Winninger, J.; and Louis, R. 1969. Role des malformations lombo-sacrees dans les sciatiques et les lombalgies: etude de 1.500 dossiers radio-cliniques dont 500 hernies discales verifiees. *La Presse Med.* 77 (23):853–55.

Pedersen, O.F.; Petersen, R.; and Staffeldt, E.S. 1975. Back pain and isometric back muscle strength of workers in a Danish factory. *Scand. J. Rehabil. Med.* 7 (3):125–28.

Pheasant, H.C.; Gilbert, D.; Goldfarb, J.; and Herron, L. 1979. The MMPI as a predictor of outcome in low back surgery. *Spine*, 4 (1):78–84.

Phillips, E.L. 1964. Some psychological characteristics associated with orthopaedic complaints. *Curr. Pract. Orthop. Surg.* 2:165–76.

Pilowski, I.; and Spence, N.D. 1975. Pattern of illness behaviour in patients with intractable pain. *J. Psychosom. Res.* 19 (4):279–87.

Pope, M.H.; Bevins, T.; Wilder, D.G.; and Frymoyer, J.W. In Press. The relationship between anthropometric, postural, muscular, and mobility characteristics of males, ages 18–55. Submitted to *Spine*.

Pope, M.H.; Rosen, J.D.; Wilder, D.G.; and Frymoyer, J.W. 1980. The relation between biomechanical and psychological factors in patients with low back pain. *Spine* 5 (2):173–78.

Redfield, J.T. 1971. The low back x-rays as a pre-employment screening tool in the forest products industry. *Journal of Occupational Medicine* 13:219–26.

Rosen, J.C.; Frymoyer, J.W.; and Clements, J.H. 1980. A further look at validity of the

MMPI with low back patients. *J. Clin. Psychol.* 36 (4):994–1000.

Rowe, M.L. 1963. Preliminary statistical study of low back pain. *J. Occup. Med.* 5 (7):336–41.

______. 1965. Disc surgery and chronic low back pain. *J. Occup. Med.* 7 (5):196–202.

______. 1969. Low back pain in industry. A position paper. *J. Occup. Med.* 11 (4):161–69.

Schultz, A.; Andersson, G.; Ortengren, R.; Haderspeck, K.; and Nachemson, A. 1982. Loads on the lumbar spine, validation and biomechanical analysis by measurements of intradiscal pressures and myoelectric signals. *J. Bone Joint Surg.* 64A (5):713–20.

Shaffer, J.W.; Nossbaum, K.; and Little, J.M. 1972. MMPI profiles of disability insurance claimants. *Amer. J. Psychiatry* 129 (4):403–407.

Sorensen, K.H. 1964. Scheuermann's juvenile kyphosis. Doctoral dissertation.

Spangfort, E.V. 1972. The lumbar disc herniation. *Acta. Orthop. Scand. Suppl.* 142:1–95.

Splithoff, C.A. 1953. Lumbosacral junction. Roentgenographic comparison of patients with and without backaches. *J.A.M.A.* 152 (7):1610–13.

Sternbach, R.A., *et al* 1973. Chronic low-back pain. The "low-back loser". *Postgrad. Med.* 53 (6):135–38.

Svensson, H.O., and Andersson, G.B.J. 1982. Low back pain in forty to forty-seven year old men. I. Frequency of occurrence and impact on medical services. *Scand. J. Rehabil. Med.* 14 (2):47–53.

Svensson, H.O.; and Andersson, G.B.J. 1983. Low back pain in forty to forty-seven year old men: Work history and work environment factors. *Spine* 8 (3):272–76.

Svensson, H.O.; Vedin, A.; Wilhelmsson, C.; and Andersson, G.B.J. 1983. Low back pain in relation to other diseases and cardiovascular risk factors. *Spine* 8 (3):277–85.

Tauber, J. 1970. An unorthodox look at backaches. *J. Occup. Med.* 12 (4):128–30.

Tilley, P. 1970. Is sacralization a significant factor in lumbar pain? *J. Am. Osteopath. Assoc.* 70:238–41.

Torgerson, B.R.; and Dotter, W.E. 1976. Comparative roentgenographic study of the asymptomatic and symptomatic lumbar spine. *J. Bone Joint Surg.* 58A (6):850–53.

Valkenburg, H.A., and Haanen, H.C.M. 1982. The epidemiology of low back pain. In *Symposium on idiopathic low back pain.* Eds. A.A. White, and S.L. Gordon, pp. 9–22. St. Louis: Mosby.

Waring, E.M.; Weisz, G.M.; and Bailey, S.I. 1976. Predictive factors in the treatment of low back pain by surgical intervention. In *Advances in pain research and therapy* (Vol. 1). Eds., J.J. Bonica and D.G. Albe-Fessard. pp. 939–42, New York: Raven Press.

Weber, H. 1983. Lumbar disc herniation. A controlled, prospective study with ten years of observation. *Spine* 8(2):131–40.

Westrin, C.G. 1970. Low back sick-listing. A nosological and medical insurance investigation. *Acta Soc. Med. Scand.* 2-3:127–34.

Westrin, C.G. 1973. Low back sick-listing. A nosological and medical insurance investigation. *Scand. J. Soc. Med. Suppl.* 7:1–116.

Wilfling, F.J.; Klonoff, H.; and Kokan, P. 1973. Psychological, demographic and orthopaedic factors associated with prediction of outcome of spinal fusion. *Clin. Orthop* 90:153–60.

Wiikeri, M.; Nummi, J.; Riihimaki, H.; and Wickstrom, G. 1978. Radiologically detectable lumbar disc degeneration in concrete reinforcement workers. *Scand. J. Work Environ. Health* 4 (Suppl. 1) 4:47–53.

Wiltse, L.L. 1971. The effect of the common anomalies of the lumbar spine upon disc degeneration and low back pain. *Orthop. Clin. North Am.* 2 (2):569–82.

Wiltse, L.L.; and Rocchio, P.D. 1975. Preoperative tests as predictors of success of chemonucleolysis in the treatment of the low-back syndrome. *J. Bone Joint Surg.* 57A (4):478–83.

Woodforde, J.M., and Merskey, H. 1972. Personality traits of patients with chronic pain. *J. Psychosomatic Res.* 16 (3):167–72.

PART IV
Patient Care

8 Evaluation of the Worker with Low Back Pain

John W. Frymoyer and Raymond L. Milhous

The evaluation of workers with low back pain (LBP) is essential to optimize initial treatment, eventual rehabilitation, and prognosis. This evaluation should include appropriate history and physical examination. Additional laboratory, radiographic, and specialized diagnostic studies may be necessary in some workers. The treatment response should also be evaluated. Although not the topic of this chapter, it should be remembered that often not only the worker but also the workplace needs to be evaluated.

STANDARD MEDICAL EVALUATIONS

1. Medical History

A complete history should include:

(a) The history of LBP and related symptoms prior to the current episode. If the patient has chronic LBP and/or prior episodes have been present, information should be sought about precipitating events, activities, or postures that aggravate or relieve pain, and the pattern of pain as well as related symptoms. The response to previous treatment and the duration of previous disability (if any) are important to an understanding of the current episode.

(b) History of current episode. If the symptoms are acute, the precise mechanism of onset should be obtained. For example, did the symptoms occur following a lifting episode (the size and weight of object being lifted and the posture of lifting), twisting episode, or fall? If the onset is gradual, changes in overall health, work requirements, recreational activities, and psychosocial factors should be elicited. The pain pattern should be clarified, including intensity of pain, localization of pain, radiation of pain, factors that accentuate pain such as body posture, coughing, sneezing, and factors that relieve pain such as rest and response to medications. If the symptoms are progressive, the temporal pattern of pain is important. For example, does the pain worsen through the day, is the pain worse when the patient first arises and then improves during the day (this suggests spondyloarthropathies), or is the pain unrelenting or even worse at rest (this suggests tumors and/or infections) (Calin 1981). Neurologic symptoms associated with the low back episode should be identified such as sensory changes (numbness, tingling), subjective sense of lower extremity muscular weakness, and changes in bladder and bowel control.

(c) Associated Symptoms and Diseases. A general inquiry should be made into the patient's overall health. This includes significant medical diseases, prior surgery, and the relationship, if any, of low back symptoms to bodily function, particularly gastrointestinal and genitourinary, as well as other general musculoskeletal complaints (joint pain, swelling), which could suggest inflammatory disease. The history should also include drug intake and any allergies.

(d) Functional Assessment. The clinician should determine to what extent the back pain incapacitates the worker. How does the worker spend his day? How does the pain affect normal activities of daily living, leisure time, and sports activities? Does emotional tension affect the pain? Does the worker understand how to care for and protect his or her back? Has he attended a back school?

2. Physical Examination

The physical examination consists of inspection, measurement, palpation, percussion, and specialized tests. It is useful to develop a system of examination, because it reduces the time involved and also ensures that all parts of the examinations are completed. A flowchart is suggested in Table 8.1. In the low back examination, one should modify the sequence of tests so that the examinations most likely to be painful are left to the end. Modifications of the examination are also required in accordance with the worker's ability to stand, move and sit.

(a) In the Standing Position. The worker's gross body movements (while undressing), gait (if walking), and his standing posture will indicate the immediate severity of symptoms and indicate functional limitations. A normal, unguarded pattern of gait and posture implies symptoms of minimal severity, whereas a carefully guarded gait and posture indicate symptoms of greater severity.

As the worker stands, one can observe sagittal posture (kyphosis, lordosis) and the presence or absence of scoliosis including muscles prominently in spasm (this can often be confirmed by palpation). Occasionally, a step deformity indicative of spondylolisthesis is observed, or can be palpated.

TABLE 8.1
Flowchart for the Examination of the LB Injured Worker

Patient Standing	Posture	Scoliosis Lordosis–kyphosis Muscle spasm
	Gait	
	Range of motion	Flexion Extension Lateral bend Axial rotation
	Screening muscle strength test	Heel-toe walk Quads test (stairs)
Patient Recumbent (supine)	Measurements	Circumferences Leg length
	Neurologic exam	Reflex Akle jerk Knee jerk Posterior tibial
		Sensation Muscle strength
	Nerve tension signs	Hip ROM Straight leg raising Confirmatory tests Abdominal exam Peripheral pulses
Recumbent (prone)	Neurologic	Complete sensation including perineum Femoral stretch test
	Palpation	Muscle spasm Spinous processes Interspinous spaces Sacroiliac joints
Seated	Observation	Seated posture
	Neurologic	"Flip test" Confusional SLR
Ancillary		Rectal Pelvic exam (female)

Source: The authors of this chapter.

The worker should be instructed to move in each plane of spine motion within his range of comfort (flexion/extension, right and left lateral bending, right and left axial rotation). Patterns of motion and restriction thereof may offer clues to diagnosis, although limitations of motion generally are nonspecific and primarily pain-related. For example, patients with sciatica may stand with a scoliosis away or toward the leg with sciatica dependent on the position of disc herniation relative to the nerve root (Fig. 8.1) (Edgar and Park 1974). In an attempt to quantify range of motion (ROM), many systems have been devised, none of which are particularly accurate. Because some of the error occurs from simultaneous motion at the hips, particularly in flexion

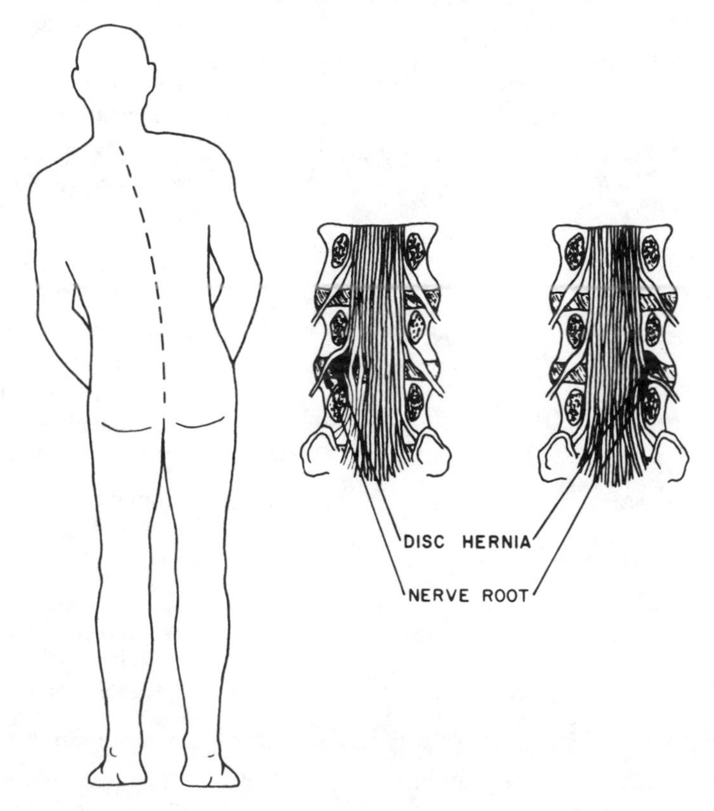

Figure 8.1
The direction of scoliosis in patients with sciatica may indicate the position of disc herniation relative to the nerve root. In this instance, the patient is showing a scoliosis concave to the left. In a patient with *left* leg sciatica, it would be suspected that the disc herniation is medial to the nerve root. Therefore, the scoliosis would tend to pull the left affected nerve root away from the herniation. In a patient with right sciatica, the position of the herniation would be more likely lateral to the root. Again the curvature to the left in the second instance will tend to pull the affected right nerve root away from the disc herniation.

and axial rotation, the pelvis should be controlled by the examiner (see Fig. 8.2). Chest expansion can also be measured. Reductions below 1.5 cm suggest spondyloarthropathies (Calin 1981).

The worker should be instructed to walk on their heels and toes, if possible. Ability to carry out these two maneuvers suggests intact function of the L5 and S1 nerve roots. Quadriceps function (i.e., the L3 and 4 nerve roots) can be tested by instructing the patient to climb up on a chair as if it were a step, using first the right and then the left leg.

(b) In the Supine Position. The individual is placed in a back-lying position. A systematic examination should include abdominal palpation for tenderness or a mass. Evaluation of the peripheral pulses (groin, back of the knee, dorsum of the foot) can also be accomplished at this time. Patients who are not responding to the treatment of acute symptoms or who have chronic LBP complaints should have a breast and pelvic examination if they are women, and a rectal examination if they are males.

Examination of the lower extremities is used to determine the presence or absence of any significant joint deformities and to assess neurologic function. Measurement of leg length and atrophy should be included. As illustrated in Figure 8.3, the authors use a simple method for determining the level of circumferential measurement of the thigh and calf. Small differences in circumference (less than 1 cm) have little meaning because of measurement error. Hip motion should be evaluated, as indicated below.

A neurological examination can be done with the patient in either the supine or sitting position. Routinely, the knee (quadriceps) and ankle (Achilles) reflexes are obtained. In addition, the posterior tibial reflex may be

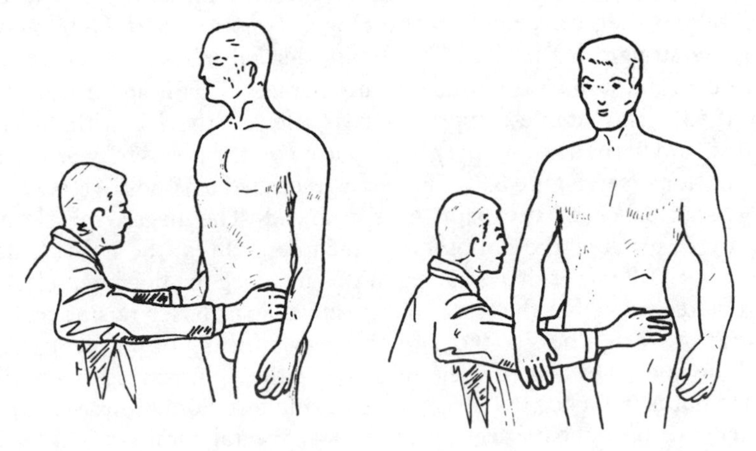

Figure 8.2
Motion may be measured in flexion, extension, lateral bend, and axial rotation. The examiner should place his hands on the pelvic girdle to control these motions, particularly axial rotation and lateral bend.

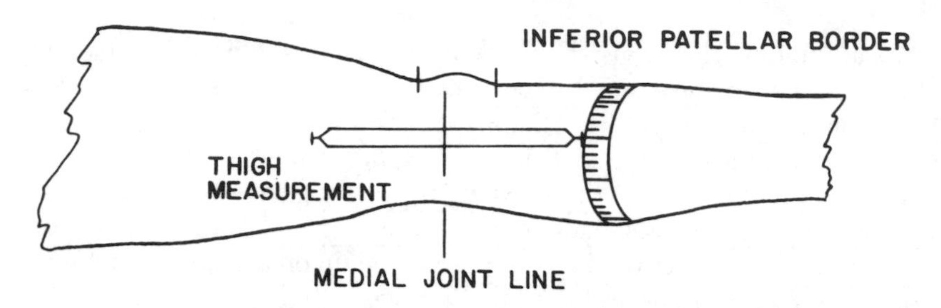

Figure 8.3
For measurement of thigh and calf circumference, a reference landmark is selected, using either the medial knee joint line, or the inferior border of the patella. A standard pen is used to determine the distance above and below that landmark, at which circumferential measurements will be made. Hence, the name the BIK pen measurement.

obtained. This reflex is present bilaterally in approximately 40 to 50 percent of normal subjects (Donaghy 1946). If the reflex is unilaterally absent, the possibility of an L5 root dysfunction is suspected; if bilaterally absent, it has no meaning. Sensitivity to pin prick and touch, and a manual muscle examination should then be carried out (Keegan 1944). In general, muscle strength measurements should include dorsiflexors, plantarflexors, long toe extensors, invertors and evertors of the foot. The Babinski sign is a measure of central nervous system involvement and should also be elicited. Figures 8.4, 8.5, and 8.6 illustrate the neurologic tests.

The measurement of nerve root tension signs is an important part of the examination. The authors prefer to log roll the limb initially to measure hip flexibility and then carry out a gentle range of motion of the hips prior to testing the straight leg raising. The straight leg raising maneuver has been described in many ways and under many names (Scham and Taylor 1971; Cram 1953). The simplest approach is to elevate the leg with the knee extended, and with the examining hand placed on the pelvis to guard against pelvic motions (see Figure 8.7). The test is recorded as a positive nerve root tension sign when sciatic pain is reproduced. The degree of elevation necessary to produce the symptoms is recorded. Often, the individual will complain of LBP rather than sciatica; in this instance the degree of elevation should be recorded, but the test is not positive. Straight leg raising may also be restricted by tightness of the hamstrings or a feeling of muscle tightness behind the knee. The degree of elevation when the tightness is felt should be recorded but is not a positive straight leg raising test. Sometimes sciatic pain will occur in the opposite leg, referred to as contralateral crossed positive straight leg raising test. Again, the degree of elevation should be recorded (Woodhall and Hayes 1950).

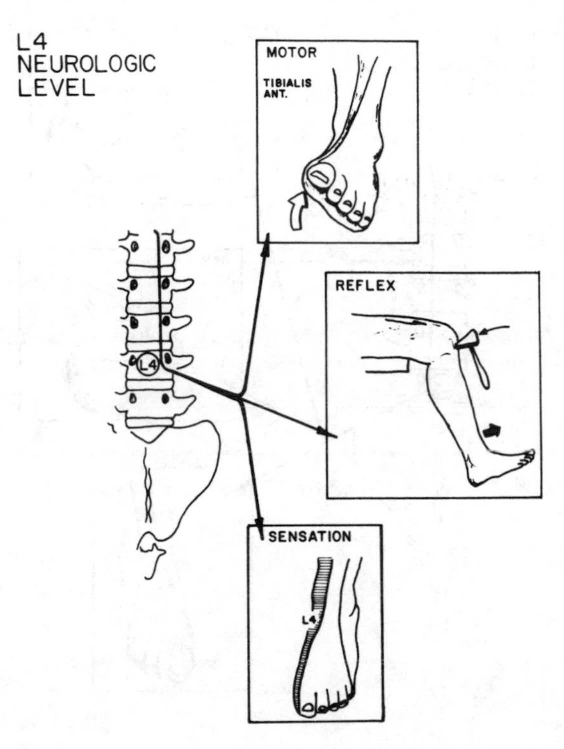

Figure 8.4
Tests for neurologic function of the L4 root include the knee jerk, quadriceps strength, and sensory areas as depicted.

A variety of confirmatory tests have been derived for straight leg raising (Scham and Taylor 1971, Troup 1981). Only three will be discussed here, since they are used quite commonly.

1. *Lasegue's sign*. The hip and knee joints are flexed 90°, and the lower limb then raised to the point where the sciatic pain is reproduced (see Figure 8.8a). The test can be further strengthened by dorsiflexing the foot. A positive response is confirmatory of the straight leg raising test.
2. *Internal rotation test.* The leg is reduced from the straight leg position which is causing sciatica to the point where pain is relieved. The hip is then internally rotated. A positive test is reproduction of the sciatic pain (Breig and Troup 1979).
3. *Femoral stretch test.* The femoral stretch test is a nerve root tension sign that selectively affects the L3 and L4 nerve roots. The patient is placed in the prone position. The examiner places his hand on the buttocks, and then, with the hip extended, the knee is flexed. A positive test is

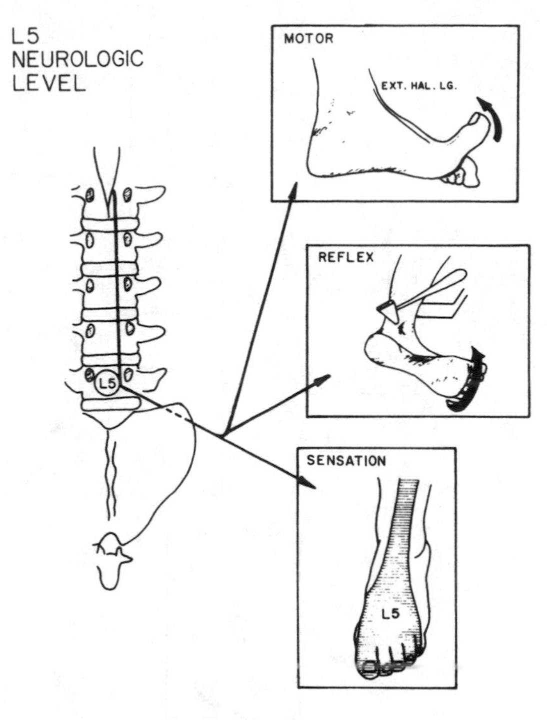

Figure 8.5
Tests for neurologic function of the L5 root include decreased strength and sensation as depicted. The posterior tibial reflex when present is a useful confirmatory test, but is bilaterally absent in 50 percent of all patients.

reproduction of anterior thigh pain, which is seen in this group of nerve root entrapments. The test's validity diminishes as soon as the pelvis begins to rock under the examiner's hand.

(c) Examination In the Prone Position. The worker is then placed on his stomach. General palpation of the spinous processes, interspinous space, paraspinal muscles, sacroiliac joints and sciatic nerve is then carried out. The examination should not be limited to the lumbar spine but should include also the thoracic spine. (One pitfall, particularly in elderly persons who have had an acute fall, is to make an incomplete examination of the spine, thereby missing local tenderness at the thoracolumbar junction which is the common site of fracture.) The course of the sciatic nerve is palpated. Tenderness of the nerve implies active "neuritis" indicative of nerve root irritation. Local swelling and tenderness can indicate a nerve tumor. The degree of forceful percussion that one then uses in re-examining the spine varies according to the degree of sensitivity elicited in more gentle palpation.

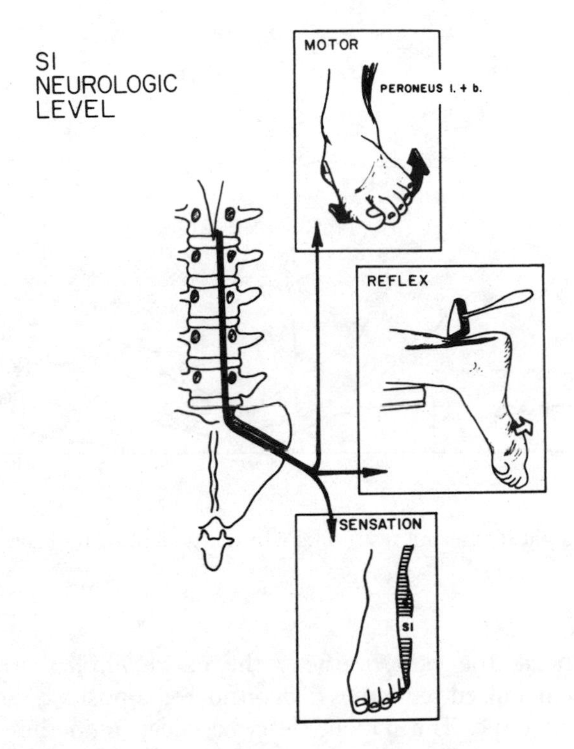

Figure 8.6
Tests for neurologic function of the S1 root is depicted.

Peroneal and gluteal cleft sensation should be evaluated for pin prick sensitivity, since diminished sensibility in this region is suggestive of an acute cauda equina lesion (see Chapter 3) (Aho, Auranen and Pesonen 1969).

(d) In the Sitting Position. The neurologic examination of the lower extrémities and some of the specialized tests described in the next section can be carried out with the worker sitting. In addition, it is helpful to observe his sitting tolerance and posture as an indication of LBP.

(e) Specialized Tests. A variety of specialized tests have evolved to test the function of various structures and joints, for example the sacroiliac joints. In routine examination of most patients, we have found these tests to be of little use with the exception of five tests that Waddell *et al.* (1979) described. Those tests should alert the examiner to the possibility of psychologic distress and/or malingering. These tests are of greater importance in evaluating the worker who is failing to respond to simple conservative measures within the expected time interval than patients with acute symptoms. Five types of physical signs are described; within each group specific tests are carried out.

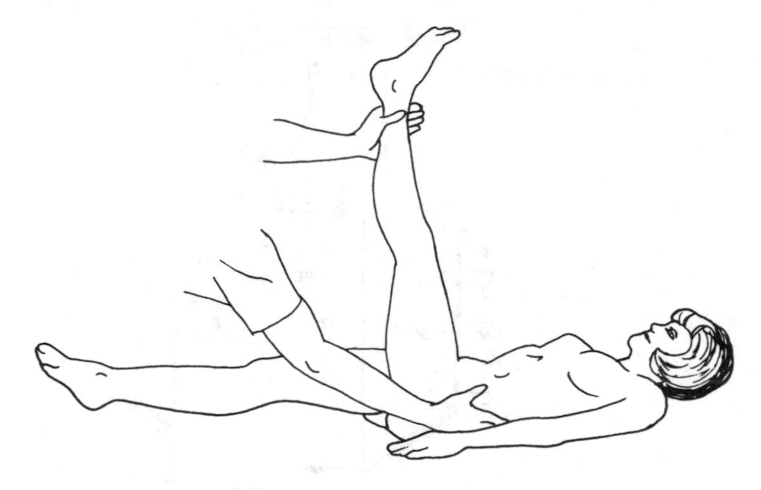

Figure 8.7
The classic straight leg raising test done with the patient recumbent.

1. *Tenderness:*

   Superficial: the skin overlying the entire lumbar area is lightly pinched. Generalized tenderness is considered a positive sign.

   Non-anatomic: The patient describes deep tenderness, when one palpates over a wide area of anatomic structures. For example, a positive test would be a patient who describes local tenderness in the thoracic spine, paraspinal muscles, pelvis and sacrum.

2. *Simulation Tests:*

   The object of these tests is to give the patient the impression that a specific test is being performed, when in fact the test is not being performed.

   Axial Loading: Compressive load is applied to the head. Persons complain of neck pain, but this maneuver rarely will reproduce back pain. Reproduction of back pain is therefore a positive test.

   Rotation: The shoulders and pelvis are passively rotated together so that no actual spine motion occurs yet the patient believes rotation is being tested. Reproduction of back pain is a positive test.

3. *Distraction Tests:*

   The object of this group of tests is to distract the person in such a fashion that a test which has been experienced as a positive under one set of circumstances now becomes negative in the distracted patient. An example is the Distraction Straight leg raising test: The usual method is to have the worker sit and perform the straight leg raising test maneuver, under the guise that the knee, foot, or ankle are being examined. If he complains of pain reproduction with the straight leg raising test being performed in the supine position and not in the sitting position, the test is considered as positive (Figure 8.8a).

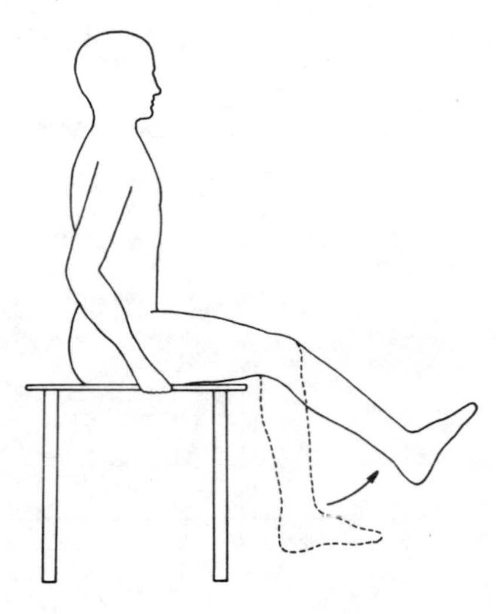

Figure 8.8a
The patient is sitting, and a straight leg raising maneuver is performed, usually by extending the leg from the knee bent to the fully extended knee position. In this exam, the patient is often told that the knee is being examined. If the patient has complained of sciatic pain on the recumbent straight leg raising, but the sitting straight leg raising (SLR) test is negative, it may be suggestive of malingering behavior.

4. *Regional Disturbances:*

   In general, pain distributions are a function of known anatomic pathways and structures. Interpretation of the examination therefore depends upon the patient giving non-anatomic, or non-physiologic responses to testing.

   Weakness: In a positive test, voluntary muscle contraction is accompanied by recurrent giving away, producing motions similar to a cog wheel.

   Sensory: Alterations in sensibility to touch and pin prick occur in a non-anatomical pattern, of which the most common varieties are the "stocking-glove" distribution, or diminished sensation over the entire half of the body or over one quadrant of the body.

5. *Overreaction:*

   During the entire examination, the person demonstrates overreactive behavior characterized by excessive verbalization of pain, facial expression signifying pain, collapsing episodes, and sweating.

Waddell and associates have demonstrated that when three of five categories of tests are positive, there is a high probability of non-organic

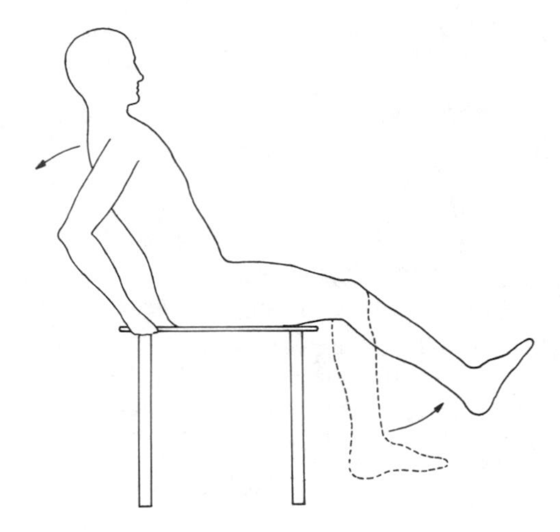

Figure 8.8b
When a SLR maneuver is performed with a seated patient, he may experience exacerbation of the sciatica, and attempt to shift into the extended spinal position, thus minimizing the degree of SLR elevation. The so-called "flip" test is one confirmatory test of nerve root entrapment.

pathology. In specific, they have used these tests to identify the individual who needs further psychologic assessment.

3. Formulation of the History and Physical Examination

At the completion of the history and physical examination, a general formulation can be made of the worker's low back problem, including (1) decisions regarding the immediate need for further diagnostic tests; (2) an initial therapeutic plan; (3) a preliminary determination of the immediate disability; and (4) an outline of reasonable expectations for return to work. The following questions are helpful in making this formulation:

(a) Has the worker had a major traumatic event (for example a fall) that is sufficient to produce structural injury? A history of major injury combined with a physical examination which shows local tenderness, swelling, and skin discoloration should alert one to this possibility. Obviously, the presence of such an injury, particularly when accompanied by any sign of neurologic impairment, should lead to careful transport of the patient to the nearest hospital or trauma center. Radiographic examination will be required in this group.

(b) Does the worker have an underlying disease process that might weaken structures to produce acute structural injury, despite the magnitude of the force being less than usually anticipated to produce such an injury?

Persons who have taken cortisone or its derivatives, those with chronic debilitating diseases (for example rheumatoid arthritis), and those with known neoplasms are examples of workers at risk. Osteoporosis can also contribute to lower resistance of tissues to trauma. In this group, spinal radiographs are warranted (Johnston and Epstein 1981).

(c) Does the worker have some other, nonmusculoskeletal causation of LBP that requires immediate attention? Although this is unlikely to occur in an industrial setting, identification of an abdominal mass that is pulsatile would be cause for immediate action (aneurysm of the aorta).

(d) Does the worker have significant neurologic dysfunction? For example, a person with acute low back and leg pain, diminished muscle strength in lower extremities, and diminished sensation in the perianal area would warrant immediate attention because of the suspicion of a cauda equina lesion.

The probability that any of these events will occur in industry is low, but the four decisions are important because of the potential preventable loss of function or the need for early intervention. In most instances, the individual undergoing evaluation will either do so in the setting of recurrent or chronic LBP, with an acute exacerbation, or their first episode of LBP with trauma insufficient to produce bony injury.

DIAGNOSTIC STUDIES

For the majority of workers, further diagnostic studies are not required. The major determinations to be made include whether the individual can continue work at that time or requires some form of therapeutic intervention, most commonly bedrest or reduced activity.

Further diagnostic studies are indicated under the following circumstances, in addition to the four indications listed previously:

(a) The worker is experiencing chronic and increasing LBP with or without associated neurologic symptoms.

(b) The worker is experiencing recurrent episodes that are increasing in frequency or intensity.

(c) The worker is not responding within the anticipated time period to treatment of the acute symptoms (usually 2 to 3 weeks).

The further evaluation of the worker is dependent on his pain patterns and the severity of his symptoms.

Laboratory Studies

1. *Blood Tests*

A variety of blood tests may be used in the evaluation of the worker with LBP. Table 8.2 lists these tests, the meaning of the measurement, and the low back conditions in which abnormalities might be detected (Calin 1981, Johnston and Epstein 1981, Galen 1979, Goldsmith 1979, Nawab and Azar 1979, Van Lente 1979).

It is apparent from reviewing Table 8.2 that abnormalities in laboratory studies will not be found commonly with LBP. Obtaining these tests therefore depends upon clinical suspicion of infection, tumor, or a spondyloarthropathy. They are not incidated as a general screening test in low back disease.

2. *Spinal Radiography*

Radiographs are warranted if the worker's pain does not improve over a period of two to four weeks (Feldman 1983). If radiographs have been done within the past two years, there is a minimal probability that interval changes will either change the diagnosis or influence treatment (Scavone *et al.* 1981). The exceptions would be workers who have had acute major trauma or who are now demonstrating symptoms suggestive of tumor or neoplasm, or spondyloarthropathy. The predictive value of radiographs will be discussed in Chapter 11, where the limitations of radiography will be considered.

The preferred method of spinal radiography is a standing anteroposterior and lateral radiograph centered at the L4-5 disc level (Pope *et al.* 1977). The radiographic findings that are not significant have largely been clarified in Chapter 3.

3. *Additional Plane Radiographic Studies*

(a) Oblique Radiography. Oblique radiographs usually are not warranted and add unnecessarily high x-ray exposure for routine use. In a patient with spondylolisthesis, the oblique may be useful in determining the integrity of the pars interarticularis. The traditional role of oblique radiographs in assessing facet arthropathy and orientation seems to be of limited value (Scavone *et al.* 1981).

(b) Sacral Radiography. When a worker is suspected of having a spondyloarthropathy, or on those rare occasions when a sacral lesion is thought to produce LBP (sacral cordoma), 30° tilt shots of the sacrum will define that structure more completely and also give greater information regarding the sacroiliac joints.

(c) Flexion/Extension Radiography. Persons who have had recurrent episodes of LBP accompanied by postural disturbance (scoliosis) or those with persistent LBP may be suspected of having a segmental instability. To demonstrate instability, the person must be relatively mobile. Obtaining adequate flexion and extension radiographs may therefore have to be delayed until acute symptoms subside or may require that pain is temporarily blocked by a facet injection (see below). In the flexion/extension radiography, the patient is either seated or stands, and is instructed to flex maximally and extend his spine. At the fully flexed and extended positions a

TABLE 8.2
Usefulness of Clinical Laboratory Studies in Diagnosing Low Back Pain

Test	*What Test Measures*	*Low Back Conditions Where Altered*
Complete Blood Count		
Hematocrit/ hemaglobin	A measure of volume of circulating red blood cells.	May be diminished in systemic diseases, i.e. neoplasm, & in chronic spinal infections.
White blood count & differential	Amount & type of circulating white blood cells.	Total white blood cell and shifts in differential may be present in spinal infections, or occasionally in spondyloarthropathies.
Sedimentation Rate	Non-specific test of inflammation.	Increased in spinal infections, may be increased in neoplasms & spondyloarthropathies.
Chemistry		
Calcium, Phosphorous	A measure of circulating calcium & phosphorous.	Calcium is elevated in hyperparathyroidism, may be elevated with primary & secondary osseous tumors, alterations in the distribution of calcium & phosphorous accompany many metabolic disorders but are normal in osteoporosis.
Alkaline Phosphatase	Enzyme associated with bone formation. Therefore elevation implies increased bone formation.	May be elevated in primary or secondary osseous neoplasms.
Acid Phosphatase	An enzyme associated with tumors metastatic to bone.	Increased in prostatic dysfunction.
Serum Proteins (albumin/globulin/ protein electro- phoresis)protein.	Measurement of amount & type of circulating	Elevations of one fraction of globulin is associated with multiplemyeloma.

(continued on next page)

TABLE 8.2 *(continued)*

Test	*What Test Measures*	*Low Back Conditions Where Altered*
HLA 27-B Antigen	A circulating antigen	Usually individuals with spondyloarthropathies are HLA 27-B positive. Note 6–8% of males have this antigen and therefore its presence is not confirmatory of a spondyloarthropathy.

Source: The authors of this chapter.

lateral radiograph is taken. Figure 8.9 illustrates a positive test, and the criterion for its interpretation (Hanley *et al.* 1976, Macnab 1971, Posner 1982, Knutsson 1944, Moran and King 1957). This method is also used to assess the integrity of a previously performed spinal fusion (Frymoyer 1979).

Some specialists add "dynamic" anteroposterior radiographs, taken in maximal right and left lateral bend. Restrictions in motion at motion segment levels are interpreted as evidence of disc herniation at that level. We think that the information to be gathered is of insufficient use to warrant the routine use of this test.

(d) Tomograms

Plane x-ray tomograms can be useful in evaluating severe osseous trauma.

4. Radionuclide Scanning

A bone scan involves the injection of a radioactive isotope that is selectively taken up by bone or other tissues (Goldstein 1983). Following injection, the entire body or a local area is scintigraphically scanned. Areas of increased uptake correlate with foci of bone formation plus or minus destruction.

A bone scan is indicated when clinical histories and pain patterns suggest a neoplasm or infection. Specifically, this test would be applied in the following circumstances:

(a) If the worker has plane radiographic evidence suggestive but not confirmatory of a neoplasm or infection.

(b) The worker has a laboratory abnormality associated with neoplasm (for example elevated alkaline phosphatase).

(c) The worker has acute LBP associated with those sports in which there is an increased risk of fatigue fracture of the neural arch. These

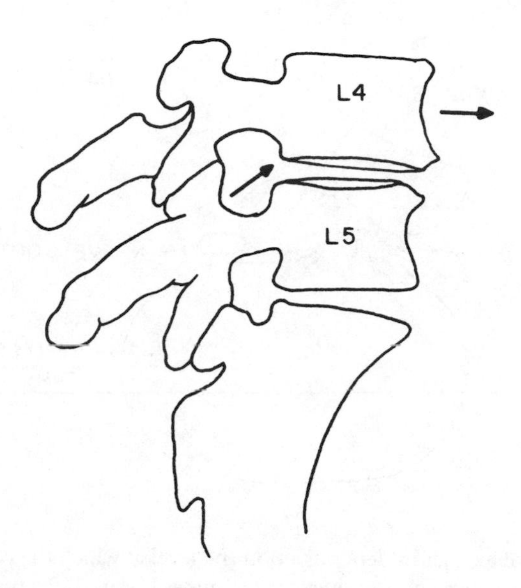

Figure 8.9
A flexion/extension radiograph is shown where there is segmental instability. Signs of this disorder include anterior traction spurs, and in forward bending films anterior diminishment of the L4-5 or other disc space height, and forward translation of the superior vertebra relative to its lower neighbor. Displacements of 3-4 mm are considered as positive evidence of segmental instability.

generally will be younger (teenage) individuals (see Chapter 4) (Wiltse 1975).

5. *CT Scan*

The CT scan is a powerful tool in the diagnosis of low back disease (Chafetz and Genant 1983, Burton *et al.* 1979). However, it involves significant radiographic exposure; many of the observed "abnormalities" probably exist in asymptomatic subjects; and the method is not necessary to determine the causation of pain in most LBP sufferers. Like any other diagnostic test, specific indications are available for CT scanning:

(a) Suspected Herniated Nucleus Pulposus. The patient presents clinical symptoms and signs of herniated nucleus pulposus (HNP). As illustrated in Figure 8.10, a very clear picture of the herniation may be obtained in CT scans.

(b) Spinal stenosis. In patients with signs and symptoms of stenosis, the CT scan offers one of the most reliable methods for assessing spinal canal dimensions (see Figure 8.11). In a few centers, persons suspected of having stenosis are first screened by ultrasonic scanning of the spinal canal, a topic beyond the scope of this chapter (Porter *et al.* 1980).

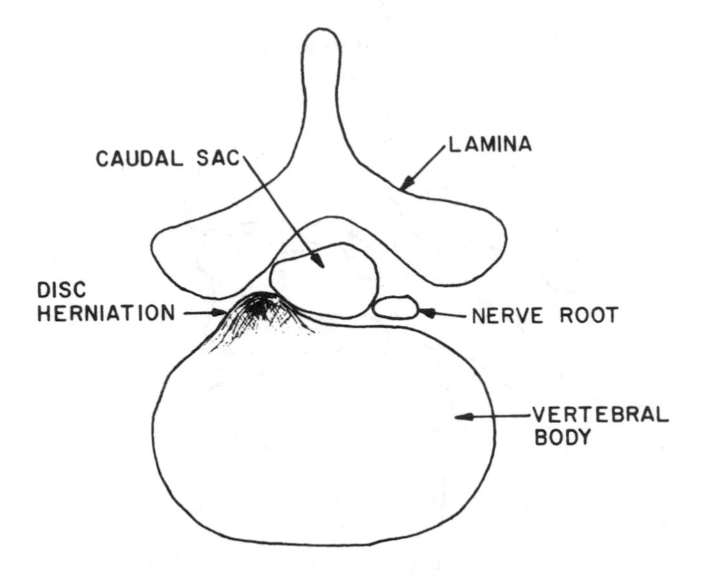

Figure 8.10
The appearance of a CT scan depends upon the level at which the radiographic cut is made. In this instance, a disc herniation appears on the left and is depicted compressing the nerve root, and distorting the caudal sac.

(c) Assessment of facets. If the worker is thought to have facet arthropathy, or relief of symptoms occurs with facet block, these joints can be most thoroughly investigated by CT scan.

In addition to these general indications, CT scan is invaluable in assessing the person with recurrent or continued symptoms after unsuccessful spinal surgery, in the preoperative planning for patients with fracture and fracture/dislocations of the spine, and to determine the extent of involvement in bony neoplasms or in soft tissue neoplasms adjacent to the spine.

6. *Myelography*

Myelography is a well established method for the evaluation of extra- and intradural space occupying lesions in the spine (Figure 8.12). The contrast media injected intrathecally may either be fat soluable (Pantopaque) or water soluable (Metrizamide) (Hudgins 1970, Rothman *et al.* 1974). Currently, the water soluable contrast media are most popular because they do not have to be removed at the completion of the procedure; they give a higher resolution with better definition of the nerve root sheaths, and therefore better definition of peripherally located extradural lesions; they also seem to have less risk of producing inflammation of the meninges

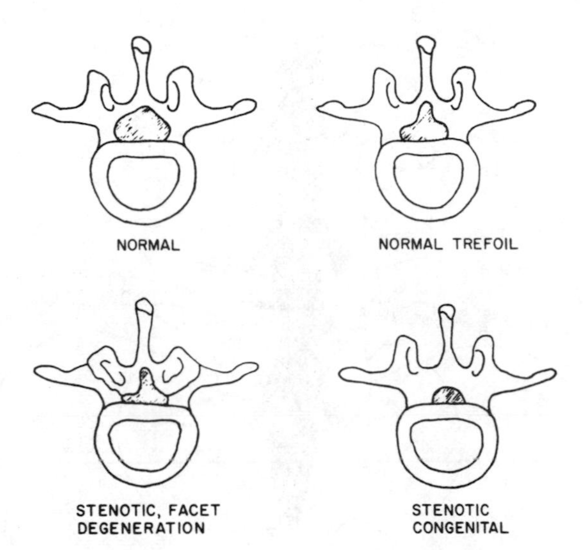

Figure 8.11
Spinal stenosis as depicted here represents one of many appearances which may be identified on CT scanning. The normal spine dimensions are shown, and a normal variant, the trefoil-shaped canal. The latter shape makes the nerve root more vulnerable to small disc herniations. In the lower two drawings, narrowing of the spinal canal is demonstrated, the first instance being related to degeneration of the facets and bony overgrowth producing a trefoil shape, the latter being on a congenital basis. Multiple combinations of spinal stenosis can occur.

(arachnoiditis). Regardless of the contrast media used, myelography is not a benign procedure, and post-myelographic complications occur in 10 to 20 percent of individuals. These complications include seizure-like activity of the lower extremities which occur in 10 percent of persons undergoing metrizamide myelography, and grand mal seizures which occur in one percent of such persons, particularly among known alcohol abusers. Of interest is that complications of myelography are more common when the examination is negative, or Workmen's Compensation is involved, which again underscores the importance that this examination should not be used in the general evaluation of patients with low back pain (Rothman 1983). Methods used to prevent these complications include the administration of Diazepam (Valium (Rx)) before examination.

The specific indications for myelography are as follows:

(a) The evaluation of persons whose clinical signs and symptoms suggest the diagnosis of herniated nucleus pulposus, and in whom surgical intervention is actively being contemplated. This focuses the appropriate attention on the myelogram as a preoperative planning tool for the precise

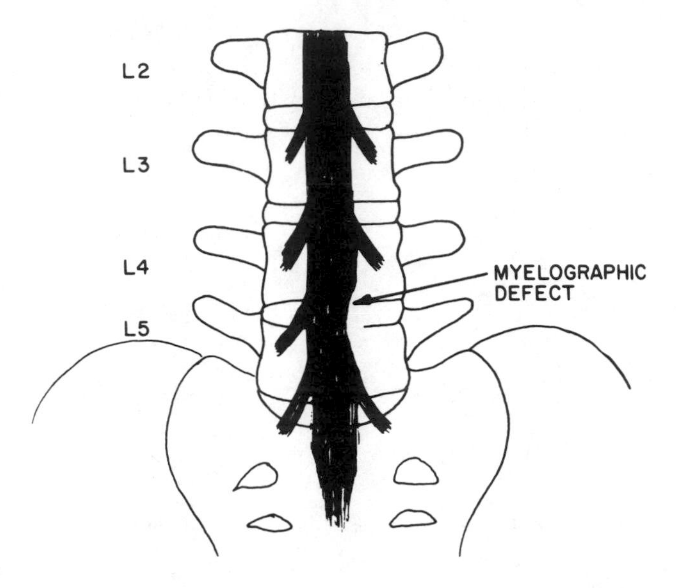

Figure 8.12
A myelogram is depicted. In actual practice, the myelographic dye appears as a whitish column on the radiograph. Note that nerve roots fill with dye, and are symmetric at each spinal disc space level, except at L4-5 on the right, where the nerve root does not appear to fill, and there is indentation into the dye column. This is a fairly typical appearance of a disc herniation.

localization of disease, rather than as a screening tool. (The test is of minimal use in defining the causation of nonspecific low back pain, be it chronic or episodic in nature).

(b) To evaluate the extent of disease in a patient with spinal stenosis in whom surgery is contemplated. For the test to be meaningful, a full column technique is desirable which means the entire lumbar canal is filled with dye.

(c) To evaluate persons with sufficient signs and symptoms to suggest a spinal cord neoplasm.

(d) To evaluate the spinal canal in persons with severe spinal injuries (fractures, fracture/dislocations) where a neurologic defect exists or is progressing, or where the patient is being considered for spinal stabilization.

The correlation of the clinical signs and symptoms with the myelogram are essential to its interpretation. In one study, 35 percent of persons without low back symptoms had defects present in their lumbar myelograms (Hitselberger and Witten 1968). In persons who previously have had spinal surgery, the myelogram is notoriously difficult to interpret because of the postsurgical defects which are present, usually because of scar formation

Combining a myelogram with a CT scan may be useful in this group or in persons with complex low back disorders accompanied by neurologic signs and symptoms. This combined examination is termed a metrizamide enhanced CT scan.

7. *Discography*

Injection of radio-opaque contrast media into the nucleus pulposus has been controversial, and its application varies widely. Proponents of the discogram feel they can demonstrate degenerative lesions and adequately define disc herniations (Collis 1962). Additionally, the amount of dye that the disc accepts and the variable reproduction of back and/or leg symptoms at the time of injection is thought to increase its reliability and yield additional diagnostic information. The negative view relates to the interpretation of the test, particularly the interpretation of a "degenerate disc." Forty percent of asymptomatic males have positive discograms (Holt 1968). Currently the test is becoming more popular in patients undergoing chymopapain treatment of disc herniations. Because the chymopapain injection requires needle placement within the nucleus pulposus, discography is easily obtained.

8. *Electrodiagnostic Studies*

Electrodiagnostic studies are most applicable to workers who have the signs and symptoms of nerve root entrapment (caused by herniated nucleus pulposus or spinal stenosis). These tests measure impairment of physiologic neuromuscular function and can be useful in further defining the level of nerve compression. Generally, the studies applied to low back problems involve electromyography which measures the electrical activity of muscles. Usually, muscles of the lower extremity as well as the paraspinal musculature are investigated. The afferent measurement of conductivity of nerves can also be measured through the evoked somatosensory potential. In this test, peripheral nerves are stimulated, and the transmission of electrical impulse is measured by electrode attached to the skull. This test is currently most useful in monitoring surgical procedures that involve the spinal cord or its adjacent structures, although there is some interest in its use in assessing spinal stenosis. In other instances, conduction of electrical impulses through nerves is measured by stimulating a peripheral nerve and measuring the time for the impulse to reach another segment of that nerve. One can therefore describe a temporal passage of electricity through the nerve as well as analyze the produced electrical signal. The measurement of nerve conduction is used to differentiate a nerve root lesion from a peripheral nerve lesion.

9. *Thermography*

Disturbances in nerve root function are sometimes accompanied by alterations in local blood flow, producing secondary localized abnormalities in tissue temperature. Some feel these temperature changes can be measured by superficial thermography. The proponents of this method cite the advantage that the test is noninvasive (Pochaczevsky 1983). However, many experts believe the test has little value and we agree.

10. *NMR*

Nuclear magnetic resonance (NMR) is a promising, new, noninvasive technique for the evaluation of the lumbar spine. The principles and underlying physics of NMR are beyond the scope of this book. It does not involve any radiographic exposure, nor does it have any known adverse effects on biologic tissues. Hence it offers an extremely exciting possibility for further understanding of the mechanisms of LBP. However, the equipment, as well as the tests, is extremely costly, and its place in the routine evaluation of the lumbar spine remains uncertain.

In Chapter 1 we stressed that, for a structure to be considered as pain-productive, one must be able to produce symptoms when the structure is noxiously stimulated and relieve symptoms when its nerve supply is blocked by local anesthesia. This principle has been applied to the facet joints, nerve roots, and less selectively through the use of intra- and extradural injections of local anesthetics.

1. *Facet Joint Blocks*

There are no uniform criteria for the use of facet joint blocks (Badgley 1941; Fairbank *et al.* 1981; Mooney and Robertson 1976; Paris 1983; Mooney 1983). This procedure appears indicated mainly in patients with chronic or recurrent LBP that is failing to respond to conservative management, particularly when CT scan has demonstrated local facet abnormalities or when the plane radiographic picture or flexion/extension films are suggestive of segmental instability. Another relative indication is to provide temporary pain relief for patients with LBP so that more complete flexion and extension radiographs can be obtained. Because definition of mechanical LBP and segmental instability is variable and the criteria imprecise, the selection of the joints to be injected is highly variable, and often is based upon suggestive plane radiographic findings such as asymmetric disc space collapse, presence of traction spurs, presence of retrospondylolisthesis, or scoliosis. Usually, the L4–5 facet joints, and/or the L5–S1 facets are injected. Twenty to twenty-two gauge spinal needles are introduced into the joint, or adjacent to it, under fluoroscopic control (see Figure 8.13). A pain provocative test may be done by injecting a small amount of hypertonic saline. Also, some clinicians attempt to outline the

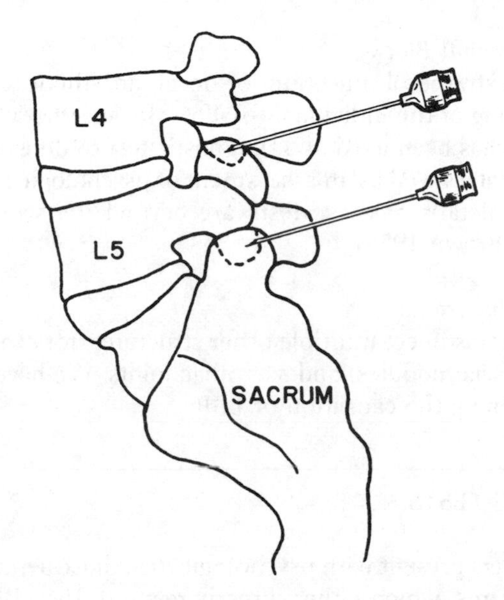

Figure 8.13
In a facet block, the facet joint capsule can be punctured, although many experts prefer to inject material adjacent to the facet joints rather than directly into it. Typically, the right and left L4-5, or L5-S1 facet joints are injected, although not uncommonly all four joints may be injected.

facet joints more completely by injecting contrast media into it thus producing a facet arthrogram (Maldague *et al.* 1981, Glover 1977). Following these procedures, one-half to one cc of local anesthetic (usually one of the longer acting anesthetics) is injected. Often, the anesthetic is accompanied by a corticosteroid. The individual's initial response and pain relief is noted, and the person is asked to report the duration of pain relief. Approximately 20 percent of persons undergoing a facet block will experience lasting relief if the initial response has been favorable. However, the person should be warned that the test is primarily for diagnostic purposes and that the therapeutic outcome is highly variable.

2. *Nerve Root Block*
Selective injection of a nerve root at its foraminal exit can be a useful test in persons with nerve root entrapment symptoms that are not easily defined. Simultaneous injection of a radio-opaque contrast media may produce an outline of the nerve root far beyond that seen in routine myelography and thus define extraforaminal sources of nerve root compression. This test is generally reserved for highly complex nerve root pain problems, particularly in patients who have undergone previous surgery (Krempen *et al.* 1975, Krempen and Smith 1974).

3. *Differential Spinal Block*

Intra- or extrathecal injection of local anesthetic accompanied by careful positioning of the individual to allow the anesthetic to slowly rise in the spinal canal has been used as a diagnostic test to determine the level of nerve involvement, as well as an assessment of psychologic malfunction. The techniques and details of these tests are beyond the scope of this book. (Mooney 1975, Brown 1975).

4. *Other Local Injections*

Some clinicians inject multiple other structures, for example ligaments, so-called myofascial nodules, and sacroiliac joints. We have found them of little use in defining the causation of LBP.

PSYCHOLOGIC TESTS

Some workers present with psychologic maladjustments or maladaptive coping mechanisms, which either directly relate to the LBP or affect their ability to deal with their low back complaints. Clinicians experienced in the evaluation of LBP may be quite accurate in identifying this group of patients. Waddell's tests, described in the section of this chapter headed "Physical Examinations" help the examiner to select those workers who are at risk. Structured psychiatric interviews are frequently used in evaluating this group, although the most commonly applied test vehicle is the Minnesota Multiphasic Personality Inventory (MMPI) (Lawlis and McCoy 1983). The psychologic profiles derived from the MMPI have been discussed in Chapter 3. Because the MMPI is both time consuming and costly, and requires specialized interpretation, other more simple screening tests have been sought. The most popular is the pain drawing (see Figure 8.14) (Ransford *et al.* 1976) which easily can be applied in an office or industrial setting. Other psychologic health inventories and modifications of the MMPI have been used. These latter tests usually tap selectively those scales of the MMPI that are most relevant to LBP and may be useful as screening devices (Rosen *et al.* 1980).

SUMMARY

Based on the natural history of LBP, its management can be divided into acute, subacute and chronic stages. Improvement in 85 to 90 percent of cases occurs in 6 to 12 weeks. Attempts should be made to reduce the duration of the disability. Bedrest, back school, traction and manipulation appear to be effective in acute LBP. For those with chronic LBP, the options include exercise, corsets and braces, surgical procedures, and comprehensive

NUMBNESS ==== PINS & NEEDLES oooo BURNING XXXX STABBING ////

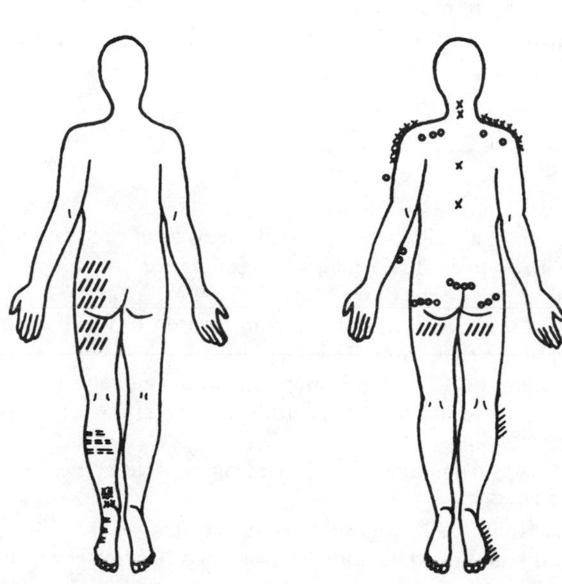

Figure 8.14
A variety of pain drawings are used. Typically the patient is told to draw the location of pain using some code for pain (XX), numbness (OO), tingling or other sensory changes. In usual instances, the drawings include both front and back views. Depicted here are two drawings of the posterior view of a patient, one from a patient with clear cut sciatica, the other from a patient with numerous psychologic problems. Note in the second case, pain is depicted as being external to the body, and located in multiple areas. (After Ransford, A.O.; Cairns, D.; and Mooney, V. 1976. The pain drawing as an aid to the psychologic evaluation of patients with low back pain. *Spine* 1:127-134.)

rehabilitation. The effectiveness of most drugs has not been proven, and narcotic analgesics should be avoided. The majority of LBP patients do not require surgery. Comprehensive rehabilitation involves the steps of accepting the impairment, goal establishment, understanding barriers, reinforcement and pain management.

The evaluation of the worker with LBP is most critically dependent on an accurate history and physical examination. Further immediate diagnostic testing would be appropriate in a small group of workers with specific pain patterns and typical findings or in those individuals who have had major acute injuries. For most persons, the response to initial treatment will dictate need for further testing. These tests should be performed only when properly indicated and must be correlated with clinical signs and symptoms. The indiscriminate use of tests in the hope of finding a cause for LBP is not only costly, but in certain instances risky, both because of direct test complications and also because over-interpretation leads to inappropriate surgical intervention.

REFERENCES

Aho, A.J.; Auranen, A; and Pesonen, K. 1969. Analysis of cauda equina symptoms in patients with lumbar disc prolapse. Preoperative and follow-up clinical and cystometric studies. *Acta Chir. Scand.* 135 (5):413–20.

Badgley, Carl E. 1941. The articular facets in relation to low back pain and sciatic radiation. *J. Bone Joint Surg.* 23 (2):481–96.

Breig, A.; and Troup, J.D.G. 1979. Biomechanical considerations in the straight-leg-raising test: Cadaveric and clinical studies of the effects of medial hip rotation. *Spine* 4 (3):242–50.

Brown, Mark D. 1975. Diagnosis of pain syndromes of the spine. *Orthop. Clin. North Am.* 6 (1):233–48.

Burton, C.V.; Heithoff, K.B.; Kirkaldy-Willis, W.; and Ray, C.D. 1979. Computed tomographic scanning in the lumbar spine. Part II. Clinical considerations. *Spine* 4 (4):356–68.

Calin, Andrei. 1981. Ankylosing spondylitis. In *Textbook of Rheumatology*, Vol. 2, pp. 1017–30. Eds., William N. Kelley, Edward D. Harris, Jr., S. Ruddy, Clement B. Sledge, Philadelphia: W.B. Saunders Co.

Chafetz, Neil; and Genant, Harry K. 1983. Computed tomography of the lumbar spine. *Orthop. Clin. North Am.* 14 (1):147–69.

Collis, John S., Jr.; and Gardner, W. James 1962. Lumbar discography. An analysis of one thousand cases. *J. Neurosurg.* 19:452–61.

Cram, R.H. 1953. A sign of sciatic nerve root pressure. *J. Bone Joint Surg.* 35B(2):192–95.

Donaghy, R.M. Peardon. 1946. The posterior tibial reflex. A reflex of some value in the localization of the protruded intervertebral disc in the lumbar region. *J. Neurosurg.* 3:457–59.

Edgar, Michael A.; and Park, William M. 1974. Induced pain patterns on passive straight-leg raising in lumbar disc protrusion. A prospective clinical, myelographic and operative study in fifty patients. *J. Bone Joint Surg.* 56B (4):658–67.

Fairbank, J.C.T.; Park, W.M.; McCall, I.W.; and O'Brien, J.P. 1981. Apophyseal injection of local anesthetic as a diagnostic aid in primary low back pain syndromes. *Spine* 6(6):598–605.

Feldman, Frieda. 1983. The symptomatic spine: Relevant and irrelevant roentgen variants and variations. *Ortho. Clin. North Am.* 14(1):119–45.

Frymoyer, J.W.; Hanley, E.N., Jr. ; Howe, J.; Kuhlmann, D.; and Matteri, R.E. 1979. A comparison of radiographic findings in fusion and nonfusion patients ten or more years following lumbar disc surgery. *Spine* 4 (5):435–40.

Galen, Robert S. 1979. The predictive value of laboratory testing. *Orthop. Clin. North Am.* 10 (2):287–97.

Glover, John R. 1977. Arthrography of the joints of the lumbar vertebral arches. *Orthop. Clin. North Am.* 8 (1):37–42.

Goldstein, Harold A. 1983. Bone scintigraphy. *Orthop. Clin. North Am.* 14 (1):243–56.

Goldsmith, Ralph S. 1979. Calcium, phosphate, and vitamin D. *Orthop. Clin. North Am.* 10 (2):319–27.

Hanley, Edward N., Jr.; Matteri, Richard E.; and Frymoyer, John W. 1976. Accurate roentgenographic determination of lumbar flexion-extension. *Clin. Orthop.* 115:145–48.

Hazlett, John W. 1975. Low back pain with femoral neuritis. *Clin. Orthop.* 108:19–26.

Hitselberger, William E.; and Witten, Richard M. 1968. Abnormal myelograms in asymptomatic patients. *J. Neurosurg.* 28 (3):204–206.

Holt, Earl P., Jr. 1968. The question of lumbar discography. *J. Bone Joint Surg.* 50A (4):720–26.

Hudgins, W. Robert. 1970. The predictive value of myelography in the diagnosis of ruptured lumbar discs. *J. Neurosurg.* 32 (2):152–62.

Johnston, C. Conrad., Jr.; and Epstein, Solomon. 1981. Clinical, biochemical, radiographic, epidemiologic, and economic features of osteoporosis. *Orthop. Clin. North Am.* 12 (3):559–69.

Keegan, J. Jay 1944. Neurosurgical interpretation of dermatome hypalgesia with herniation of the lumbar intervertebral disc. *J. Bone Joint Surg.* 26 (2):238–48.

Knutsson, R. 1944. The instability associated with disk degeneration in the lumbar spine. *Acta Radiol.* 25:593–609.

Krempen, John F.; and Smith, Buel S. 1974. Nerve-root injection. A method for evaluating the etiology of sciatica. *J. Bone Joint Surg.* 56A (7):1435–44.

Krempen, John F.; Smith, Buel S.; and DeFreest, Lynn J. 1975. Selective nerve root infiltration for the evaluation of sciatica. *Orthop. Clin. North Am.* 6 (1):311–15.

Lawlis, G. Frank; and McCoy, C. Edward. 1983. Psychological evaluation: Patients with chronic pain. *Orthop. Clin. North Am.* 14 (3):527–38.

Macnab, Ian 1971. The traction spur. An indicator of segmental instability. *J. Bone Joint Surg.* 53A (4):663–70.

Maldague, Baudouin; Mathurin, P.; and Malghem, J. 1981. Facet joint arthrography in lumbar spondylolysis. *Radiology* 140 (1):29–36.

Mooney, Vert. 1975. Alternative approaches to the patient beyond the help of surgery. *Orthop. Clin. North Am.* 6 (1):331–34.

Mooney, Vert. 1983. The syndromes of low back disease. *Orthop. Clin. North Am.* 14 (3):505–15.

Mooney, Vert; and Robertson, James. 1976. The facet syndrome. *Clin. Orthop.* 115:149–56.

Moran, Francis P.; and King, Thomas. 1957. Primary instability of lumbar vertebrae as a common cause of low back pain. *J. Bone Joint Surg.* 39B (1):6–22.

Nawab, Rehana A.; and Azar, Henry A. 1979. The laboratory diagnosis of plasma cell myeloma and related disorders. *Orthop. Clin. North Am.* 10 (2):391–404.

Paris, Stanley V. 1983. Anatomy as related to function and pain. *Orthop. Clin. North Am.* 14 (3):475–89.

Pochaczevsky, Ruben. 1983. The value of liquid crystal thermography in the diagnosis of spinal root compression syndromes. *Orthop. Clin. North Am.* 14 (1): 271–88.

Pope, M.H.; Hanley, E.N.; Matteri, R.E.; Wilder, D.G.; and Frymoyer, J.W. 1977. Measurement of intervertebral disc space height. *Spine* 2:282–86.

Porter, R.W.; Hibbert, C.; and Wellman, P. 1980. Backache and the lumbar spinal canal. *Spine* 5 (2):99–105.

Posner, I.; White, A.A.,3rd.; Edwards, W.T.; and Hayes, W.C. 1982. A biomechanical analysis of the clinical stability of the lumbar and lumbosacral spine. *Spine* 7 (4):374–89.

Pykett, Ian.L. 1982. NMR imaging in medicine. *Scientific American* 246 (5):78–88.

Ransford, A.O.; Cairns, D.; and Mooney, V. 1976. The pain drawing as an aid to the psychologic evaluation of patients with low back pain. *Spine* 1:127–34.

Rosen, James C.; Frymoyer, John W.; and Clements, Janice H. 1980. A further look at validity of the MMPI with low back patients. *J. Clin. Psychol.* 36 (4):994–1000.

Rothman, R.H. 1974. Patterns in lumbar disk degeneration. *Clin. Orthop.* 99:18–29.

Rothman, R. 1983. *Personal communication.*

Scavone, Jerome G.; Latshaw, Robert F.; and Rohrer, G. Victor. 1981. Use of lumbar spine films. Statistical evaluation at a university teaching hospital. *JAMA* 246 (10):1105–08.

Scham, Stewart M.; and Taylor, T.K.F. 1971. Tension signs in lumbar disc prolapse. *Clin. Orthop.* 75:195–204.

Troup, J.D.G. 1981. Straight-leg-raising (SLR) and the qualifying tests for increased root tension: Their predictive value after back and sciatic pain. *Spine* 6 (5):526–27.

Van Lente, Frederik. 1979. Alkaline and acid phosphatase determinations in bone disease. *Orthop. Clin. North Am.* 10 (2):437–50.

Waddell, G.; McColloch, J.A.; Kummel, E.; and Venner, R.M. 1979. Nonorganic physical signs in low back pain. *Spine* 5 (2):117–25.

Wiltse, Leon L; Widell, Eric H., Jr.; and Jackson, Douglas W. 1975. Fatigue fracture: The basic lesion in isthmic spondylolisthesis. *J. Bone Joint Surg.*, 57-A (1):17–22. 17–22.

Woodhall, Barnes; and Hayes, George J. 1950. The well-leg-raising test of Fajersztajn in the diagnosis of ruptured intervertebral disc. *J. Bone Joint Surg.* 32A (4):786–92.

9 Treatment, Education and Rehabilitation

Thomas R. Lehmann, John W. Frymoyer, and Raymond L. Milhous

INTRODUCTION

The management of the low back injured worker can be categorized according to the duration of symptoms: acute (0 to 3 months in duration; immediate onset); subacute (0 to 3 months; slow onset); and chronic (symptoms greater than three months duration (Nachemson and Andersson 1982). This division is based on the natural history of low back episodes, the costs attendant to treatment, and the special problem posed by the low back injured worker as the duration of the episode increases.

The natural history of LBP is remarkably constant. White (1966) and Bergquist-Ullman and Larsson (1977) both found 85 to 90 percent of low back injured workers improved within 6 to 12 weeks. The remaining 10 to 15 percent of patients whose symptoms were greater than three months in duration posed the single largest problem in management and also incurred the greatest costs.

The overall lesson to be derived is that health professionals involved in the care of low back injured workers should have a clear plan to manage the patient at the onset of symptoms. If the condition is not improving, early appropriate referrals should be encouraged. All effort should be made to reduce the duration of the disability. The physician/surgeon is important in controlling the "tempo" of initial diagnosis, prescription, and need for further consultation (See Figure 9.1). The industrial physician/nurse or physical therapist is faced with a special challenge because of his unique function. At times he is the primary physician and therefore will require skills in the diagnosis and initial treatment of LBP. He will also determine those workers who require further special evaluation and treatment. At other times, the industrial health professional is responsible for monitoring the costs of medical care and disability incurred by the injured worker. If the

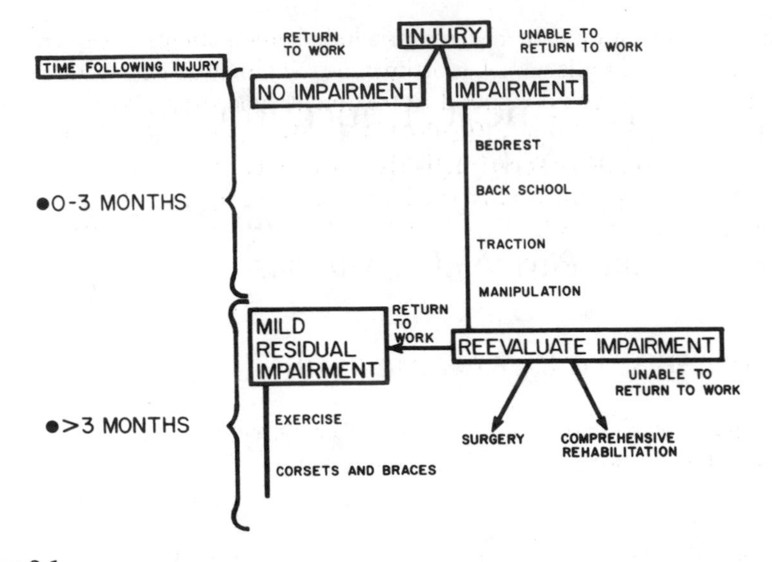

Figure 9.1
This outlines the general algorithm following industrial LBP injury. The treating physician is in the ideal position to govern the tempo of these interventions. Note that in the first three months following injury there is temporary impairment, the majority of workers returning activity within the three-month interval with only mild residual impairment. If the usual treatments are not leading to functional restoration, then re-evaluation of impairment is necessary followed by the appropriate intervention.

condition becomes more chronic and disabling, the industrial health professional may at times have to disrupt the established patient/therapist relationship in hopes of increasing the rate of recovery and minimizing the costs.

LBP is a symptom, and the structural cause often cannot be determined. It is often difficult to determine how effective various treatment methods are in reducing the duration of symptoms and their attendant disability. The natural history of acute LBP is generally favorable, and apparent favorable response to a drug or treatment may simply result from the natural history, rather than the specific treatment. There is also a significant possibility of the placebo effect. In most studies of pain management, 35 to 40 percent of subjects receiving a placebo, i.e., an inactive drug or treatment, will report symptomatic relief.

Although the patient and the physician may enthusiastically endorse a treatment method, the scientific evidence to establish that treatment's efficacy must rest on confirmed experimental analysis. In low back treatment, these experimental problems are compounded by the methodological difficulties including the fact that outcome measures of success are crude.

Frequently, the studies by necessity include some patients with acute pain and some with chronic pain. Unfortunately, there may be situations in which the acute patient would respond to treatment, whereas the chronic patient would not. Despite these difficulties, there is a fairly substantial body of evidence that supports some methods of treatment as being beneficial. We will discuss these efficacious methods according to the classification by duration of symptoms.

ACUTE AND SUBACUTE LOW BACK PAIN

Those treatments that have proven to be effective in acute LBP include bedrest, back school, traction, and manipulation.

Bedrest

The most significant study on this form of treatment for acute LBP was performed on military basic combat trainees experiencing their first episode of LBP (Wiesel *et al.* 1980). Patients treated with bedrest were able to return to full duty in a mean of 6.6 days compared to a mean 11.8 days for the ambulatory group. Based on these findings, the authors recommend bedrest as the treatment of choice for acute LBP, although the need for bedrest in patients with mild acute symptoms might be questioned. Perhaps immediate return to lighter work might be acceptable, thereby saving lost man-hours at work. In Wiesel's study the patients were classified into mild back pain (subjective complaints with no objective findings), moderate (limited range of spinal motion and paravertebral muscle spasm), and severe (inability to straighten spine and difficulty walking). Regardless of the symptom severity, bedrest was proven to be superior to the ambulatory treatment.

Wiesel also analyzed the effects of drugs on acute LBP. No significant shortening of the disability was demonstrable with the use of anti-inflammatory drugs (aspirin, phenylbutazone) or narcotic analgesics (codeine, oxycodone). However, the analgesic medications were effective in reducing the amount of pain. There are no studies that demonstrate reduced disability from the use of medication. Thus, analgesics should be kept in their proper context. Although they are capable of reducing pain complaints, analgesics do not have any specific healing properties. The same is true of most "muscle relaxants."

The majority of acute LBP patients will be able to return to function by two weeks, and therefore further diagnostic evaluation will be unnecessary. After the resolution of symptoms, a few days of graded increase in function may be advisable prior to returning to work. As will be seen, education and muscular rehabilitation may play an important role in the secondary prevention of further LBP episodes.

In those cases in which symptoms persist for greater than fourteen days, a thorough evaluation becomes indicated. At this time, the clinical picture may be clearly indicative of a specific pathologic or anatomic diagnosis. For example, an acute LBP episode may herald a herniated nucleus pulposus (HNP), but the signs of radiculopathy only present later. If the condition is one that has a known likelihood for chronic disability, such as HNP, then further diagnosis and management may be specifically directed. When the condition remains nonspecific, several management options are available. There are three major categories of therapy that may be beneficial based on controlled studies: 1) back school, 2) traction, and 3) manipulation. The experimental studies of these treatment modalities have often included acute as well as subacute patients. Therefore, institution of these treatments should not necessarily be delayed. The judgment of the clinician will be required to determine when a patient can tolerate the physical demands of a back school or is suitably comfortable and relaxed for a manipulation.

Back School

Low back schools are one of the most important methods for the treatment of individuals who have low back pain. We have deliberately placed in extensive discussion of this topic in Chapter 12.

Traction

Bedrest with traction is a traditional method for the treatment of LBP. Traction with five or ten pounds applied to each lower extremity has little mechanical effect other than reinforcing bedrest. However, traction utilizing forces in the range of 50 percent of body weight has been shown to decrease intradiscal pressures by 25 percent. The methods currently popular are inversion and gravity lumbar reduction traction. Inversion is a treatment where the patient hangs upside down, usually with boot fixation. Gravity lumbar reduction is a technique that has the patient upright and reduces loads with the fixation of a vest. Controlled studies of these specific modalities have yet to be published for nonspecific LBP. Lidstrom and Zachrisson (1970) were able to demonstrate that traction (at approximately 50 percent of body weight), in combination with postural instructions and isometric exercises, was beneficial in a randomized controlled trial of patients with LBP and sciatica greater than one month's duration. It is impossible to know from their study whether the benefit derived was from the traction or the isometric exercise and postural instructions.

Manipulation

Mobilization, manual therapy, or manipulative therapy is being enthusiastically endorsed by a growing number of medical doctors and physical therapists in addition to the traditional users, chiropractors and osteopathic physicians. There is a wide variation in the actual manipulative techniques applied or the indications for their use. Although there are many theories as to how and why manipulation works, its actual mechanism of action lacks scientific validation. The most complete study of the efficacy of manipulation has been conducted by Jayson *et al.* (1981) who analyzed two groups: outpatients with milder forms of LBP and hospitalized patients with more severe LBP. The hospitalized group did not respond to mobilization by the Maitland technique. However, the outpatients showed slight but statistically significant symptom improvement. The beneficial effect was not lasting, and no difference was observed between the two groups at the three-month and one-year interval after treatment. Thus, mobilization and manipulation have a limited capacity to hasten the resolution of symptoms although it is known that these modalities do not affect the long-term prognosis. It is also clear that manipulation does not change the position of a disc herniation. Studies of myelograms done before and after rotatory manipulation show no change in the degree of protrusion (Chrisman *et al.* 1964).

CHRONIC LOW BACK PAIN

When LBP persists for more than three months and is disabling, more comprehensive and costly treatments are indicated. If a patient is unable to work because of pain, then the cost of disability payments is likely to exceed the costs of further medical management. The management of chronic pain therefore depends on a careful assessment of the impairment and severity of disability. This discussion of chronic LBP management considers two categories of patients, those with continued work-absence and those still working. The principles of treatment for chronic low back pain are not significantly different for disabled and working patients. Options for management of these chronic low back LBP patients include exercise, corsets and braces, surgical procedures, and comprehensive rehabilitation.

Exercise

In a controlled trial of patients with "long-standing," nonspecific backache, Kendall and Jenkins (1968) studied three exercise programs: Group A—exercises to mobilize and strengthen the trunk with flexion type predominating, Group B—isometric flexion exercises, and Group C—extensor muscle strengthening exercises. All groups received postural instruction and groups A and C also received lifting instructions. The best results three months later were in Group B of which seven were symptom-free, four improved, three had no change, and none were worse. By contrast, 14 percent of Group A and 35 percent of Group B were worsened by their treatment. These data provide additional support for Lidstrom and Zachrisson's study which showed superior results from isometric flexion exercises when given in conjunction with traction (Lidstrom and Zachrisson 1970). Exercise should enhance the stability of the spine and improve dynamic control. Furthermore, the improvement of trunk strength and control is one of the few ways the patient can affect his condition. We usually advise simple, safe isometric abdominal exercises to chronic patients as a matter of daily hygiene (Fig. 9.2). Proper instruction in the program is essential and may involve more than one visit to the physical therapist to reinforce the proper method and importance of the exercise program.

A quite different program which teaches a patient to assume a posture of greater extension or lordosis is gaining popularity (McKenzie 1981). Experienced therapists may be able to discern those patients who will respond to an extension program. Because flexion is known to increase intradiscal pressure (i.e., sitting and bending) patients with discogenic LBP may be improved by increasing the lordosis which reduces the intradiscal pressures. Although the extensor exercise program has some theoretical appeal, currently there are no controlled trials to validate this approach.

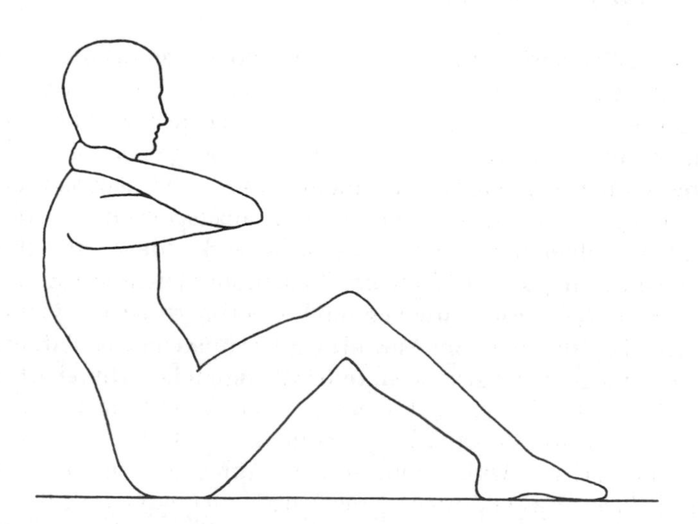

Figure 9.2
The current method for doing a sit-up.

Corsets and Braces

There are no established indications for the use of a lumbar orthoses except in patients who have undergone surgery to correct spinal deformities (for example scoliosis), have had severe spinal trauma, or are convalescent from a spinal fusion. Because of the favorable natural history of acute low back pain, the extensive use of orthoses is not justified in most cases of nonspecific back pain. However, the use of orthoses in LBP is widespread (Fig. 9.3). In a survey of 3,410 orthopaedic surgeons, Perry (1970) found that 99 percent had prescribed an orthosis for low back pain patients. The respondents to that survey emphasized the importance for each patient to eliminate the need for external support in the future and that external supports are mainly for short-term use. Therefore, treatment with a lumbar orthosis should be accompanied by a trunk exercise program.

To determine whether patients used their orthoses, Ahlgren and Hansen (1978) conducted a survey of 260 randomly selected patients for whom a corset had been prescribed four years previously. Approximately 50 percent of the patients were still using their corsets. These results indicate that patients do comply with the prescription of an orthosis but also suggest patients may be using the device too long. The effectiveness of corsets in chronic LBP has been studied only minimally.

Million *et al.* (1982) demonstrated that a corset with lumbar support was superior to a corset without the support on several subjective measures of functional ability (standing, sitting, etc.) in patients with low back pain of greater than six months' duration. This randomly controlled trial would suggest that prescription of an orthotic device can be efficacious for chronic low back pain patients.

Very little is known about how rigid braces and orthoses function. It has been shown that low back braces do not immobilize the lumbar spine or individual motion segments (Norton and Brown 1957; Waters and Morris 1970) leading to the suggestion that brace effectiveness is related to increased intra-abdominal pressure rather than immobilization. This concept is discussed in Chapter 1. Others have felt postural control was the major desired effect and have created newer orthotics made of thermoplastic material, with pelvic modules to enhance pelvic fixation and maintain an antilordosis posture. Micheli, Hall and Miller (1980) found that athletes responded well to this type of bracing in an uncontrolled study. These athletes were able to compete while wearing their orthoses. Best results were obtained in patients with spondylolysis. Although this study was an uncontrolled trial, it would support the concept that more rigid bracing may be effective for low back pain with specific etiologies such as spondylolysis and/or spondylolisthesis. For nonspecific chronic low back pain, a trial of a corset with a lumbar support or stays may be helpful.

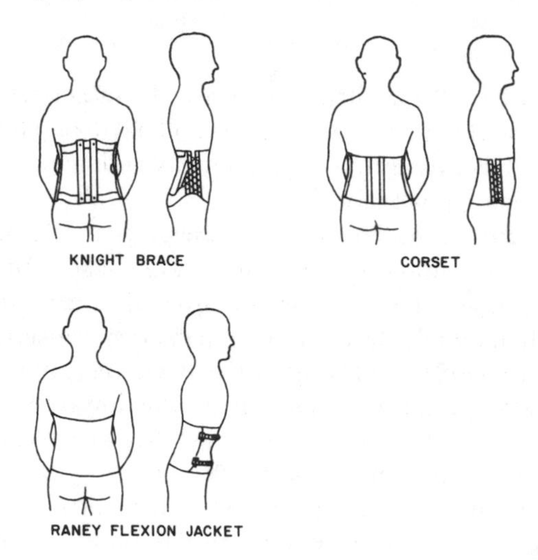

Figure 9.3
Spinal orthoses in common use. Knight brace, Camp corset, Raney jacket.

Other Treatments

A variety of medications are used in the treatment of chronic LBP. Pain medication, nonsteroidal acute inflammatory drugs, and "muscle relaxants" are heavily prescribed. There are no studies to show the effectiveness of most of these drugs, although patients with multilevel degenerative diseases seem to improve with anti-inflammatory drugs. It is critical to avoid narcotic analgesics such as morphine, demerol, codeine, and percodan. One of the major problems in chronic LBP is drug abuse.

SURGICAL PROCEDURES

The majority of patients with LBP do not require surgery and should not be considered as candidates for that treatment. The number of surgical procedures for LBP is relatively small. In patients under 50 years of age, three procedures are most likely to be indicated: 1) discectomy for HNP; 2) chymopapain injection for HNP; and 3) decompression and/or fusion for patients with the isthmic spondylolisthesis.

In patients over 50 years of age, two syndromes are most commonly treated by surgery: 1) spinal stenosis; and 2) degenerative types of spondylolisthesis. In both age groups, spinal fusion for undiagnosable LBP has poor results. Other indications for surgery (tumor, infection, fracture) are only rarely indicated for the injured worker.

Disc Herniation

It is generally recognized that the presence of a disc herniation alone does not indicate the need for surgery and is dependent upon careful assessment of the history, physical findings, neurologic deficits, and degree of straight leg raising positivity.

The patient with a HNP proven by radiographic (CT scan, myelogram) and clinical findings (straight leg raise, neural deficit, electromyographic abnormality) has three major treatment options: 1) open lumbar discectomy (Fig. 9.4); 2) chemonucleolysis; and 3) palliative conservative treatment. These three treatment methods appear to have relatively equivalent long-term results. Weber performed a prospective randomized controlled trial on patients with a HNP proven by myelography. Patients were assigned to open discectomy group or to six weeks of in-patient conservative treatment. One year later (Weber 1978a), the surgical group had results that were superior to the nonoperated group as measured by pain relief and function. Four years

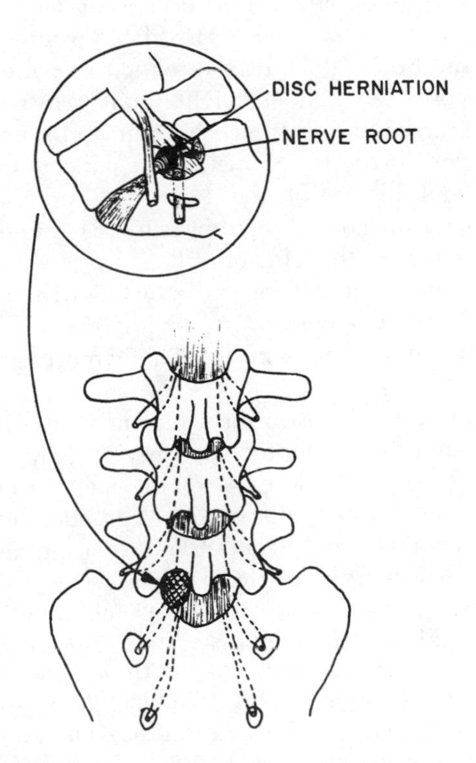

Figure 9.4
Open lumbar discectomy. This demonstrates the position of the herniation with respect to the nerve root.

later, the tendency for better results still persisted in the surgical group but was not statistically significant (Weber 1978b). Ten years later both groups were essentially the same (Weber 1983). Because the long-term results are similar with and without surgery, the major advantage of surgery is an earlier resolution of the symptoms. Although there is a current wave of enthusiasm for "microsurgical discectomy," this approach probably has little impact on shortening disability or altering long-term results. For patients with little neurologic involvement, conservative treatment is definitely warranted.

The effect of lumbar discectomy on disability costs is not entirely clear. Surin (1977) studied the total duration of disability in 116 patients who had undergone lumbar discectomy. Many of his patients were observed for a ten-year period prior to and a ten-year period after the surgery. Work-absence for the total group increased continuously from the beginning of the observation period to the end and in general was unaffected by surgical intervention. However, early surgery (short period of preoperative sciatica), diminished total preoperative period of sick leave, and immediate relief of sciatic pain were factors predictive of a short total period of postoperative disability. For surgery to be effective in decreasing work absence, careful selection of patients without unnecessary delay would seem profitable. Leavitt, Johnston and Beyer (1972, Part 3) found that patients who received surgery were high cost cases, although eight of nine returned to work. In the cases for which surgery was mentioned but not performed, less than half returned to work. One can therefore conclude: 1) patients should understand the favorable natural history of the HNP syndrome without surgery; 2) surgery does not restore normal anatomy (cure); 3) the goals of surgery are the relief of sciatica and not the relief of LBP; 4) surgery may be followed by the need for repeat lumbar surgery in approximately 10 percent of cases; and 5) surgery offers them the opportunity to be improved earlier but does not assure that result (Frymoyer and Hanley 1978; Frymoyer and Matteri 1978; Frymoyer 1981).

The second surgical alternative for a patient with a herniated disc is chemonucleolysis, in which a proteolytic enzyme (chymopapain or collagenase) is injected to digest portions of the nucleus pulposus (Wiltse 1983). The mechanism by which this treatment produces its clinical results is not totally clear. However, in well controlled clinical trials these enzymes have been shown to be more effective than placebo in patients who have classic signs and symptoms of sciatica. They are not an effective means of relieving undiagnosed LBP. Approximately 70 to 75 percent of patients can be expected to obtain good results with these methods. Whether chemonucleolysis is superior or inferior to open surgical discectomy is debatable. The major advantage of this procedure is that it is less invasive and avoids postoperative surgical scarring. The disadvantages are a risk of severe anaphylactic reaction in 1/100 cases; death has been reported in 1/10,000 patients so treated, although newer tests probably allow for the allergic patient to be identified. The cost of chemonucleolysis is comparable to open surgery. The indications therefore should be the same as for open surgical discectomy. Approximately 20 percent of patients will be sufficiently dissatisfied after chemonucleolysis to request open surgical discectomy.

Less is known about the medical efficacy and cost benefit of surgical treatment of other low back pain syndromes (spondylolisthesis, "instability," spinal stenosis), and indications for surgery are considerably less objective (Selby 1983). Consider, for example, the adult patient with spondylolisthesis and chronic low back and/or leg pain. The patient with a herniated disc has multiple objective tests (myelography, CT scanning, EMG, straight leg raising, neurological findings) which, when positive and correlating with each other, can predict surgical success. In contrast the patient with spondylolisthesis has only a plane radiographic finding and a report of pain. Since one to three percent of the population has this radiographic change and frequently the lesion is painless, surgery may not be warranted. For patients with disabling pain, surgery yields 80 to 85 percent good results (Wiltse 1983, Frymoyer and Matteri 1978). In general, spondylolitis and segmental instability are treated by spinal fusion (Fig. 9.5).

Patients with spinal stenosis may be considered as candidates for surgical intervention if they have increasing and progressively disabling neural claudicatory leg pain as previously described. The decision to operate is based on the same indications used in disc surgery, namely, the patient has had adequate conservative management, symptoms fail to respond or increase in the face of that treatment, and there are objective signs, symptoms and confirmatory tests, for example CT scan and myelogram. When surgery is performed for spinal stenosis, the goal is to decompress all of the levels where there is significant stenosis as measured by the myelogram. This may require extensive removal of the laminae and even the facet joints (Fig. 9.6). It remains controversial as to who should be fused if the facet joints are sacrificed. In general, the older the patient, the less likely fusion is to be required. Conversely, the younger the patient and the more extensive the decompression or facet sacrifice, the higher the probability for fusion. The success rate for such operations is lower than that for HNP, probably because of the greater extent of the surgery and the fact that scarring is a significant probability after such extensive surgery.

Although surgery may produce the desired pain relief, the patient's return to work may be influenced by many other significant factors: 1) age, 2) education, 3) job skills, 4) patient's motivation, 5) employer's motivation, 6) availability of jobs within the patient's ability, and 7) other psychological and social factors. Failure to appreciate these confounding factors may also be associated with poor results (Wiltse 1975, Wilfling *et al.* 1973).

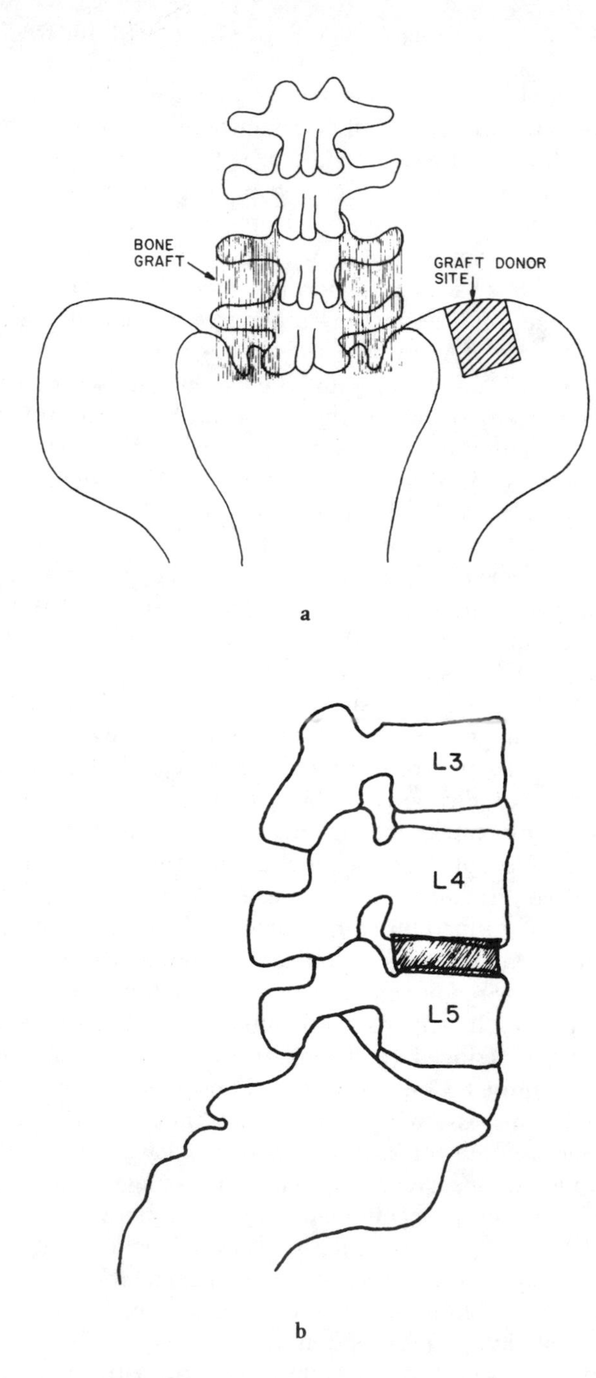

Figure 9.5
Spinal fusion showing:

a) bone graft along transverse process, and;
b) graft between adjacent bodies.

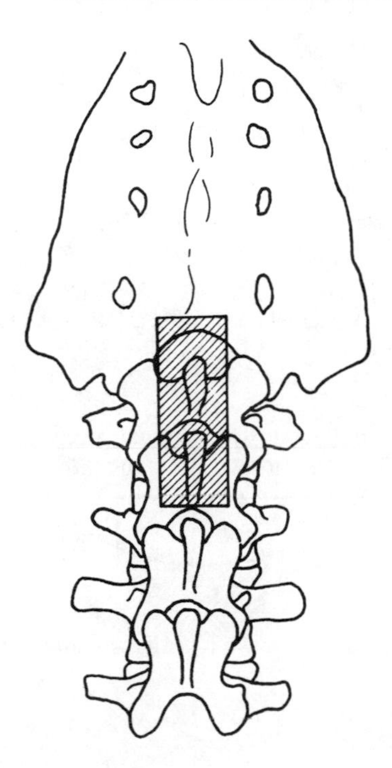

Figure 9.6
Removal of laminae.

COMPREHENSIVE REHABILITATION

The rehabilitation of patients with chronic LBP is analogous to the rehabilitation of patients with other musculoskeletal handicaps. The key to the process is that the patient rehabilitates himself while the therapist serves as an educator, counselor, and coach. The handicap represents a loss of function to the extent that normal everyday living, work, and recreational activities are limited. Our environment, created for the unimpaired, presents the handicapped with barriers that impede function. The rehabilitation process is one of providing the handicapped person with the knowledge and skills to overcome barriers.

Steps in Comprehensive Rehabilitation

Five steps taken in the rehabilitation process will be discussed (Fig. 9.7). Since each step takes time, one does not wait necessarily for mastery or achievement of each step before proceeding to the next. These steps include accepting the impairment, goal establishment, understanding barriers, reinforcement, and pain management.

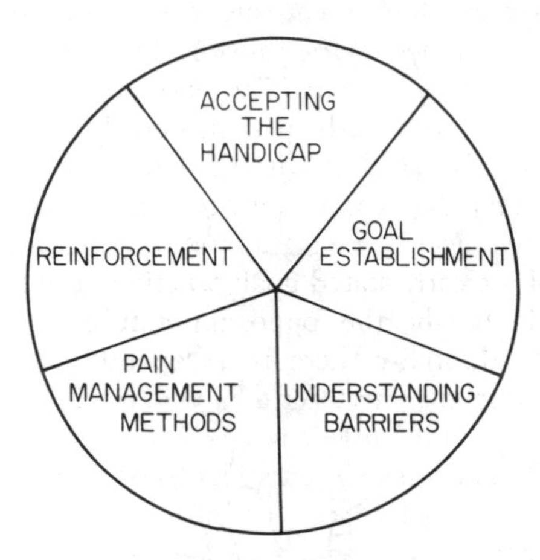

Figure 9.7
The rehabilitative process can be viewed as five equally important steps.

Accepting the Impairment

The first step for the patient is accepting the handicap. After the loss of any bodily function there is a period of mourning and anger and an unwillingness to accept the handicap. Some patients never get past this stage. The successful process is two-fold: 1) accepting the reality of the loss, and 2) accepting the self-responsibility of adaptation. The disabled LBP patient may have difficulty because the handicap is invisible and not finalized, in contrast to other impairments. The LBP-impaired cannot see the handicap, and is often confused by the lack of a clear explanation for its cause.

Others around the patient (family, physicians, employers, neighbors) also have a difficult time seeing, understanding and accepting that the LBP patient has a handicap. The patient's failure to function is seen as a weakness, contrary to the work ethic and questionably motivated. The total energies of the patient may be consumed in the search for an explanation that will restore credibility to his complaints. This often includes visits to many health professionals.

Another typical reaction is anger. It is one order of anger to be hurt in the first place, but a much higher order of anger to be hurt and be thought of as a malingerer. This anger also misdirects the patient's energy. To get even rather than get well becomes the goal.

The therapist must understand these uncertainties and feelings and be able to sense their presence in the patient. A major therapeutic effort must be made. The therapist must also explain the nature of the impairment. If possible, the nature of that condition must be clearly explained. Words that can potentially frighten the patient such as "degeneration" or "disintegration" are carefully avoided. If the cause is unknown, the patient must be taught the possible sources of pain. The permanent nature of the impairment must be explained so that the patient's energy is no longer devoted to finding the cure but rather dealing with the handicap. The patient must not view the rehabilitation process as a treatment trial. Surgery cannot be held in the mind as an ultimate cure. The rehabilitation process has to be viewed as the achievement of a clearly stated goal: *function despite the handicap*.

The patient is taught the opposing nature of the disability and rehabilitation. With disability there is a dependency on others: doctors, family, insurance company, employer, and welfare. In contrast, the purpose of rehabilitation is to become independent. The patient should have trust in his ability to be self-supportive. The patient must be his own 24-hour-a-day therapist who is knowledgeable and dependable in his own treatment. Once the problem is accepted as real and as one requiring self-responsibility, the next step is establishing goals.

Goal Establishment

The therapist teaches that the terms handicap and impairment are not the same as the term disability. The handicap or impairment is that body part which does not function normally. The disability is the loss of function as a result of the handicap. The patient is taught that the presence of the handicap does not necessarily have to incur disability. Total disability and total rehabilitation are viewed as two ends of a yardstick (Fig. 9.8). The major goal for the patient is to achieve the maximal rehabilitation. Short- and long-term goals should be identified. The long-term goal of total independence, and financial and emotional security may appear unattainable, and therefore short-term, attainable goals must be reached. Each short-term goal is considered a plateau. Examples of short-term goals are 1) achieving a certain level of repetitions in an exercise; 2) acquiring certification of high school equivalence; 3) solving a family problem; 4) working as a volunteer; 5) taking a lower paying job and accepting lower socioeconomic status until return to a better position becomes possible. Achievement of each short-term goal reinforces positive behavior and helps restore confidence.

Understanding Barriers

The next step is learning more about the barriers that low back pain patients must overcome; these barriers may be invisible or "hidden." The therapists will teach the patients about the barriers and alternative ways to deal with them.

The barriers for the disabled low back pain patient may include fear, anger, work disincentives, misunderstanding and limited education, stressful

Figure 9.8
Patients with chronic low back conditions fall somewhere along a yardstick ranging from total disability with all of its concomitant problems, to total rehabilitation. Typically patients are in the middle, where the alternatives are either to prove they are disabled, or to "pretend they are well". Comprehensive rehabilitation through all of its various steps attempts to aid patients in moving towards total rehabilitation, rather than falling into total disability.

life situations, the physical impairment, vocational limitations, and emotional instability.

Fear and anger have adverse effects on patients with chronic pain. For example, patients may not attempt new activities or remain drug dependent because they fear pain. They may be angry about past treatment failures and the ability of the rehabilitation team to cure these.

The obstacles of work disincentives, misunderstanding, and stressful life situations often overlap. For example, they may be paid for not working. The average patient's source of advice may be from a misguided neighbor, union representative, or attorney. In addition, they may have significant financial or family problems. They often believe their only security is their Worker's Compensation benefits. For every disincentive that the patient expresses, the creative therapist can usually counter with an equally effective incentive.

The patient also learns how stress and emotions play a role in the pain behavior. Patients are taught that it is impossible to experience pain without having some emotional reaction such as anger, depression and frustration. Patients with more incapacitating emotional reactions may need intense pyschological and/or psychiatric therapy

Patients learn about the nature of their physical impairment through classes in anatomy and function of the spine. The nature of this educational process and alternatives to achieving their educational goals is discussed fully in the Low Back School section of this chapter.

The exercise program in the rehabilitation process is more intensive than in Low Back School and focuses on psychomotor skills. This learning is directed in two exercise modalities: 1) trunk strength and flexibility, and 2) cardiovascular or aerobic fitness (bicycle ergometers, walking, and swimming). The exercises are reinforced through the understanding of how loading the spine in either flexion or extension can control the strain and presumably pain of soft tissue origin. The aerobic fitness program is reinforced by the concepts that: 1) to exercise control of the spine all day long requires overall fitness and 2) that the increased activity with aerobic exercise may stimulate the naturally occurring pain blockers (endorphins) and modulate the pain experience.

To facilitate transfer of this information patients may be trained on an obstacle course. The course is set up so that patients have to perform activities of daily living and work activities. Patients are then taught specific ways to improve the use of their backs.

Vocational counselors evaluate each patient on vocational tasks based on prior job, job skills and educational background. If the old job will not be available, this must be established. In these instances, valuable time in additional counseling and remedial education is lost. Although the patient's best short-term goals might be return to the old job, the best long-term goals may be additional training for a more skilled job several years later.

Reinforcement

The fourth step in the rehabilitation process is reinforcement of all the concepts and incorporation into the patient's life. This step represents the *affective component* of the learning experience. This affective learning is enhanced through the team or multidisciplinary approach, by peer pressure or patient group interaction, and by involving the family members. Regular team conferences enhance communications within the team and with the patient. Regular team conferences allow all members to coordinate their approach and reinforce the learning experience.

The team director or coordinator has the role of delivering to the patient his report of progress or failure. In Worker's Compensation cases, the director will determine the end of the healing period, date of return to work, and work limitations. We have found that therapists must be compassionate and understanding but also must be firm in demanding achievement of reasonable goals. For example, at the commencement of the Iowa rehabilitation program, patients were told to set and achieve their own goals (Lehmann 1983). However, of the first 26 patients treated, none returned to work. A policy was developed in which patients were given a termination date for their healing period and a release to return to work within a few weeks of discharge. In the subsequent 51 patients treated by this policy: 17 returned to work and were still working at six-month return; 5 returned to work but lost their job because of the poor economy; 4 returned but were unable to function because of their backs; 3 were in vocational training; and 22 never attempted to return to work or retraining.

It is typical for most patients to improve while in the rehabilitation environment but some regress after discharge. By teaching the family the concept of behavior modification, the home environment can become a rehabilitation environment. The family often controls the success or failure of the rehabilitation effort. Patients with severe disability have returned to work because of a strong affective role played by the spouse. The opposite may occur in other families.

Utilization of Pain management Modalities

The fifth step in the rehabilitation process is the utilization of pain management techniques such as acupuncture and transcutaneous nerve stimulation. Additionally, psychoactive modalities such as hypnosis, relaxation therapy, biofeedback, operant conditioning, and cognitive behavioral therapy may also be useful. Psychotherapy is usually necessary for management of chronic pain. For further information about these modalities, the reader is referred to excellent reviews by Keefe (1982) and Turner and Chapman (1982, Part I and Part II).

Several authors have studied the role of transcutaneous electrical nerve stimulation for pain control in LBP patients. It was concluded that electroacupuncture demonstrated reduced pain. Fox and Melzack (1976) have found high intensity stimulation (similar to electroacupuncture) to be efficacious in chronic LBP patients. However, in a controlled study of chronic LBP patients, Mendelson *et al.* (1983) found *manual* acupuncture was no more effective than a placebo.

Efficacy of Rehabilitation Programs

Multidisciplinary pain management programs have been justified by the belief that a team approach is necessary for persons with multiple physical, psychological, social and vocational problems. This concept was supported by A.W.M. White's study (1966) of the Ontario Workmen's Compensation Board. All patients admitted to the study had been disabled with LBP for a period of six weeks to 12 months from the time of their injury. After the initial evaluation, patients were randomly assigned to treatment groups. One group was discharged to the care of their own doctors. The other group was treated for six weeks on an inpatient basis by rehabilitation. Patients in both groups were evaluated after six weeks and after three months to determine which patients had a "satisfactory" return to work. Of the outpatient group only 15.8 percent had a "satisfactory" return to work. Of the inpatients, 42.4 percent had a "satisfactory" return to work. White reported that the time, effort, and expense was justified because of the two and a half times greater success rate with inpatient treatment. However, the 42 percent success rate with inpatient treatment was felt to be inadequate and indicated the need for better rehabilitation methods. Beals and Hickman (1972) have shown similar results. In an attempt to understand the cause of rehabilitation failures, predictors of success have been devised.

In White's study from the Workmen's Compensation Board of Ontario, demographic factors were analyzed but were not shown to be predictive of treatment outcome (White 1966). Beals and Hickman (1972) found that a combined score based on the physician's and psychologist's clinical judgment was the best predictor of rehabilitation outcome. They also noted a steady, gradual decline in the percentage of patients returning to work as a function of longer duration of disability. Maruta, Swanson and Svensson (1979) have developed a scale for the prediction of outcome with chronic pain management utilizing the variables of duration of pain, duration of disability, number of pain-related surgeries, severity of pain, dependency on drugs and hysteria and hypochondriasis scores on the MMPI.

Leavitt and Garron (1979) have developed the Back Pain Classification Scale which consists of thirteen words describing pain. They have cross-validated this instrument's ability to distinguish between "organic" and "functional" LBP patients. They have shown that the scale can predict treatment response. Patients classified as "organic" improved more than those classified as "functional" in response to conservative medical treatment. Other methods of pyschological testing may improve our ability to predict rehabilitation outcome. Lawliss *et al.* (1982) have used motivational indices measured by the Sixteen Personality Factors (a 105-question self-report test) to predict outcome from a multidisciplinary spine pain rehabilitation program. They implied that these motivational indices predict success, whereas certain scores on the Minnesota Multiphasic Personality Inventory (MMPI) predict failure. Used in combination, scores have been developed which correlate with several measures of successful rehabilitation. The outcome measures used were consistent with return to work.

Painter, Seres and Newman (1980) compared two groups of 25 chronic pain patients each of whom had either improved or regressed following treatment. The amount of improvement during the pain center treatment did not predict later regression. Incentives such as availability of reemployment and a supportive family environment seemed to correlate with success while disincentives such as financial compensation correlated with failure. Newman *et al.* (1978) found that all patients improved during pain center treatment but that fully-employed patients maintained the greatest gains over an extended period of time. Roberts and Reinhardt (1980) reported that

77 percent of patients treated in their chronic pain management program were living normal lives one to eight years later. Their high success rate appears to be largely attributable to their stringent admission criteria. The most common cause for rejection of treatment was when patients appeared to be unwilling to return to work or their families appeared to be unwilling to cooperate or change.

The continued development of predictors of success or failure in chronic LBP rehabilitation ultimately should lead to selective admission for those likely to benefit from new regimens, and a realistic approach for those who have little or no chance of success.

SUMMARY

The treatment of low back pain depends upon the duration of symptoms and the presumed or documented clinical diagnosis. Ultimately, the only scientific measure of the effectiveness of any treatment program rests upon a controlled, randomized trial in which the effectiveness of a treatment program is compared to a placebo or some other established form of treatment. In this regard, the evidence strongly favors the age-old remedy of bedrest and pain relievers as the most effective and efficient means of dealing with most forms of acute back pain. There is also fairly strong evidence that the addition of certain muscle relaxants will reduce the severity of acute symptoms, although not necessarily their duration. Among the other treatment methods for acute conditions, manipulative therapy appears to offer temporary rather than lasting relief. In the majority of low back injured workers, a simple measure of bedrest or even reduced stresses will suffice. A key element appears to be the presence of some form of educational program, currently best organized in the low back school concept, which is most efficient and reliable. Organized physical therapy seems to have little role in reducing the duration of acute episodes or the short-term disability effects. However, the overall principle of maintaining strong supportive musculature seems logical. It is totally unclear, despite strongly held opinions, whether flexion or extension exercises are most beneficial. We favor an eclectic approach whereby the therapist, working with a particular patient, chooses that form of exercise best tolerated by the patient. Such therapy should not be instituted until the acute and disabling symptoms have subsided to the degree that would allow the program to progress.

When symptoms are persistent, it becomes essential to establish a functional diagnosis if possible, based on the clinical history, signs, and, when needed, confirmatory radiographic and laboratory tests. The purpose of such studies is to determine what form of treatment will best serve the individual patient's overall condition. If sciatica is the predominant symptom, and a disc herniation the confirmed diagnosis, the options include continued conservative management, epidural block, chymopapain injections, or various surgical types of disc excision. For most, but not all, patients with sciatica, it should be impressed upon the patient and doctor that the results of treatment, surgical or conservative, will ultimately yield equivalent rates of success. The exceptions are those patients with cauter bowel dysfunction (Cauter equinus syndrome) and significant muscle weakness. It does seem clear that if the decision to operate is made, this should be done within the time period of six weeks to three months from the onset of symptoms.

When low back pain predominates, the structural diagnosis is often more difficult to reach, and therapy therefore may be less specific. Among conditions that may require specific treatment, one can include certain forms of segmental instability and, obviously, patients who have primary or metastatic tumors or infections. The latter group will require prompt surgical or nonsurgical treatment depending upon their specific condition. Most patients with low back pain will not require surgery, but in these structural conditions, such as spondylolisthesis, the treatment option may include spinal fusion. It is totally clear that nonspecific spinal fusion, when there is not a clear structural diagnosis, has a poor prognosis for ultimate symptom relief. Fortunately, the majority of patients will improve with simple conservative measures that include the addition of back supports, ongoing supervised physical therapy, a protected work environment, and time. There remains a small, hard core of high-cost patients who, despite simple conservative or surgical treatment, do not improve. It is this group which poses the greatest challenge. We emphasize the importance of identifying those patients who are not responding to their course of treatment, within three months if possible. It is this group of patients which requires a clear decision from the clinical team about whether or not surgery can help. For this patient group, a multidisciplinary approach is essential before a pattern of disability behavior becomes established.

REFERENCES

Ahlgren, S.A.; and Hansen, T. 1978. The use of lumbosacral corsets prescribed for low back pain. *Prosthet. Orthot. Int.* 2 (2):101–04.

Beals, R.K.; and Hickman, N.W. 1972. Industrial injuries of the back and extremities. Comprehensive evaluation – an aid in prognosis and management: A study of one hundred and eighty patients. *J. Bone Joint Surg.* 54A (8):1593–1611.

Bergquist-Ullman, M.; and Larsson, U. 1977. Acute low back pain in industry. A controlled prospective study with special reference to therapy and confounding factors. *Acta Orthop. Scand.* (Suppl.) 170:1–117.

Chrisman, O. D.; Mittnacht, A.; and Snook, G.A. 1964. A study of the results following rotatory manipulation in the lumbar intervertebral-disc syndrome. *J. Bone Joint Surg.* 46A (3):517–24.

Fox, E.; and Melzack, R. 1976. Comparison of transcutaneous electrical stimulation and acupuncture in the treatment of chronic pain. *Advances in Pain Research and Therapy.* Volume 1, Proceedings of The First World Congress on Pain. Bonica, John J., *et al.* (eds.) pp. 797–801. N.Y.: Raven Press.

Frymoyer, J.W.; Hanley, E.; Howe H; Kuhlmann, D.; and Matteri, R. 1978. Disc excision and spine fusion in the management of lumbar disc disease. A minimum ten-year followup. *Spine.* 3 (1):1–6.

Frymoyer, J.W.; Matteri, R.E.; Hanley, E.N.; Kuhlmann, D.; and Howe, J. 1978. Failed lumbar disc surgery requiring second operation. A long-term follow-up study. *Spine.* 3 (1):7–11.

Frymoyer, J.W. 1981. The role of spine fusion. Question 3. *Spine* 6 (3):284–90.

Hudgins, W. Robert. 1983. The role of microdiscectomy. *Orthop. Clin. North Am.* 14 (3):589–603.

Jayson, M.I.V.; Sims-Williams, H.; Young, S.; Baddeley, H.; and Collins, E. 1981. Mobilization and manipulation for low-back pain. *Spine* 6 (4): 409–416.

Kendall, P.H.; and Jenkins, J.M. 1968. Exercise for backache: A double blind controlled trial. *Physiotherapy* 54:154–57.

Keefe, F.J. 1982. Behavioral assessment and treatment of chronic pain: Current status and future directions. *J. of Consulting and Clinical Psychology* 50 (6):896–911.

Lawliss, G.F.; Mooney, V.; Selby, D.K.; and McCoy, C.E. 1982. A motivational scoring system for outcome prediction with spinal pain rehabilitation patients. *Spine* 7 (2):163–67.

Leavitt, S.S.; Johnston, T.L.; and Beyer, R.D. 1972. The process of recovery: Patterns in industrial back injury. Part 3. Mapping the health care process. *Industrial Medicine and Surgery* 41:7–11.

Leavitt, F.; and Garron, D.C. 1979. Validity of a back pain classification scale among patients with low back pain not associated with demonstrable organic disease. *J. of Psychosomatic Res.* 23 (5):301–306.

Lehmann, T. 1983. Personal communication.

Lidstrom, A.; and Zachrisson, M.I. 1970. Physical therapy on low back pain and sciatica. An attempt at evaluation. *Scand. J. Rehabilitation Med.* 2 (1):37–42.

McKenzie, R.A. 1981. Lumbar spine. Mechanical diagnosis and therapy. Waikanae, New Zealand: Spinal Publications, Ltd.

Maitland, D.G. 1977. *Vertebral manipulation.* 4th ed. London 1:1 Butterworths.

Maruta, T.; Swanson, D.W.; and Svensson, W.M. 1979. Chronic pain: Which patients may a pain-management program help?" *Pain* 7 (3):321–29.

Mattmiller, A.W. 1980. The California back school. *Physiotherapy* 66 (4): 118–21.

Mendelson, G.; Selwood, T.S.; Kranz, H.; Loh, T.S.; Kidson, M.A.; and Scott, D.S. 1983. Acupuncture treatment of chronic low back pain. A double-blind placebo-controlled trial. *Am. J. Med.* 74 (1):49–55.

Micheli, L.J.; Hall, J.E.; and Miller, M.E. 1980. Use of modified Boston brace for back injuries in athletes. *Am. J. Sports Med.* 8 (5):351–56.

Million, R.: Hall, W.; Nilsen, K.H.; Baker, R.D.; and Jayson, M.I.V. 1982. Assessment of the progress of the back-pain patient. *Spine* 7 (3):204–21. 1981 Volvo Award in Clinical Science.

Morris, J.M. 1974. Low back bracing. *Clin. Orthop.* 102:126–32.

Nachemson, A.L.; and Andersson, G.B.J. 1982. Classification of low-back pain. *Scand. J. Work Environ. Health* 8 (2):134–36.

Newman, R.I.; Seres, J.L.; Yospe, L.P.; and Garlington, B. 1978. Multidisciplinary treatment of chronic pain: Long-term follow-up of low-back pain patients. *Pain* 4 (3): 283–92.

Norton, P.L.; and Brown, T. 1957. The immobilizing efficiency of back braces. Their effect on the posture and motion of the lumbosacral spine. *J. Bone Joint Surg.* 39A (1): 111–39.

Painter, J.R.; Seres, J.L.; and Newman, R.I. 1980. Assessing benefits of the pain center: Why some patients regress. *Pain* 8 (1):101–113.

Perry, J. 1970. The use of external support in the treatment of low-back pain. Report of the Subcommittee on Orthotics of the Committee on Prosthetic-Orthotic Education, National Academy of Sciences, National Research Council. *J. Bone Joint Surg.* 52A (7):1440–42.

Roberts, A.H.; and Reinhardt, L. 1980. The behavioral management of chronic pain: Long-term follow-up with comparison groups. *Pain* 8 (2): 151–62.

Selby, D.K. 1983. When to operate and what to operate upon. *Orthop. Clin. North Am.* 14 (3):577–88.

Surin, V.V. 1977. Duration of disability following lumbar disc surgery. *Acta Orthop. Scand.* 48:466–71.

Turner, J.A.; and Chapman, C. R. 1982. Psychological interventions for chronic pain: A critical review. I. Relaxation training and biofeedback. *Pain* 12 (1):1–21.

Turner, J.A.;and Chapman, C.R. 1982. Psychological interventions for chronic pain: A critical review. II. Operant conditioning, hypnosis and cognitive-behavioral therapy. *Pain* 12:23–46.

Waters, R.L.; and Morris, J.M. 1970. Effect of spinal supports on the electrical activity of muscles of the trunk. *J. Bone Joint Surg.* 52A:51.

Weber, H. 1978a. Lumbar disc herniation: A prospective study of prognostic factors including a controlled trial. Part I. *J. Oslo City Hosp.* 28 (3/4): 33–64.

Weber, H. 1978b. Lumbar disc herniation: A prospective study of prognostic factors including a controlled trial. Part II. *J. Oslo City Hosp.* 28 (7/8): 89–120.

Weber, H. 1983. Lumbar disc herniation. A controlled, prospective study with ten years of observation. *Spine* 8 (2):131–40.

White, A.W.M. 1966. Low back pain in men receiving Workmen's Compensation. *Canad. Med. Assoc. J.* 95:50–56.

Wiesel, S.W.; Cuckler, J.M.; DeLuca, F.; Jones. F.; Zeide, M.S.; and Rothman, R.H. 1980. Acute low-back pain: An objective analysis of conservative therapy. *Spine* 5:324–30.

Wilfling, F.J.; Klonoff, H.; and Kokan, P. 1973. Psychological, demographic and orthopaedic factors associated with prediction of outcome of spinal fusion. *Clin. Orthop.* 90:153–60.

Wiltse, L.L. 1983. Chemonucleolysis in the treatment of lumbar disc disease. *Orthop. Clin. North Am.* 14 (3):605–22.

Wiltse, Leon L., and Rocchio, Patrick D. 1975. Preoperative psychological tests as predictors of success of chemonucleolysis in the treatment of the low-back syndrome. *J. Bone Joint Surg.* 57A (4):478–83.

NOT CITED

Leavitt, Stephen S.; Johnston, Tom L.; and Beyer, Robert D. 1972. The process of recovery: Patterns in industrial back injury Part 4. Mapping the health care process. *Industrial Medicine and Surgery* 41 (2):5–9.

Leavitt, Stephen S.; Johnston, Tom L.; and Beyer, Robert D. 1971. The process of recovery: Patterns in industrial back injury. Part 1. Costs and other quantitative measures of effort." *Industrial Medicine and Surgery* 40 (8):7–14.

Leavitt, Stephen S.; Johnston, Tom L.; and Beyer, Robert D. 1971. The process of recovery: Patterns in industrial back injury. Part 2. Predicting outcomes from early case data." *Industrial Medicine and Surgery* 40 (9):7–15.

McGill, Charles M. 1968. Industrial back problems. A control program. *J. Occ. Med.* 10 (4):174–178.

PART V
Prevention

10 Concepts in Prevention

Gunnar B. J. Andersson

There is widespread agreement that the problem of low back pain (LBP) is of such magnitude that it should be addressed with general preventive measures. As will be obvious from this chapter, there are several alternative models to use, none of which has been properly evaluated. The three main preventive approaches are:

1. designing the job for the worker;
2. selecting the right worker for the right job; and
3. teaching the worker the right work method.

Although they are each discussed in separate chapters in the book, this chapter serves as an introduction.

DEFINITIONS

Classical preventive medicine divides prevention into three categories:

1. *Primary prevention*: measures taken to prevent the clinical manifestation of a disease before it occurs, for example, immunizations for childhood infections.
2. *Secondary prevention*: measures taken to arrest the development of a disease while it is still in the early, asymptomatic stage, for example, treatment of asymptomatic hypertension to prevent stroke.
3. *Tertiary prevention*: measures taken to minimize the consequences of a disease (or injury) once it has become clinically manifest, for example, coronary bypass surgery for a patient with intractable angina. Tertiary

prevention includes most of medical and surgical therapy. The prevention of treatment complications also falls into this category.

The three classes of prevention can be illustrated as in Figure 10.1. It should be clear that the process of prevention, generally, becomes more difficult and expensive as one moves from primary towards tertiary prevention. It should also be clear that primary and secondary prevention in practical terms tend to overlap as they relate to LBP.

When discussing prevention, it is also useful to differentiate between approaches applied to society in general, for example, fluoridation of drinking water, and more individual approaches, for example, advice against smoking. The first does not require the participation of the individual, i.e., it is passive, whereas the second clearly does, i.e., it is active. Changes at the workplace can be considered intermediate, requiring the good intentions of management and the acceptance of the worker.

PRIMARY PREVENTION

To institute true primary prevention of low back pain, we should know what constitutes the disease risk. Unfortunately, our knowledge lacks detail. LBP is a symptom and not a disease or an injury. The association of LBP with workplace and individual factors is discussed in Chapters 5 and 6. In spite of our superficial knowledge, there is considerable scientific evidence and common sense knowledge that permits specific suggestions for primary prevention programs. For example, we know that low back pain can occur from mechanical trauma–isolated or repeated. Although we do not know the precise limits at which physical loads become harmful or how individual factors influence those limits, we have developed a great deal of useful information. That knowledge can be used to reduce the physical work load on the lower back. The main approaches to be used are:

1. Pre-employment screening programs, as outlined in Chapter 11. Screening is clearly an area where primary and secondary prevention overlap and indeed, some consider it to be entirely the latter (Anderson, 1980). General aspects on screening procedures will be given later in the chapter.

2. Changes in work habits as outlined in Chapter 12. These changes would include both direct work-related instruction and general information (schooling) about the back and its load capacity. The importance of load moment on the spine and how it is influenced by object weight, location, size and density as well as by posture, dynamic forces, asymmetries in loading and different time factors (duration, frequency) need to be understood and not simply memorized. It is well known that, in spite of the fact that many workers learn how to handle objects at work and which postures to avoid, it

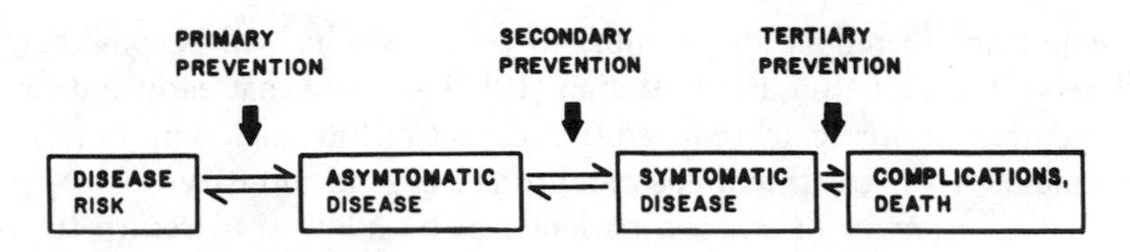

Figure 10.1
The potential course of a disease, and the opportunities for prevention.

is difficult to make them use that knowledge in the stress of the working environment. None of the back schools available at present are aimed at primary prevention nor are they designed for that purpose. They have not been evaluated from the perspective of whether the information is clearly understood or just temporarily memorized. Practical instruction at the workplace cannot be replaced by sound–slide programs and posters. Perhaps, there is a need for a "Back Man" supervisor throughout industry.

3. Changes at the workplace, as outlined in Chapter 14. These changes include reductions of permissible loads (by recommendation or regulations), optimization of postures, and ergonomic design of the workplace and work methods.

SECONDARY PREVENTION

As with primary prevention, the problem basically is to know who is at risk. Our ability to identify asymptomatic "disease" which places the industrial low back at risk is poor. Screening programs on a regular basis therefore have limited value.

Secondary prevention and tertiary prevention tend to overlap, particularly with respect to recurrences. Previous low back pain, for example, increases the risk of a subsequent episode, and this knowledge is useful in both secondary and tertiary prevention. Most low back school programs should be considered as secondary (and tertiary) preventive programs because the person enrolled has already had low back pain at one time or another, and the schools were developed with that target group in mind.

TERTIARY PREVENTION

Selection of the right therapy to reduce the chronicity of LBP and prevent recurrence may be the most immediate, practical, and preventive possibility, but it is of course totally redundant if primary prevention has been taken. Evaluation and therapy are discussed in Chapters 8 and 9 where we emphasize that early recognition of the problem and proper management are important to reduce the number of patients who will become chronic sufferers. Wiesel, Feffer and Rothman (1982) showed that early and careful management dramatically reduced the impact of low back pain in terms of lost time at work, treatment cost, and direct cost to industry.

The assessment of when a patient can be allowed to return to work safely is based on a poorly developed technology. Also to be explored is for whom and at what time modifications of the workplace and work routine can aid in reducing LBP chronicity or recurrence. Even when chronic symptoms have developed, it is important to encourage work activities in some form to avoid all of the problems of chronic disability and its treatments discussed in Chapter 9.

GENERAL CONSIDERATIONS IN SCREENING PATIENTS

Different options in worker selection, as well as their advantages and limitations are discussed in Chapter 11. Some important definitions will be given here. These include 1) accuracy, 2) sensitivity, 3) specificity, and 4) predictive value.

Accuracy is a measure of a screening test's ability to measure a quantity or quality. While a low hemoglobin value is an accurate measure of anemia, range of spinal motion measured clinically is far from accurate.

Sensitivity is a measure of a test's accuracy in correctly identifying persons with a certain condition. It can be expressed as the fraction or percent of all persons with a condition who will have a positive test:

$$\frac{\text{True Positive}}{\text{True Positive} + \text{False Negative}} \times 100 \text{ percent}$$

This means that if the sensitivity is 95 percent, then 95 of a group of 100 with a certain condition will have a positive test (true positive), and 5 a negative (false negative).

Specificity is a measure of a test's accuracy in correctly identifying persons who do not have the condition. It can be expressed as the fraction or percent of all persons free of a condition who will have a negative test (true negatives).

$$\frac{\text{True Negatives}}{\text{True Negatives} + \text{False Positives}} \times 100 \text{ percent}$$

A test with a 95 percent specificity applied to a group of 100 persons who do *not* have a disease will be negative in 95 (true negative) and positive in 5 (false positive).

The *predictive value* of a test is its ability to predict the presence of a condition. It can be expressed as:

$$\frac{\text{True Positives}}{\text{True Negatives} + \text{False Positives}} \times 100 \text{ percent}$$

The predictive value depends on the actual frequency of a condition in the population. An example used by Rockey, Fantel and Omenn (1979) illustrates this: The exercise electrocardiogram (ECG) test is about 95 percent sensitive and about 95 percent specific for coronary artery disease (CAD). If this test is given to 1,000 patients with angina pectoris, a group in which the frequency of CAD is 80 percent of the 800 persons with CAD, 95 percent or 760 will have positive exercise ECGs. Of the 200 persons without CAD, 95 percent or 190 will have negative ECGs, but 10 will have positive tests. Therefore, the (positive) predictive value of the test is 760 divided by (760 + 10) or 98.7 percent.

However, what if the test were applied to healthy eighteen to thirty-year-old job applicants, a group in which the frequency of CAD is about two percent? If 1,000 applicants were screened, 19 of the 20 (95 percent) with CAD would have positive tests. Of the 980 applicants with CAD, 931 (95 percent) would have negative tests, but 49 would have positive tests. Therefore, the predictive value of the test is 19 divided by (19 + 49) or 28 percent. This means that only 28 percent of those with a positive test actually have CAD (true positive) and the other 72 percent with a positive test are free of the disease (false positives). If such a test were used as prerequisite for employment, 72 percent of those denied employment because of a positive test would have been misclassified.

Table 10.1 illustrates the dependence of the predictive value on the disease prevalence. When the prevalence is 50 percent or higher, the predictive value is marginally influenced.

HOW TO DEVELOP A PREVENTION PROGRAM

When actually initiating the program, *the first phase* includes planning and education. Information is fundamentally important at this stage. The management, organized labor and employees should be informed about the methods, goals, risk factors and the possibility of prevention. This may require brochures, posters and presentations. Once the support of management and labor is ensured, *the second phase* can be initiated. This stage involves the actual screening–in the case of back injuries, screening of both worker and workplace. Specific subject and work-related risk factors are defined and analyzed before *the third phase* is started. The third phase

TABLE 10.1
Predictive Value of a Positive Test as a Function of the Disease Prevalence. Laboratory Test with 95 percent Sensitivity and 95 percent Specificity

Prevalance Disease (percent)	*Predictive Value of Positive Test*
1	16.1
2	21.2
5	50.0
10	67.9
15	77.0
20	82.6
25	86.4
50	95.0

Source: Galen 1979.

includes implementation of the findings of phase two to alter the workplace, reassign the workers, and institute education. A *fourth phase* should always include evaluations of the phase three interventions and reinforcement of the principles of back prevention.

SUMMARY

In spite of the less than concrete scientific evidence on the ultimate effectiveness of any of the proposed preventive routes, all avenues should be explored in light of the magnitude of the LBP problem. Combined prevention programs have proved effective in reducing the frequency of low back complaints, at least in the shorter perspective. Prevention programs applied to the society in general could learn from the successful caries-prevention programs, and should be introduced early, before "bad habits" have developed, and repeated throughout life. Workers should be made aware that they can themselves influence their LBP situation and should be charged with that responsibility. The workplace should be improved as needed, the work methods optimized, and workers should not have to exert themselves at or above their actual strength limits. If injured, a program of diagnosis, care, and rehabilitation should be put into immediate effect. Unrealistic—perhaps. Optimistic—certainly. We have to make a dent in the low back problem. However, we should also be aware that these approaches will not lead to a total extinction of LBP but will help workers understand its cause(s) and improve our treatment programs.

REFERENCES

Anderson, J.A.D. 1980. Occupational aspects of low back pain. *Clinics in Rheumatic Diseases* 6:17–35.

Galen, R.S. 1979. Selection of appropriate laboratory tests. In *Clinician and chemist: the relationship of the laboratory to the physician.* Donald S. Young *et al.* (eds.) Washington, D.C.: Assn. for Clinical Chemistry.

Rockey, P.H.; Fantel, J.; and Omenn, G.S. 1979. Discriminatory aspects of pre-employment screening: Low-back x-ray examinations in the railroad industry. *American Journal of Law and Medicine* 5(3):197–214.

Wiesel, S.W.; Feffer, H.L.; and Rothman, R.H. 1982. Industrial low back pain: A prospective evaluation of a standardized diagnostic and treatment protocol. *The George Washington University Medical Center*, Washington, D.C., unpublished manuscript.

11 Worker Selection

Gunnar B. J. Andersson and
Thomas R. Lehmann

INTRODUCTION

It is the purpose of worker selection to match the right worker to the right job. Ideally, the requirements of all jobs should be such that worker selection and restricted placements are not required. However, this is not the case today and will not realistically be so for some time because of the large variability in the performance capability of the population and the fact that many jobs cannot be redesigned. Thus, selection is one method to reduce possible harmful physical effects of work created by a mismatch of the worker and the job.

This chapter reviews several options in and general criteria for worker selection. Job classification schemes to match subject capabilities are discussed in Chapter 13.

Before a preemployment screening procedure is recommended, the general principles discussed in Chapter 10 should always be considered. The most pertinent questions are:

- Is the screening procedure safe?
- Does the screening procedure have a good probability of predicting the risk of future low back pain or injury?
- Is the screening procedure practical?
- Is the screening procedure ethical and legal?

Safety

A screening program aimed at reducing the hazards of work must clearly be safe. The risks vary from none (in obtaining a medical history) to more substantial risks when radiographs or strength testing programs are used.

Predictive Value

Three factors must be considered in this respect:

- Is the test sensitive?
- Is the test specific?
- What is the frequency of positive results in the population?

These aspects have been discussed in Chapter 10. For reference, the most important definitions are repeated in abbreviated form in Table 11.1.

Practicality

The practicality of a given screening test varies considerably and has a different meaning to different people involved in the screening procedure. A medical history can always be obtained, with little effort and time involved, whereas a strength testing program can be quite cumbersome, expensive and difficult to implement correctly. From the management point of view, practicality also means cost-effective in terms of decreasing compensation claims, number of lost work days, and protection of the worker.

TABLE 11.1
Concepts in Screening Tests

Accuracy:	The ability of a test to provide a true measure of the quantity or quality in question.
Sensitivity:	The ability of a test to recognize disease, that is the fraction of positive results in patients who have a particular disease.
Specificity:	The capacity of a test to recognize non-disease, that is, the fraction of negative results in patients free of the disease.
Predictive Value:	The ability of a test to predict the presence or absence of disease. This value is determined by the sensitivity and specificity of a disease, and by the prevalence of the disease in question.

SPECIFIC SCREENING TESTS

Medical History and Clinical Physical Examination

One purpose of the screening medical history and the physical examination is to identify people who are at *high risk* in terms of future susceptibility to back problems.

The most important part of the screening *medical history* is the identification of previous back problems. In fact, some physicians maintain that this is the only useful predictor in the medical history (Taylor 1968; Glover 1980). Recurrent episodes of LBP appear to be almost part of the natural history LBP. Rowe (1963) found that 83 percent of those with LBP had recurrent attacks. Patients with sciatica had a recurrence rate of 75 percent (Rowe 1965). Similar findings have been reported by Dillane, Fry and Kalton (1966); Horal (1969); Hirsch, Johnsson and Lewin (1969); Leavitt, Johnston, and Beyer (1971); Gyntelberg (1974); Pedersen (1981); Troup, Martin and Lloyd (1981); and Bierring-Sorensen (1983). Recurrence is a different issue from that of having the first attack, as is discussed in Chapter 4.

A few studies specifically consider the risk of having a recurrence. Chaffin and Park (1973) found a three-fold increase in risk, and Bergquist-Ullman and Larsson (1977) report that 62 percent of their group of 217 workers with acute LBP had recurrences within a year, and another 18 percent had recurrences within two years. Pedersen (1981), Troup, Martin and Lloyd (1981) and Bierring-Sorensen (1983) found a history of sciatica to be a risk indicator, in contrast to an earlier report by Dillane, Fry and Kalton (1966). Magora and Taustein (1969) found that persons who had had sciatica had more and longer sickness absence periods than those with pain only in the lower back, a finding confirmed by Andersson, Svensson and Oden (1983).

Bierring-Sorensen (1983) performed a cross-sectional survey of 558 men and 583 women, 30 to 60 years of age. He found that there was a significantly increased risk of LBP in the year following examination in subjects who, in the year immediately before the study, had either 1) many episodes of back pain, 2) many sickness absence days because of LBP, 3) short intervals between episodes, or 4) an aggravated course of LBP.

Thus, information not only about the occurrence of previous LBP but also about its severity has predictive value. Pedersen (1981) found that a history of more than three previous episodes of LBP was ominous because new episodes were both longer and more severe. Acute onset of LBP, irrespective of cause, is associated with a longer duration of the pain episode (Bergquist-Ullman and Larsson 1977; Pedersen 1981; Bierring-Sorensen 1983). In contrast, Troup, Martin and Lloyd (1981) found that truly accident-related previous LBP did not predict recurrences. This may seem paradoxical, but the patient with a fracture may return to work with the problem.

Lloyd and Troup (1983), in a prospective study, found four factors to be of predictive value in patients returning to work after LBP. Those were residual leg pain on return to work, falls as a cause for back pain, sickness absence for five weeks or more, and a history of two or more previous attacks. The greater the number of these factors, the higher the risk of a recurrent attack.

These studies would all seem to suggest that it is useful to obtain a medical history as part of the preemployment program. Snook, Campanelli and Hart (1978) found that simply obtaining a medical history for prior back pain was not enough to reduce back injury rates and advocate a more comprehensive approach (including workplace changes and personnel training).

The purpose of the screening *physical examination* is the detection of signs of dysfunction indicative of a risk of future problems. General observation of posture, ranges of motion and similar tests, do not seem to be reliable prognosticators, as discussed in Chapter 7. Age and sex have some direct influence on susceptibility to LBP; the risk of LBP increases with increasing age and, as previously discussed, there are differences between women and men. The individual variability, however, is so great within age and gender groups that to deny a person employment based solely on age or sex would be insupportable and illegal in the United States. If the patient has a history of ongoing back pain, of neoplastic lesions (including a history of such), of spondylolisthesis, and of severe postural deformities, this should lead to more extensive medical investigations.

Anthropometric data and measurements of the range of motion of the spine are easily obtained at the clinical examination but often have limited value to the selection procedure. It should be obvious from the preceding chapters that the worker must be matched to the work task according to reach and space requirements. If a specific reach requirement is documented for a particular job, it is necessary to evaluate the applicant's capability in that respect. The job requirement also should be adjusted to the subject's range of motion. He should not have to stretch or bend to the normal limits of motion. A lack of joint mobility can cause comparatively greater stresses on another joint; for example, limited ranges of motion of the hips and knee can require more motion of the spine. Also, a rigid joint is less well adapted to absorb shock. It is a well-known fact that joint motion is impaired following injury. Bierring-Sorensen (1983) found that reduced spine motion in subjects with previous LBP indicated an increased risk of a future pain episode. In contrast, previously healthy subjects with less spine mobility had a decreased risk for future LBP.

There are other physical findings that may have predictive value. Lloyd and Troup (1983) found five physical signs with significant predictive value in patients returning to work after a back pain episode. These are restriction of the pain-free range of straight-leg raising (SLR); reproduction of pain caused by SLR; inability to sit up from the supine (trunk flexor weakness); pain or weakness on resisted hip flexion tests with the patient sitting; and back pain on lumbar extension induced by passive flexion of the knees with the patient in the prone position. When more than one of these signs was present, the risk of recurrence increased further. Although Lloyd and Troup's data cannot be directly transferred to a general population, the data may have a role in secondary prevention and when reassigning workers.

General Fitness

Individuals with a good state of general fitness have a lower risk of chronic LBP and also recover more rapidly after LBP. Cady *et al.* (1979) measured the flexibility, isometric lifting strength, and cardiopulmonary tests of 1,652 firefighters in Los Angeles from 1971 to 1974. The firefighters were divided into three groups. The most fit had the least back pain. Although these results are not sufficiently conclusive to permit exclusion of workers with poor physical fitness, these data should stimulate workers to upgrade their physical status. This topic is important in the design of back schools as is discussed in Chapter 12. Svensson and Andersson (1983) found LBP to be more common in men who were physically less active in their leisure time. This relative inactivity can, of course, be as a result of existent back problems and not its causation.

Radiographic Screening

The use of preemployment radiographic assessments of the lumbar spine has been and is widespread. Of the many papers published on this controversial topic, a few are presented here for the purpose of further discussion. For a complete review of the literature see Montgomery (1976). As with any preemployment screening procedure, predictive and safety factors must be considered.

Safety Aspects

A single lumbosacral x-ray probably does not carry a significant health risk. Repeated radiographs, however, can result in considerable irradiation of the gonads and the bone marrow. The addition of oblique x-rays to the standard AP more than doubles this exposure. Low back x-rays are the largest single contributor to gonadal irradiation in the United States (Rockey *et al.* 1978). Gonadal shielding is clearly possible in males but can be difficult in females. In women with unsuspected pregnancy, there is also a significant risk for irradiation of the fetus, which can double the risk of subsequent childhood leukemia. For highly susceptible fetuses the risk can be increased ten-fold (Bross and Nataragan 1972). It has also been estimated that low-level gonadal radiation could increase the risk of irradiation-caused diseases by 2.5 to 25 percent (National Research Council [U.S.] 1972). Therefore, low back x-rays should always be considered a potential health risk, and any unnecessary use should be discouraged.

The Predictive Value of Low-Back Radiographs

The ultimate answer to the question of predictive value of spinal radiography would require a prospective, well controlled, longitudinal study. No such study exists at the present time. Most investigations on the subject have used one of two approaches: searching for an increased prevalence of radiographic abnormalities in patients who have developed back pain, or searching for decreased incidence of LBP after an x-ray screening program has been implemented.

The first method requires a control group without LBP. In the past, either a control group has been selected for that purpose, or a general population sample has been divided into a group with and one without LBP and these two groups compared. Examples of studies in which control groups have been selected are by Splithoff (1953), Horal (1969), Torgerson and Dotter (1976), Fischer, Friedman and van Denmark (1958), and LaRocca and Macnab (1969). Examples of population studies are by Bistrom (1954), Hult (1954), Lawrence (1955), Connell (1968), Hirsch, Johnsson and Lewin (1969), Rowe (1963, 1965, 1969), Magora and Schwartz (1976, 1978, 1980), Wiikeri *et al.* (1978) and Frymoyer *et al.* (1983). The difficulty of obtaining adequate control populations is obvious from most of these studies.

The second approach has been to measure alterations in LBP after preemployment x-rays. Unfortunately, all the published studies have used other screening techniques as well, and it is difficult to assess the value of x-rays *per se*. Examples of studies using this approach are by Kelly (1965), McGill (1968), Leggo and Mathiasen (1973), Redfield (1971), Kosiak, Aurelius and Hartfiel (1966), and Crookshank and Warshaw (1961).

Based on all of these studies, it is clear that so-called radiographic abnormalities are quite frequent and do not always indicate pathology. Common x-ray findings and their relationship to LBP are tabulated in Table 11.2. Tumors, fractures, and infectious and inflammatory diseases clearly were related to present or previous LBP. Spondylolisthesis and osteoporosis occur frequently in patients with no LBP.

In recent years, other approaches have also been attempted to determine the predictive value of the spinal radiograph. Rockey *et al.* (1978) reviewed

TABLE 11.2
Importance of Radiographic Abnormality in Determination of Cause for or Risk of LBP Based on Literature Review

1. Definitive Importance

    Neoplasm of destruction of bone
    Fractures
    Ankylosing spondylitis
    Infections
    Osteoporosis
    Spondylolisthesis
    Severe kyphosis
    Severe osteoarthritis apophyseal joints

2. Definitive Importance Requiring Myelogram or CT

    Disk herniation
    Spinal stenosis

3. Probable Importance

    Osteoarthritis apophyseal joints
    Marked disk degeneration (several levels)
    Disk resorption
    Spondylolysis

4. Possible Importance

    Severe scoliosis
    Lumbar kyphosis
    Transitional vertebrae

5. Minor Importance

    Disc degeneration
    Scoliosis
    Spina bifida
    Facet joint tropism
    Schmorl's nodes
    Lordosis

the treatment and outcome of treatment in 440 patients with back pain. One hundred and six (24 percent) of those had spinal x-rays. It was determined that those spinal radiographs contributed minimally to the diagnosis, little effect on the therapeutic decisions, and had no appreciable effect on the therapeutic outcome. But, patients in whom back x-rays were taken were more likely to be satisfied with their care. Brolin (1975) reviewed 68,000 consecutive low back examinations done over a ten-year period at a Swedish hospital. In only 1 of 2500 radiographs were there findings previously unsuspected on the basis of clinical evaluation. These findings were also confirmed by Scavone (1981) who also found that oblique radiographs yielded additional information in only 2.4 percent of patients. He estimates 7 million lumbar spine examinations are performed annually in the United Kingdom. The annual cost is $500 million. Repeat spinal radiographs were also of little value. Scavone (1981) found virtually no use in the repeat radiograph except in tumors and monitoring fracture healing.

Another study design was used by Gibson *et al.* (1980) who compared two retrospective cohorts (of steel workers); one hired before and one after back x-rays were introduced into an otherwise identical preplacement medical program. The inclusion of low back x-ray screening did not significantly influence the subsequent incidence of LBP in this carefully conducted retrospective study.

It should be clear from the preceding discussion that low back x-rays have low sensitivity and low specificity, and consequently also low predictive value. Rockey, Fantel and Omenn (1979) have made some interesting calculations on a hypothetical cohort of workers clearly showing how misclassification can occur for this reason and the effects of those miscalculations.

The Frequency of 'Abnormal' Radiographs in the Population

Although most radiographic abnormalities or diagnoses are quite rare (with the exception of degenerative changes), they are so many that together a large proportion of the population will have at least one such change. Foote (1982) summarized eleven papers on 40,017 subjects for whom preemployment radiographs were obtained. From 22 to 70 percent of the x-rays were considered abnormal, and from 9.8 to 28 percent of the workers were rejected or restricted. Ethical and moral concerns are obvious when such large parts of the work force are affected. Legal aspects must be considered also, as summarized by Hadler (1978) and by us in Chapter 16.

Consensus on Preemployment Back X-rays

The enthusiastic endorsements of pre-employment screening radiographs in its early days have changed gradually over the years. In a review article, Montgomery (1976) questioned the evidence and advised against the use of preemployment back x-rays for determining future risk. LaRocca and Macnab (1969), Hadler (1978), and Rockey, Fantel and Omenn (1979) came to a similar conclusion. In 1973 the American College of Radiology, the American Academy of Orthopaedic Surgeons and the American Occupational Medical Association joined together in a conference to evaluate the effectiveness of such x-rays when compared to the potential radiation hazard. They concluded that the use of x-rays as the sole criterion for selection of workers was not justified and that more concern was needed to protect workers from unnecessary radiation in such examinations (Amer. Coll. of Radiol. 1973). Foote (1982) published a dissenting opinion in response to that report.

Houston (1977) listed "three compelling reasons" that radiologists should avoid complicity in routine preemployment spine radiographs in an editorial in the *Journal of the Canadian Association of Radiologists*. Those were:

1. There is no predictive value from such examination.
2. The individual examined receives the slight potential harm of gonadal irradiation without benefit to health.
3. An asymptomatic condition, which would not impede effective work, might nevertheless prevent an applicant from getting a particular job infringing on his civil right of equal opportunity for employment and possibly even affecting his future insurability.

Predictably, there were letters in response to his article expressing different opinions.

We generally concur with the opinions of those taking a negative stand on the issue of preemployment low back x-rays. The overwhelming bulk of evidence indicates that there are few degenerative or developmental criteria that can be applied to an individual to reasonably predict the risk of low back disability. When a previous history of back pain is present, or clinical signs indicate possible disease, or in clinical signs such as spondylolisthesis suspected from clinical examination, radiographs can be a valuable aid in determining further risk.

Laboratory Screening Procedures

There are no direct laboratory screening tests for LBP. Clinical interest, however, has been generated by the statistically significant association found between ankylosing spondylitis and a histocompatibility complex antigen HLA B-27. While the sensitivity of the test is high (95 percent), the specificity is low (Calin 1980), making its use as a screening procedure of little value (Sandstrom and Andersson 1983).

Strength Testing

The use of preemployment strength testing to reduce the incidence and severity of musculoskeletal problems has been met with considerable enthusiasm over the past decade. The obvious goal is to assure that only people with sufficient strength to perform a job safely will be assigned to that job. It is clear from previous studies that there is an enormous variation in strength among people. Many jobs in industry do indeed require exertions that approach or exceed individual strengths. It is beyond the scope of this chapter to review different muscle strength evaluation methods and equipment. The reader is referred to Chaffin (1975), Kroemer (1970), Garg, Mitel and Asfour (1980), Stobbe (1982), and Chaffin and Andersson (1983). There are static (isometric) strength testing methods available (Chaffin, Herrin and Keyserling 1978; Keyserling, Herrin and Chaffin 1980; Stobbe 1982), whereas dynamic strength tests are yet to be further developed (Pytel and Kamon 1981). Psychophysical approaches also can be used, as discussed by Snook (1978) and Ayoub *et al.* (1980).

Safety

Various types of physical strength and endurance tests must be carefully evaluated to assure that they are safe. Previous history of musculoskeletal problems or cardiovascular problems may contraindicate testing of a specific person or at least require a modified test procedure.

Predictive Value

There is some epidemiologic support that strength testing is a useful means of reducing back injury rates. Chaffin and Park (1973) found a sharp increase in the low back injury rates in subjects performing jobs requiring strength that was greater or equal to their isometric strength-test values. In fact, the risk was three times higher for the weaker subjects. A second longitudinal study was performed by Chaffin *et al.* (1977) to determine further the value of strength testing, and again the back injury incidence rate was found to be almost three times higher in the over-stressed group. In a later paper, Chaffin, Herrin and Keyserling (1978) suggested that specific placement and selection programs should be undertaken by industry based on strength performance criteria.

Another study by the Michigan group involved the application of strength tests and a simultaneous biomechanical job analysis in the rubber industry (Keyserling, Herrin and Chaffin 1980). Subjects were strength tested and jobs were biomechanically classified. The subjects were then assigned to jobs in such a way that some were overstressed and others were understressed. The medical records were followed for one year to determine m''sculoskeletal problems over that period. Although the follow-up interval was short, job matching based on strength criteria appeared to be beneficial. beneficial.

Although none of these studies allows absolute predictive values to be computed for strength testing, there appears to be some beneficial value. Troup, Martin and Lloyd (1981), in a prospective study, found reduced dynamic strength of back flexor muscles to be a consistent predictor for recurrence or persistence of back pain.

Practicality

Because strength testing requires equipment specific to a given work situation and familiarity with strength test procedures, strength testing may not always be practical. Chaffin (1983) has suggested the following three general guidelines for equipment and procedure: hardware capable of simulation of different work situations; minimal administration time; minimal time of instruction and learning.

In addition, a careful analysis must be made of the workplace to allow for the strength test to truly assess those strength capabilities that are important in a given job.

SUMMARY

The most important part of the screening medical history is the identification of previous back problems. General observations of posture and range of motion are not useful prognosticators of future LBP but may help to match the worker to the job. Workers should not be excluded from work on the basis of poor fitness but physical fitness should be encouraged. Preemployment radiographs are not justified unless the clinical signs suggest possible disease.

Strength testing is in its infancy, and there is little consensus based on experience and predictive value. It appears that strength testing can be preventive. The testing procedure should be specific to the job being sought by the worker. It remains unclear whether a well-matched worker will develop a problem from a heavy physical job should the exposure period be extended beyond one year. Muscle strength and tissue resistance to future stress may not be directly related. In general, a worker should at least have the demonstrable strength required to perform a job for which he is employed prior to having to perform the job.

REFERENCES

American College of Radiology. 1973. Conference on low-back x-rays in pre-employment physical examinations, Tucson, 1973. *Proceedings of Meeting*, January 11–14, sponsored by NIOSH, contract HSM–00–72–153.

Andersson, G.B.J.; Svensson H.O.; and Oden, O. 1983. The intensity of work recovery in low back pain. *Spine*. In press.

Ayoub, M.M.; Mital, A.; Bakken, G.M.; Asfour, S.S.; and Bethea N.J. 1980. Development of strength and capacity norms for manual materials handling activities. The state of the art. *Human Factors* 22 (3):271–83.

Bergquist-Ullman, M.; and Larsson, U. 1977. Acute low back pain in industry. A controlled prospective study with special reference to therapy and confounding factors. *Acta Orthop. Scand.* (Suppl.) 170:117.

Bierring-Sorensen, F. 1983. The prognostic value of the low back history and physical measurements. Unpublished doctoral dissertation. University of Copenhagen, Copenhagen.

Bistrom, O. 1954. Need degenerative changes in the spinal column entail back pain? *Ann. Chir. Gynaec. Fenn.* 43:29–44.

Brolin, I. 1975. Produktkontroll av rontgenundersokningar av landryggraden. *Lakartidningen* 72 (7):1793–95.

Bross, I.D.J.; and Nataragan, N. 1972. Leukemia from low-level radiation. Identification of susceptible children. *New Engl. J. Med.* 287 (3):107–110.

Cady, L.D.; Bischoff, D.P.; O'Connell, E.R.; Thomas, P.C.; and Allan, J.H. 1979. Strength and fitness and subsequent back injuries in fire-fighters. *J. Occup. M.* 21 (4):269–72.

Calin, A.; Kay, B.; Sternberg, M.; Antell, B.; and Chan, M. 1980. The prevalence and nature of back pain in an industrial complex: Questionnaire and radiographic and HLA analysis. *Spine* 5:201–205.

Chaffin, D.B. 1982. Functional assessment for heavy physical labor. M.H. Alderman and M.J. Hanley (eds.) *Clinical medicine for the occupational physician.* New York: Marcel Dekker.

Chaffin, D.B. 1975. Ergonomics guide for the assessment of human static strength. *Am. Indus. Hyg. Assoc. J.* 36 (7):505–511.

Chaffin, D.B.; and Andersson G.B.J. 1983. *Occupational biomechanics.* New York: Wiley. In press.

Chaffin, D.B.; and Park, K.Y.S. 1973. A longitudinal study of low-back pain as associated with occupational weight lifting factors. *Am. Indus. Hyg. Assoc. J.* 34 (12):513–25.

Chaffin, D.B.; Herrin, G.D.; Keyserling, W.M.; and Foulke, J.A. 1977. Pre-employment strength testing in selecting workers for materials handling jobs. *NIOSH Technical Report*, NIOSH Physiology and Ergonomics Branch, Cincinnati, Ohio. NIOSH Contract No. CDC–99–74–62, May 1977.

Chaffin, D.B.; Herrin, G.D.; and Keyserling, W.M. 1978. Preemployment strength testing. An updated position. *J. Occup. Med.* 20 (6):403–408.

Connell, M.A. 1968. Bony anomalies of the low back in relation to back injury. *Southern Med. J.* 61:482–86.

Crookshank, J.W.; and Warshaw, L.M. 1961. The lumbar spine in the workman. *Southern Med. J.* 54:636–38.

Dillane, J.B.; Fry, John; and Kalton, G. 1966. Acute back syndrome–A study from general practice. *Br. Med. J.* 2 (5505):82–84.

Fischer, F.J.; Friedman, M.M.; and van Demark, R.E. 1958. Roentgenographic abnormalities in soldiers with low back pain. A comparative study. *Am. J. Roentgenol.* 79 (4):673–76.

Foote, G.A. 1982. Pre-employment radiography of the lumbosacral spine. Radiology in health screening. *Australas. Radiol.* 26 (1):25–29.

Frymoyer, J.W.; Newberg, A.; Pope, M.H.; Wilder, D.G.; Clements, J.; and MacPherson, B. 1983. The relationship between spinal radiographs and LBP severity in males 18-55. Pending publication *J. Bone Joint Surg.*

Garg, A.; Mital, A.; and Asfour, S.S. 1980. A comparison of isometric strength and dynamic lifting capability. *Ergonomics* 23 (1):13–27.

Gibson E.S.; Martin, R.H.; and Terry, C.W. 1980. Incidence of low back pain and pre-employment X-ray screening. *J. Occup. Med.* 22(8):515–19.

Glover, J.R. 1980. Prevention of back pain. The lumbar spine and back pain. 2nd ed. Malcolm Jayson (ed.), Tunbridge Wells (Eng.), Pitman Medical.

Gyntelberg, F. 1974. One year incidence of low back pain among male residents of Copenhagen aged 40-59. *Dan. Med. Bull.* 21 (1):30–36.

Hadler, N.M. 1978. Legal ramifications of the medical definition of back disease. *Ann. Intern. Med.* 89 (6):992–99.

Hirsch, C.; Jonsson, B.; and Lewin, T. 1969. Low back symptoms in a Swedish female population. *Clin. Orthop.* 63:171–76.

Horal, J. 1969. The clinical appearance of low back disorders in the city of Gothenburg, Sweden. Comparisons of incapacitated probands with matched controls. *Acta Orthop. Scand.* (Suppl.) 118:1–109.

Houston, C.S. 1977. Pre-employment radiographs of lumbar spine, editorial. *J. Canad. Assoc. Radiol.* 28 (3):170.

Hult, L 1954. Cervical, dorsal, and lumbar spinal syndromes. *Acta Orthop. Scand.* (Suppl.) 17:1–102.

Kelly, F.J. 1965. Pre-employment medical examinations including back x-rays. *J. Occup. Med.* 7 (4):132–36.

Keyserling, W.M.; Herrin, G.D.; and Chaffin, D.B. 1980. Isometric strength testing as a means of controlling medical incidents on strenuous jobs. *J. Occup. Med.* 22 (5):332–36.

Kosiak, M.; Aurelius, J.R.; and Hartfiel, W.F. 1966. Backache in industry. *J. Occup. Med.* 8 (2):51–58.

Kroemer, K.H.E. 1970. Human strength. Terminology, measurement and interpretation of data. *Human Factors* 12 (3):297–313.

LaRocca, H.; and Macnab, I. 1969. Value of pre-employment radiographic assessment of the lumbar spine. *Can. Med. Assoc. J.* 101 (7):383–88.

Lawrence, J.S. 1955. Rheumatism in coal miners; occupational factors. *Brit. J. Indus. Med.* 12:249–61.

______. 1969. Disc degeneration. Its frequency and relationship to symptoms. *Ann. Rheum. Dis.* 28 (2):121–38.

Leavitt, S.S.; Johnston, T.L.; and Beyer, R.D. 1971. The process of recovery: Patterns in industrial back injury. Part 1. Costs and other quantitative measures of effort. *Ind. Med. Surg.* 40 (8):7–14.

Leggo, C.; and Mathiasen, H. 1973. Preliminary results of a preemployment back x-ray program for state traffic officers. *J. Occup. Med.* 15 (12):973–74.

Lloyd, D.C.E.F.; and Troup, J.D.G. 1983. Recurrent back pain and its prediction. *J. Soc. Occup. Med.* 33 (2):66–74.

Magora, A.; and Schwartz, A. 1976. Relation between the low back pain syndrome and x-ray findings. 1. Degenerative osteoarthritis. *Scand. J. Rehabil. Med.* 8 (304):115–25.

Magora, A.; and Schwartz, A. 1978. Relation between the low back pain syndrome and x-ray findings. 2. Transitional vertebrae (mainly sacralization). *Scand. J. Rehabil. Med.* 10 (3):135–45.

Magora, A.; and Schwartz, A. 1980. Relation between low back pain and X-ray changes. 4. Lysis and olisthesis. *Scand. J. Rehabil. Med.* 12 (2):47–52.

Magora, A.; and Taustein, I. 1969. An investigation of the problem of sick-leave in the patient suffering from low back pain. *Industr. Med. Surg.* 38 (11):398–408.

McGill, C.M. 1968. Industrial back problems. A control program. *J. Occup. Med.* 10 (4):174–78.

Montgomery, C. H. 1976. Preemployment back x-rays. *J. Occup. Med.* 18 (7):495–98.

National Research Council (U.S.). 1972. Advisory Committee on the Biological Effects of Ionizing Radiation, Washington. The effects on populations of exposure to low levels of ionizing radiation. *National Academy of Sciences*, p. 55.

Nordgren, B.: Schele, R.; and Linroth, K. 1980. Evaluation and prediction of back pain during military field service. *Scand. J. Rehab. Med.* 12 (1):1–8.

Pedersen, P.A. 1981. Prognostic indicators in low back pain. *J. Royal Coll. Gen. Pract.* 31 (225):209–16.

Pytel, J.L.; and Kamon, E. 1981. Dynamic strength test as a predictor for maximal and acceptable lifting. *Ergonomics* 24 (9):663–72.

Redfield, J.T. 1971. The low back x-ray as a pre-employment screening tool in the forest products industry. *J. Occup. Med.* 13 (5):219–26.

Rockey, P.H.; Fantel, J.; and Omenn, G.S. 1979. Discriminatory aspects of pre-employment screening: Low back x-ray examinations in the railroad industry. *Am. J. Law and Med.* 5 (3):197–214.

Rockey, P.H.; Tompkins, R.K.; Wood, R.W.; and Wolcott, B.W. 1978. The usefulness of x-ray examinations in the evaluation of patients with back pain. *The Journal of Family Practice* 7 (3):455–65.

Rowe, M.L. 1963. Preliminary statistical study of low back pain. *J. Occup. Med.* 5 (7):336–41.

______. 1965. Disc surgery and chronic low back pain. *J. Occup. Med.* 7:196–202.

______. 1969. Low back pain in industry. A position paper. *J. Occup. Med.* 11:161–69.

Runge, C.F. 1958. Pre-existing structural defects and severity of compensation back injuries. *Ind. Med. Surg.* 27 (4):249–52.

Sandstrom, J.; and Andersson G.B.J. 1983. HLA-B27 as a diagnostic screening tool in chronic low back pain. *Scand. J. Rehab. Med.* In press.

Scavone, J.G.; Latshaw, R.F.; and Weidner, W.A. 1981. Anteroposterior and lateral radiographs: An adequate lumbar spine examination. *A.J.R.* 136 (4): 715–17.

Snook, S.H. 1978. The design of manual handling tasks. *Ergonomics* 21 (12):963–85.

Snook, S.H.; Campanelli, R.A.; and Hart, J.W. 1978. A study of three preventive approaches to low back injury. *J. Occup Med.* 20 (7):478–481.

Splithoff, C.A. 1953. Lumbosacral junction. Roentgenographic comparison of patients with and without backaches. *JAMA* 152 (17):1610–13.

Svensson, H.O.; and Andersson, G.B.J. 1983. Low back pain in forty to forty-seven year old men: Work history and work environment factors. *Spine* 8 (3):272–76.

Stobbe, T. 1982. Unpublished dissertation. (Industr. Eng.) Univ. Michigan, Ann Arbor.

Taylor, P.J. 1968. Personal factors associated with sickness absence. A study of 194 men with contrasting sickness absence experience in a refinery population. *Brit. J. Ind. Med.* 25 (2):106–118.

Torgerson, W.R.; and Dotter, W.E. 1976. Comparative roentgenographic study of the asymptomatic and symptomatic lumbar spine. *J. Bone Joint Surg.* 58A (6):850–53.

Troup, J.D.G.; Martin, J.W.; and Lloyd, D.C.E.F. 1981. Back pain in industry. A prospective survey. *Spine* 6 (1):61–69.

Wiikeri, Matti; Nummi, Juhani; Riihimaki, Hilkka; and Wickstrom, Gustav. 1978. Radiologically detectable lumbar disc degeneration in concrete reinforcement workers. *Scand. J. Work Environ. and Health* 4 (Suppl. 1) 47–53.

12 Education and Training

Stover H. Snook and Arthur H. White

INTRODUCTION

Education and training have been the most common approaches used in industry to prevent low back pain (LBP). Safety and personnel departments typically have had responsibility for instructing employees in proper methods and work procedures. Training in safe lifting procedures has been a part of safety programs in industry for over 50 years. In recent years there has been increased emphasis on strength and fitness training, and the use of back schools. The training of supervisors and other management personnel to respond appropriately to the LBP injured worker is a relatively new use of education and training to minimize the impact of low back disorders. In this chapter, four strategies for education and training are reviewed.

TRAINING IN SAFE LIFTING

Several studies have shown that approximately one-half of all compensable LBP episodes are associated with manual lifting tasks (Cady *et al.* 1979; Snook *et al.* 1980; Snook *et al.* 1978). Consequently, there has been a major industrial effort to train workers to lift safely; however, the rationale, particularly the effects of the educational programs, and the compliance with them are controversial. The most common type of manual lift in industry is from the floor or close to the floor (Snook *et al.* 1978). The original concept of safe lifting required the worker to maintain a straight back, bending the knees to lower the body, and then lifting with the leg muscles. It was recognized that the intervertebral disc could better withstand the compressive forces of straight back lifting than the shear forces of bent back lifting

(Hickey and Hukins 1980). However, as we have learned in Chapter 3, the bent knee lift results in greater load moments if the load is bulky.

Additional principles of safe lifting have been added to this original concept. Holding the object close to the body is as important, if not more important, than keeping a straight back (Andersson *et al.* 1976). Slow, smooth lifting without jerking tends to minimize the effects of acceleration on the lower back. Turning with the feet instead of twisting the trunk reduces the torsional loads on intervertebral discs (Hickey and Hukins 1980). Many compensable LBP episodes are associated with twisting motions.

The correct positioning of the feet, chin, arms and hands is emphasized by "kinetic lifting" (Anderson and McClurg 1970) (currently recommended by the International Labour Office [Himbury 1967] and the National Safety Council in the United States) [National Safety Council 1971]). Kinetic lifting is defined as a lift whereby the worker adds kinetic energy either to a horizontally moving load or to the vertically moving body before actually lifting the load vertically. This is done by imparting some velocity to the object. However, there is not always a concensus about the elements of a safe lifting program. For example, some authorities emphasize bending the knees (Ring 1982), while others place greater importance upon maintaining a straight back (Hall 1973).

Lifting with straight back and bent knees places the quadriceps muscles at a severe mechanical disadvantage and requires greater energy expenditure than lifting with bent back and straight knees. Perhaps this is why workers seldom use straight back and bent knee lifting in industry, even after programs in safe lifting are conducted (Brown 1971, Garg 1979). Most workers use a combination of bent back/bent knees, depending upon the characteristics of the lifting task. Keeping the back straight to prevent shear forces appears to have merits, but it is difficult to enforce. Workers who have or have had LBP have been found to be more likely to keep their backs straight when lifting. Consequently, it may be more appropriate to teach the straight back lift to those workers who have had LBP, since they may accept the tradeoff of shifting the load to the back rather than the knees. In this case, the purpose of the training would be to prevent recurrences of LBP, instead of the initial episode.

The National Institute of Occupational Safety and Health (NIOSH) observes that many safe lifting programs have tended to rely on a dogmatic style of instruction and sets of 'rules' for safe lifting are common. The drawback is that literal application of some of these rules has led to quite unsafe lifting practices. For example, to insist on the rule to always 'Lift with the knees' is often impractical and may in fact be dangerous in some circumstances. In some people the quadriceps femoris muscle may be insufficient for the task at hand. Moreover, this lifting posture applied nonselectively can lead to lifting bulky objects at arm's length in front of the knees, thus creating more stress on the spine than the straight-legged stoop posture. Any rules used as memory aids should at least teach a basic aim or principle. What matters is that the trainee is led to a proper understanding of the problem and not merely expected to remember a set of "catch-phrases" (U.S. Dept. Health and Human Services 1981b). To achieve these goals, NIOSH recommends that training in safe lifting should include the following information (Fahrni 1975):

1. the risks to health of unskilled lifting;
2. the basic physics of lifting;
3. the effects of lifting on the body;
4. individual awareness of the body's strengths and weaknesses;
5. how to avoid the unexpected physical factors that might contribute to LBP;
6. the development of handling skills;
7. the use of handling aids.

Despite these practical and common sense guidelines, the effect of training in safe lifting on the reduction of LBP and disability has not yet been adequately studied. Several studies have reported a 65 to 70 percent reduction in low back disability (Glover 1976; Miller 1977), while other studies have reported little or no effect (Brown 1971; Dehlin 1976; Snook 1978). Unfortunately, these studies have not been controlled nor are they of sufficiently long duration. Many authorities recommend a minimum experimental observation of three years because of the cyclic nature of low back disability in industry and the "Hawthorne"* effect of many training programs. Two studies have investigated the biomechanical aspects of lifting with straight back and bent knees versus lifting with bent back and straight knees. One study concluded that straight back/bent knees should be recommended only for small, compact objects that can be lifted between the knees and close to the body (Park 1974). The other study could find no clear biomechanical rationale for deciding between the two lifting postures (Garg 1979).

NIOSH agrees that the value of any training in safe lifting is open to question because there have been no controlled studies showing a consequent drop in LBP (U.S. Dept. Health and Human Services 1981b). "Yet so long as it is a legal duty for employers to provide such training or for as long as the employer is liable to a claim of negligence for failing to train workers in safe methods of . . . [lifting], the practice is likely to continue despite the

---

*The Hawthorne effect is improvement merely because there has been an intervention of some kind.

lack of evidence to support it" (U.S. Dept. Health and Human Services 1981b).

STRENGTH AND FITNESS TRAINING

Strength and fitness training variably emphasizes different factors such as musculoskeletal strength, cardiovascular fitness, aerobic capacity, endurance and flexibility. Exercises to increase strength and fitness have been a part of LBP treatment programs for many years, but only recently have strength and fitness programs been advocated in industry to reduce or prevent the onset of LBP. The efficacy of strength and fitness training as a technique for prevention is debatable, and further study is needed.

One of the better studies of the effects of strength and fitness on compensable LBP was conducted with Los Angeles firefighters (Cady 1979). A five point fitness index was used, consisting of endurance, strength, flexibility, exercise blood pressure and recovery heart rate. Firefighters were evaluated and grouped into three categories of most fit, middle fit, and least fit. Over a period of four years, 7.1 percent of the least fit group, 3.2 percent of the middle fit group, and 0.8 percent of the most fit group suffered compensable LBP. Although these are very encouraging results, the cost per case did not exhibit the same trend. The cost per case for the least fit group was $5,622; for the middle fit group, $4,972; and for the most fit group, $73,700. However, the sample size in the most fit group was not large enough to allow definite conclusions. It is possible that the most fit group incurred the most serious injuries, accounting for the high cost. A similar study of physical fitness emphasizing flexibility and strengthening exercises has been shown to reduce the injury rates in Canadian ambulance drivers (Imrie 1983).

Several other studies have indicated a positive role for strength and fitness training in reducing the likelihood of an LBP episode. A study at the Eastman Kodak Company investigated the relationship between LBP and abdominal muscle weakness, and between LBP and trunk stiffness (Rowe 1969). Although a definite relationship was found, it could not be determined which came first–the muscle weakness/trunk stiffness or the LBP. A similar conclusion was derived from the epidemiologic studies of Pope *et al.* (1983).

The relationship between preemployment strength testing and musculoskeletal injuries also has been investigated. The results indicate that the incidence of musculoskeletal injuries among weaker employees is up to three times greater than the incidence among stronger employees (Chaffin 1978; Keyserling *et al.* 1980). The YMCA has established a fitness program for individuals with LBP, entitled the "Y's Way to a Healthy Back." Preliminary data indicate excellent results for 29 percent of the participants, good results for 36 percent, fair results for 25 percent, and poor results for 9 percent (Krause *et al.* 1977). However, the study was not controlled, and since most back pain is of a sporadic nature, it is not known how many participants would have improved without the program. Two other studies have investigated the effects of muscle strengthening exercises in recovering from LBP, and both studies concluded that isometric abdominal exercises were significantly superior in aiding recovery (Kendall and Jenkins 1968; Lidstrom and Zachrisson 1970).

On the other hand, not all investigations of strength and fitness have demonstrated its relationship to LBP. Isometric strength measurements of back muscles were not related to the incidence of LBP in Danish factory workers (Pedersen *et al.* 1975). An epidemiological study of nearly 5000 people in Copenhagen could not relate the degree of physical fitness or activity to LBP (Gyntelberg 1974). Measurements of muscle strength in adults with and without LBP yielded little or no difference between the two groups, leading the authors to conclude that the strength of spinal and abdominal muscles are of doubtful importance for the prevention of LBP (Berkson *et al.* 1977; Nachemson and Lindh 1969).

In addition to the conflicting data on strength and fitness training, it must be recognized that the enforcement of such training is difficult in industry. In most cases, industry must rely on voluntary compliance. However, the growing popularity of physical fitness worldwide represents a good opportunity that should not be missed. Industry should encourage and promote physical fitness through the development of programs and facilities for its employees. There are many health benefits associated with physical fitness, including a possible reduction of LBP.

BACK SCHOOLS

The concept of the back school can be traced to the teachings of Delpech in Toulouse (Peltier, 1983). In 1958, Fahrni began using the concepts of education to control LBP in his patients on a one-to-one basis (Fahrni 1966, 1975). In 1969, Zachrisson-Forssell organized a back school in Stockholm (Zachrisson-Forssel 1981). Early back schools were also developed in San Francisco (Mattmiller 1980), and Downey, California (Mooney 1975), Dallas (Melton 1983), Atlanta (O'Donnell 1981), and Toronto (Hall 1980). The original concept of the back school was to educate patients who were already suffering (or had recently suffered) from LBP, i.e., it was a form of treatment. A more recent use of the back school is to educate workers in industry on how to prevent or reduce LBP.

The back school concept emphasizes the importance of involvement of the patient in his own care. Since the prognosis of most LBP episodes is favorable despite the treatment prescribed, the most likely determinant of outcome may be how the back is used and cared for 24 hours a day rather than single physical therapy treatments. The patient as his own therapist may be a more efficient mechanism to apply this principle. Although patients can be educated in the principles of back care by the physicians, pressures of time realistically may interfere. The formalized curriculum of the low back school overcomes this problem. Learning is likely enhanced by the present notion of a school. That is, the patients attending a low back school understand from the outset that *their* treatment and *their* outcome depend on learning. The low back school assures that all pertinent aspects of a curriculum be covered. Ideally, the patient would be tested for comprehension prior to and after the educational program.

The present back school is an attempt to educate the worker in all aspects of back care; it represents a much more comprehensive approach to back care and includes instruction in safe lifting, strength, and physical fitness. Most back schools include discussions on anatomy and physiology of the lower back, body mechanics, posture and nutrition. Some back schools will also offer advice on stress management, coping with pain, relaxation, drug use and abuse, first aid and acute care, epidemiology, activities of daily living, and vocational guidance. The style of teaching can range from informal demonstrations and discussions to structured classroom settings. Job simulation or actual visits to the workplace can also be a part of the back school curriculum.

In addition to the educational material, back schools attempt to convey two important messages. The first message is that the responsibility for recovery from LBP is shared by the patient and the practitioner. The patient has very definite responsibilities, and the most successful recovery results from meeting these responsibilities. The second message is that recovery may not mean complete absence of pain. Residual pain may remain, or pain may recur. The overall goal is to teach the student how to deal, how to cope, and how to function with LBP.

The format and curriculum of low back schools differ from center to center. The type of learning that is emphasized may have one or more attributes of classic learning: cognitive, psychomotor, and affective experiences. Cognitive learning is purely information, presented in a didactic, classroom-like atmosphere. The presented material may include printed brochures, audiovisual programs, or live professional instruction. Currently, there are many prepackaged low back schools available. Usually, anatomy and function of the spine, posture, home exercises, epidemiology, and psychological aspects of the pain experience are included in these instructional units.

Psychomotor learning is a strong component of the California Back School, where the concept of the "obstacle course" has been introduced (Mattmiller 1980). Patients are presented with information about the spine, but are also observed and physically instructed in the correct way to perform activities of daily living and work tasks. Inclusion of a psychomotor component merits consideration because many patients may not be able to incorporate the knowledge into their lives' activities.

The affective component of the learning takes into consideration the importance of motivation in patient education. For most patients, medical therapy should have obvious relevance to the LBP (e.g., a pill, a manipulation, a corset). Low back schools may not meet that goal unless specifically structured to effect the desired changes in the patient's use of his back. Live, professional education is generally considered the most effective way to achieve that goal. The back schools that have enjoyed the greatest success have done so largely because their evangelistic "deans" have the ability to sell their concepts to patients and motivate them to use the methods taught.

Another consideration in establishing a back school is whether to design a curriculum that fits most patients' needs or to design a flexible curriculum. A flexible curriculum might individualize postures most likely to benefit that particular patient, i.e., flexion (antilordosis) or extension (increased lordosis), mobilization or immobilization of the spine. This concept is most highly structured in the California Back School where a specific back school prescription is written based on examination and diagnosis. The therapist then instructs the patient in a specifically tailored program.

The faculty of the back school consists of professional practitioners such as physical therapists, physicians, occupational therapists, and psychologists. Professional practitioners can answer questions, lead discussions and create an atmosphere in which LBP is considered a real and legitimate disorder. Instructors are sympathetic toward the students and sensitive to the many problems that can accompany LBP.

The size of the back school class can range from one to twenty-five or more. Most classes, however, range from five to fifteen; and they represent a form of group therapy by which patients can learn from each other as well as from the instructor. Since most back schools deal with injured workers, the furniture is important. Therefore, mats, pillows, and reclining lounges are often provided in addition to well designed chairs. Students are encouraged to change their posture as often as necessary to remain comfortable.

There has only been one controlled study of the effectiveness of low back schools. Using the Swedish back school developed by Zachrisson-Forssell, Bergquist-Ullman and Larsson investigated the effect on acute LBP patients from the Volvo Company in Goteborg, Sweden (Bergquist-Ullman and Larsson 1977). Patients were randomly assigned to one of three types of treatment: back school, physical therapy, or placebo. The back school consisted of four sessions, with six to eight patients in each class. Physical therapy was conducted on a one-to-one basis. The placebo group was given treatment with "short waves of the lowest possible intensity." Seventy patients completed the back school; 72 patients, physical therapy; and 75 patients, placebo. The number of days necessary for relief from pain was 14.8 for the back school, 15.8 for physical therapy, and 28.7 for the placebo group. There was no statistically significant difference between the back school and physical therapy; both required significantly fewer days (*p* less than .05) than the placebo group. The number of days necessary for return to work was 20.5 for the back school, 26.5 for physical therapy, and 26.5 for the placebo group. The back school required significantly fewer days (*p* less than .05) than either the physical therapy or placebo groups. The authors concluded that the back school is superior to the placebo treatment in time required for pain relief and return to work. Although there was no significant difference between the back school and physical therapy in time required for relief from pain, the authors observed that the back school is economically superior because one therapist can treat more than one patient at a time. The study did not investigate the effect of the back school on patients with chronic pain or the effect of the back school as a preventive technique for industrial workers. For example, Hall (1980) reviewed his experience with 6418 participants in the Canadian back education units. Significant improvement occurred in 69 percent. The rate of success was influenced by duration of LBP, higher levels of education and a recognition of the emotional levels of pain. Poorer results occurred in those with greater than six months duration of LBP, sciatica, and workers' compensation.

Other studies have been reported for preventive types of back schools in industry, but they have not been controlled studies. For example, the Southern Pacific Company used the California Back School as a preventive program for 30,000 workers; they report a 22 percent reduction in low back cases and a 43 percent reduction in lost work days after two years (Pilcher 1979). PPG Industries used the Atlanta Back School as a preventive program for 2000 workers and report between 70 and 90 percent reduction in the number of injuries and the cost of injuries after two years (Johnson 1981). It is difficult to assess the results of these studies, however, because of the lack of a control group and because other types of preventive techniques were used concurrently.

MANAGEMENT TRAINING

The use of education and training to prevent or reduce LBP in industry has been almost exclusively directed toward the worker. Equally important is the training of management. It must be recognized that within the current state of the art, LBP cannot be entirely prevented. "Secondary prevention" is a technique that is used by management to prevent poor results in workers with LBP. Poor results are often associated with adversary situations, litigation, hospitalization, and lack of follow-up and concern. Management can prevent many of these situations by training foremen, supervisors, managers, engineers and medical personnel.

Fitzler and Berger describe a program at American Biltrite in which management was trained in the positive acceptance of LBP (Fitzler and Berger 1982, 1983). An atmosphere was created in which workers were encouraged to report all episodes of LBP (even minor episodes) to the company clinic. An immediate, conservative, in-house treatment was provided by the company nurse, including worker education. Attempts were made to keep the worker on the job, often with modified duties that were consistent with the worker's condition. If necessary, referrals were made to the company doctor, where treatment and progress were closely monitored. Over a three-year period, workers'compensation costs for low back claims were reduced from over $200,000 per year to less than $20,000 per year—a decrease of ten-fold. Although this was not a controlled study, the results are impressive.

Data from the Weyerhaeuser Company indicate that workers with back injuries who are off work more than six months have only a 50 percent possibility of ever returning to productive employment; for more than one year's absence, the possibility is 25 percent; for more than two year's absence, the possibility of returning is almost nil (McGill 1968). Another study concludes that, with the passage of time following injury, patients increasingly elaborate and exaggerate their symptoms but are less truly depressed by their predicament (Beals and Hickman 1972). Even if the physical disability remains constant, the patient's psychological posture and the likelihood of returning to work changes. These studies emphasize the importance of providing modified work as a means of returning the worker to the job as quickly as possible. Unfortunately, modified work is not well accepted in industry, and management often refuses to return an employee to work unless he is 100 percent recovered. The reluctance to accept modified work is costly to industry, unhealthy for the worker, and should be the focus of intensive management training.

Company medical personnel should also be trained in the benefits of early intervention, initial conservative treatment, patient follow-up, and job placement techniques. Medical personnel should be familiar with recent literature, such as presented in this book, that objectively evaluates various types of treatment for LBP (Deyo 1983, Quinet and Hadler 1979, Rowe 1969). Medical personnel should also be familiar with the physical demands of jobs performed in the company, in order to adequately place injured employees and new employees.

Industrial engineers should be trained in the ergonomic principles of good job design. New techniques are available for identifying high risk jobs and for evaluating suggested job modifications (Snook 1978, U.S. Dept. Health and Human Services 1981b); many are given in Chapters 13 and 14.

Good job design reduces the onset of low back episodes, allows the worker to stay on the job longer, and permits the injured worker to return to the job sooner. Data from the Liberty Mutual Insurance Company indicates that job redesign can prevent up to one-third of all compensable LBP (Snook *et al.* 1978). Although job redesign usually involves some cost, it is frequently less than the average cost of a single case of compensable LBP.

Management should also be aware that no one approach will solve the problem of LBP. All of the approaches described here are necessary for an effective program in controlling LBP. Innovative and dedicated education and training can produce beneficial results for both the employee and employer–but only through a total commitment by management at the highest levels.

SUMMARY

Fifty percent of compensable LBP is associated with lifting but instructions for proper technique have frequently been faulty or incomplete. Most workers use a combination of bent back and bent knees. It is probable that those most likely to use safe lifting techniques are those who have had LBP. Strength and fitness training to prevent LBP does not have proven benefit. However, the general health benefits of improved physical fitness should be encouraged by industry. Recently, back schools have been employed to discuss such topics as "how the back functions," lifting techniques, and general emotional and physical health. There are many models of the back school, but they all stress that there is a shared responsibility for recovery; recovery may not mean complete absence of pain. Controlled study shows that the back school is better than placebo and economically superior to physical therapy. There is some indication that preventive schools can be helpful but controlled studies are lacking. Management training is also helpful. Company physicians should be trained in the benefits of early intervention and conservative treatment, follow-up, and careful job placement. Industrial engineers should be trained in ergonomic principles since job redesign can prevent 30 percent of compensable LBP. All of these approaches require the enthusiastic cooperation of management.

REFERENCES

Anderson, T.N. 1970. Human kinetics in strain prevention. *Brit. J. Occup. Safety* 8 (93):248–50.

Andersson, G.B.J.; Ortengren, R.; and Nachemson, A. 1976. Quantitative studies of back loads in lifting. *Spine* 1:178–85.

Beals, R.K.; and Hickman, N.W. 1972. Industrial injuries of the back and extremities. *J. Bone Joint Surg.* 51A:1593–1611.

Bergquist-Ullman, M.; and Larsson, U. 1977. Acute low back pain in industry. *Acta Orthop. Scand.* Suppl 170, 1977.

Berkson, M.; Schultz, A.; Nachemson, A.; and Andersson, G. 1977. Voluntary strengths of male adults with acute low back syndromes. *Clin. Orthop.* 129:84–95.

Brown, J.R. 1971. *Lifting as an industrial hazard.* Labour Safety Council of Ontario, Ontario Department of Labour, Toronto.

Cady, L.D.; Bischoff, D.P.; O'Connell, E.R.; Thomas, P.C.; and Allen, J.H. 1979. Strength and fitness and subsequent back injuries in firefighters. *J. Occup. Med.* 21:169–272.

Cady, L.D., Bischoff, D.P., O'Connell, E.R., Thomas, P.C.; and Allen, J.H. 1979. Letters to the editor: Authors' response. *J. Occup. Med.* 21:720–25.

Chaffin, D.B.; Herrin, G.D.; and Keyserling, W.M. 1978. Preemployment strength testing: An updated position. *J. Occup. Med.* 20:403–408.

Dehlin, O.; Hedenrud, B.; and Horal, J. 1976. Back symptoms in nursing aides in a geriatric hospital. *Scand. J. Rehabil. Med.* 8:47–53.

Deyo, R.A. 1983. Conservative therapy for low back pain. *J. Am. Med. Assoc.* 250:1057–62.

Fahrni, W.H. 1966. Backache relieved through new concepts of posture. Springfield, Il: Charles C. Thomas.

______. 1975. Conservative treatment of lumbar disc degeneration: Our primary responsibility. *Orthop. Clin. North Am.* 6:93–103.

Fitzler, S.L.; and Berger, R.A. 1982. Attitudinal change: The Chelsea back program. *Occup. Health Safety* :24–26, February.

Fitzler, S.L.; and Berger, R.A. 1983. Chelsea back program: One year later. *Occup. Health Safety*, 52–54, July.

Garg, A.; and Herrin, G.D. 1979. Stoop or squat: A biomechanical and metabolic evaluation. *AIIE Transactions* 11:293–302.

Glover, J.R. 1976. Prevention of back pain. In Jayson M. (ed.), *The lumbar spine and back pain.* New York: Grune and Stratton.

Gyntelberg, F. 1974. One year incidence of low back pain among male residents of Copenhagen aged 40-59. *Danish Med. Bul.* 21:30–36.

Hall, H.W., Sr. 1973. "Clean" vs "dirty" leaning, an academic subject for youth. *ASSE J.* 20–25, October.

Hall, H. 1980. The Canadian back education units. *Physiotherapy* 66:115–17.

Hall, H; and Iieta, J.A. 1983. Back schools. An overview of the Canadian back education units. *Clin. Orthop* .179:10–17.

Hickey, D.S.; and Hukins, D.W.L. 1980. Relation between the structure of the annulus fibrosus and the function and failure of the intervertebral disc. *Spine* 5:106–116.

Himbury, S. 1967. Kinetic methods of manual handling in industry. Occupational Safety and Health Series No. 10, International Labour Office, Geneva.

Imrie, D. 1983. Personal communication.

Johnson, C.D. 1981. Safety Forum. *Ind. Safety Prod. News.* 21 June.

Kendall, P.H.; and Jenkins, J.M. 1968. Exercises for backache: A double-blind controlled trial. *Physiotherapy* 54:154–57.

Keyserling, W.M.; Herrin, G.D.; Chaffin, D.B.; Armstrong, T.J., and Foss, M.L. 1980. Establishing an industrial strength testing program. *Am. Ind. Hyg. Assoc. J.* 41:730–36.

Kraus, H.; Melleby, A.; and Gaston, S. 1977. Back pain correction and prevention. *N.Y. State J. Med.* 77:1335–38.

Lidstrom, A.; and Zachrisson, M. 1970. Physical therapy on low back pain and sciatica: An attempt at evaluation. *Scand. J. Rehabil. Med.* 2:37–42.

Mattmiller, A.W. 1980. The California back school. *Physiotherapy* 66:118–21.

McGill, C.M. 1968. Industrial back problems: A control program. *J. Occup. Med.* 10:174–78.

Melton, B. 1983. Back injury prevention means education. *Occup. Health Safety* 20–23, July.

Miller, R.L. 1977. Bend your knees! *Nat. Safety News*, 57–58, May.

Mooney, V. 1975. Alternative approaches for the patient beyond the help of surgery. *Orthop. Clin. North Am.* 6:331–34.

Nachemson, A.; and Lindh, M. 1969. Measurement of abdominal and back muscle strength with and without low back pain. *Scand. J. Rehab. Med.* 1:60–65.

National Safety Council. 1971. Human kinetics . . . and lifting. *National Safety News*, 44–47, June.

Nordin, M.; Frankel, V.; and Spengler, D.M. 1981. A preventive back care program for industry (Abstract). International Lumbar Spine Meeting, Paris, May 17–20.

O'Donnell, R.J. 1981. Prevention of back injury. The Back School. Atlanta, Ga.

Park, K.S.; and Chaffin, D.B. 1974. A biomechanical evaluation of two methods of manual load lifting. *AIIE Transactions* 6:105–113.

Pedersen, O.F.; Petersen, R.; and Staffeldt, E.S. 1975. Back pain and isometric back muscle strength of workers in a Danish factory. *Scand. J. Rehabil. Med.* 7:125–28.

Peltier, L. 1983. The back school of Delpech in Montpellier. *Clin. Orthop.* 179:4–9.

Pilcher, O.J. 1979. Personal communication.

Pope, M.H.; Bevins, T.; Wilder, D.G.; and Frymoyer, J.F. 1983. The relationship between anthropometric, postural, muscular, and mobility characteristics of males, ages 18–55. Submitted to *Spine*.

Quinet, R.J.; and Hadler, N.M. 1979. Diagnosis and treatment of backache. *Semin. Arthritis Rheum.* 8:261–87.

Ring, L. 1981. Facts on backs–A simplified approach to back injury prevention and control. Loganville, Ga: Institute Press.

Rowe, M.L. 1969. Low back pain in industry: A position paper. *J. Occup. Med.* 11:161–69.

Rowe, M.L. 1983. Backache at work. Fairport, NY: Perinton Press.

Snook, S.H. 1978. The design of manual handling tasks. *Ergonomics* 21:963–85.

Snook, S.H.; Campanelli, R.A.; and Ford, R.J. 1980. A study of back injuries at Pratt and Whitney Aircraft. Liberty Mutual Insurance Company, Research Center, Hopkinton, Ma.

Snook, S.H.; Campanelli, R.A.; and Hart, J.W. 1978. A study of three preventive approaches to low back injury. *J. Occup. Med.* 20:478–81.

U.S. Department of Health and Human Services. 1981b. Work practices guide for manual lifting. DHHS (NIOSH) Publication No. 81–122, March.

Zachrisson-Forssell, M. 1981. The back school. *Spine* 6:104–106.

13 Workplace Evaluation

Gary D. Herrin, Don B. Chaffin, Gunnar B. J. Andersson, and Malcolm H. Pope

RATIONALE FOR WORKPLACE EVALUATION

The need to describe manual work activities quantitatively was first recognized in the late 19th century. Since that time a number of systems have been developed to meet a variety of objectives with respect to organizing workplaces and predicting the labor requirements to accomplish any job. Only recently, however, have the available methods attempted to describe comprehensively the physically stressful aspects of work. This chapter describes several of the current workplace evaluation systems. A special emphasis is placed on systems that concentrate on those aspects believed to contribute to the risk of low back pain (LBP).

One reason for the paucity of such systems in the available literature is, as one might suspect, a general disagreement on what, precisely, should be measured. One survey of the available research literature disclosed the following physical workplace factors as being important (Herrin 1978):

1. *Loads*—measures of the vector forces and moments (lifting/pushing/pulling, etc.) acting on the body during materials handling;
2. *Dimensions*—measures of size/shape and form of objects handled;
3. *Distribution of Loads*—measures of the location of the object center of gravity with respect to the worker;
4. *Couplings*—measures of the interfacing between the worker and the load (e.g., handle design parameters such as size, shape, location, coating and texture);
5. *Stability of Load*—measures of the consistency of the load's center of mass (as in handling liquids and bulky materials);
6. *Workplace Geometry*—measures of the spatial properties of the task such as movement distances, directions and extent of motion paths;

obstacles; nature of the destination (each of which affect worker posture);

7. *Temporal Factors*—measures of the frequency, duration, and pace of work activities over the short and long term;
8. *Complexity*—measures of manipulation requirements; objective of activity, tolerances for motion error;
9. *Environment*—measures such as temperature, humidity, lighting, noise, vibration, foot traction, toxic agents, etc.;
10. *Organization*—measures of administrative factors such as use of teamwork, machine pacing, work incentives, extended work shifts, job rotations, personal protective devices, etc.

Although this list is far from complete, it does suggest that the etiology of LBP is quite complex and involves the interaction of many factors.

A practical difficulty with evaluation of workplaces for potential risks of physical trauma is that the available literature is unclear about whether the process is, indeed, attributable to *acute* or *cumulative trauma*. If the cause is normally a long-term, degenerative, "wear and tear" process, then presumably the workplace evaluation should include the temporal aspects of work such as frequencies and durations of exertions, fatigue, work-rest regimes, and so forth.

If, on the other hand, the etiology is predominantly acute trauma, as in slipping and falling or being struck, then infrequent, "unpredictable", atypical aspects of work, as in emergency procedures, maintenance and machine setup, and perhaps leisure time activities, etc. should receive primary attention.

From the preceding chapters it is apparent that the risks of *both* acute and cumulative trauma are major concerns in industry; hence it will be necessary to apply *multiple* evaluation procedures.

HISTORICAL PERSPECTIVE

The organization of efforts of groups of workers has traditionally involved a sequence of four steps (Chaffin and Andersson 1984):

Development of a Preferred Method

Assuming that a task (or job) can be performed using any number of different methods, it is incumbent that a preferred method be defined. To accomplish this, one must be able to:

- State the objective of the operation or activity.
- Identify those methods that meet or exceed the objective.

- Implement feasible methods on a trial basis.
- Select the best method that meets the objective.

Preparation of a Standard Practice

This requires that the preferred method be stated formally. This includes:

- Sketching of the workplace (including relevant dimensions and locations of all machines and tools used).
- Listing of abnormal working conditions that could affect work performance (e.g., illumination, dust, and heat).
- Tabulating the sequence of motions required of a worker to complete the operation.

Determination of a Time Standard

This is accomplished either by having a skilled worker perform the operation while being timed or by consulting normative time data available in the literature (e.g., synthetic time systems). Overall operation standard times can be predicted by cumulating the elemental standard times assuming an experienced, well-trained worker.

Training of the Worker

Instruction sheets, verbal directions, and training aids must be developed and used to assure that workers understand the job requirements and that preferred work methods are followed.

These four steps usually evolve from a concern on the part of management that labor is not well utilized and that time is being wasted. As the cost of labor increases, more emphasis is invariably given to assuring that each of the above steps is fully implemented by the industrial or production engineering functions of a well managed plant or industry.

Within the last decade, however, an additional concern has been imposed by top managements: namely, that any operation must be accomplished *safely*. It is this relatively new awareness and concern for worker safety and health that makes job and workplace evaluation such an important function today. To understand better the role of conventional workplace evaluation in injury prevention and to give some perspective to the current practice of workplace evaluation, a brief history of traditional job methods and motion analysis techniques is in order.

Many individuals contributed to the early developments of motion and methods analysis. For an extensive discussion of the techniques, a number of excellent textbooks are available, for example, Barnes (1968), Niebel (1972), and Mundel (1978). Perhaps the most influential of the early contributors to the field was Frederick W. Taylor.

Taylor's "Principles of Scientific Management" (1929) indicated the need for standardization of work methods, time study, instruction of workers, selection of workers, and offering incentives for completing work correctly and on time. Taylor's study of shoveling rice, coal and iron ore in 1898 was a rigorous and comprehensive investigation of the optimal design for a shovel in terms of the size, shoveling frequency, and rest allocation necessary to maximize the total output of a group of workers (Copley 1923). Taylor showed that a shovel that would allow about 22 pounds (10.3 kg) of material to be lifted each time resulted in maximal total daily output for a group of physically strong workers. By a combination of careful selection and training of workers, provision of special shovels for different weight materials, and paying bonuses for above average output, he was able to demonstrate that 140 men could perform the same work as was previously done by 400 to 600 workers (Copley 1923).

At about the same time, Frank and Lillian Gilbreth were advocating that manual activities in industry needed to be carefully categorized to minimize fatigue and monotony while maximizing productivity (Barnes 1968). One of the earliest and most quoted studies by Frank Gilbreth was of bricklaying (Gilbreth 1911). By using photographs of the bricklayers in his construction business he was able to demonstrate that fatigue and wasted motions could be minimized and high output achieved if bricks were first sorted for good and bad bricks; bricks were oriented with the best side facing out; and bricks were delivered to the bricklayer at a comfortable working height by means of an adjustable scaffold. His methods resulted in a doubling of total output. His study also demonstrated the concept of "division of labor." Since bricklaying tasks could be divided into skilled and unskilled portions, the necessary labor could also be divided between different workers (with different skills and presumably different wage rates).

Two specific methodological contributions to the study of manual labor must be attributed to the Gilbreths. First is the use of *micromotion study* (Gilbreth 1912) which allows great attention to be given to categorizing individual body motions, referred to as "Therbligs" (an anagram derived from their name). These basic motions were carefully timed and later became the basis for one of the first predetermined time systems.

Second, the Gilbreths are credited with the first cycleograph (or chronocycleograph) used to compare alternative motion sequences. A set of blinking lights (flashing at a known rate) was attached to the arm and hand and a photograph was made with the camera shutter open for a prolonged period. The motion path was transcribed as a sequence of dots onto film. By measuring the numbers and displacements of the dots, the Gilbreths were able to measure and compare different job motion requirements. Today this technique has been considerably enhanced by modern video and computer techniques to allow whole body motion studies, but the concept is virtually unchanged.

It should be clear that the general approach and investigative methods of work analysis developed by Taylor and the Gilbreths early in this century continue to be directly applicable to workplace evaluation today. Without these mechanisms and procedures of systematic work classification, time study, motion study, and motion analysis, it will be impossible to reduce job-induced musculoskeletal stress in industry.

Several conventional work analysis systems derived from these earlier efforts are briefly described, followed by several new work analysis systems that have been specifically directed at defining musculoskeletal stress in manual work.

CONVENTIONAL WORK ANALYSIS SYSTEMS

In 1922, Asa Segur began his classic development of a set of motion-time prediction equations for various classes of motions. His film analyses of industrial operations during World War I were originally designed to determine ways of training blind and other handicapped workers to perform useful work in support of the war (Brisley and Eady 1982). Similar data were soon published by others during the 1930s.

What these authors collectively demonstrated was that the time required of people to perform certain basic or *elementary motions* is about the same for different people. Thus the time to perform a complex collection of motions could be predicted by accumulating elementary motion times. The resulting elemental time values have become known as predetermined elemental motion times. Because different groups of experts performed the analyses several times, different systems have emerged. Table 13.1, developed by Barnes (1968), describes the background of several popular predetermined motion-time systems in use today.

These synthetic, predetermined motion-time systems are most useful in planning staffing requirements for new workplaces before they become operational. Once a workplace is operational, then direct observation techniques (such as time study) can be used to refine the work methods and procedures.

A survey of the use of such techniques reported by Karger and Bayha (1966) disclosed that about two-thirds of firms sampled use some form of work measurement system, and that about 56 percent of these used a predetermined motion-time system. They also found that of the many

TABLE 13.1
Some Conventional Motion-Time Data Systems

Name of System	*First Publication Describing System*	*How Data Were Originally Obtained*
Motion-Time Analysis (MTA)	Data not published, but information concerning MTA published in *Motion-Time Analysis Bulletin*, a publication of A.B. Segur & Co., 1924	Motion pictures, micromotion analysis, kymograph
Body Member Movements	*Applied Time and Motion Study* by W.G. Holmes, Ronald Press Co., New York, 1938	Not known
Motion-Time Data for Assembly Work (Get and Place)	*Motion and Time Study*, 2nd ed., by Ralph M. Barnes, John Wiley & Sons, New York, 1940, Chs. 22 and 23	Time study, motion pictures of factory operations, laboratory studies
The Work-Factor System	"Motion-Time Standards" by J.H. Quick, W.J. Shea, and R.E. Koehler, *Factory Management and Maintenance*, Vol. 103, No. 5, pp. 97-108, May, 1945	Time study, motion pictures of factory operations, study of motions with stroboscopic light unit
Elemental Time Standard for Basic Manual Work	"Establishing Time Values by Elementary Motion Analysis" by M.G. Schaefer, *Proceedings Tenth Time and Motion Study Clinic*, IMS, Chicago, pp. 21-27, November 1946	Kymograph studies, motion pictures of industrial operations) and electric time-recorder studies (time measured to 0.0001 minute)
Methods-Time Measurement (MTM)	*Methods-Time Measurement* by H.B. Maynard, G.J. Stegemerten, and J.L. Schwab, McGraw-Hill Book Co., New York, 1948	Time study, motion pictures of factory operations
Basic Motion Timestudy (BMT)	Manuals by J.D. Woods & Gordon, Ltd., Toronto, Canada, 1950	Laboratory studies
Dimensional Motion Times (DMT)	"New Motion Time Method Defined" by H.C. Geppinger, *Iron Age*, Vol. 171, No. 2, pp. 106-108, January 8, 1953	Time study, motion pictures, laboratory studies
Predetermined Human Work Times	"A System of Predetermined Human Work Times" by Irwin P. Lazarus, Ph.D. thesis, Purdue University, 1952	Motion pictures of factory operations

Source: Barnes 1968.

predetermined motion-time systems available today, the Methods-Time Measurement (MTM) system is used about twice as often as other systems. An international organization assists its members in learning and implementing this system.

MTM-1: An Example System

One of the original and most popular systems, used by thousands of practitioners throughout the world, is referred to as MTM-1. To use this particular motion classification system with accuracy and consistency requires training in a special course of 24 to 80 classroom hours (offered and certified by the MTM Assn.). For a detailed discussion of the use of such systems the reader should consult textbooks by Barnes (1968), Niebel (1972), Karger and Bayha (1966), Karger and Hancock (1982), and Konz (1979).

The methods-time measurement system developed by Maynard, Stegemerten, and Schwab in the late 1940s is based on a modified collection of "Therbligs" as shown in Table 13.2. The Therbligs with superscripts are ineffective elements of movement. Whenever they occur, an attempt should be made to eliminate them if possible.

The resulting time values for each elemental motion (Therblig) in the MTM procedure are given in units of one hundred thousandths of an hour (0.00001 hour) which is referred to as one TMU (Time Measurement Unit). Use of such a fraction of an hour was chosen as the basic unit of measure because production rates in industry are often expressed in units produced per hour. However, for more general reference, one TMU equals 0.0006 minutes of 36 milliseconds.

The time for Therbligs such as Move, Reach, Grasp, Position, Regrasp, Turn, Apply Pressure, Release or Disengage can be predicted. Additional allowances are necessary for Eye Focus/Travel time (required to search for an object within the field of view), and for Motion of the Body, Legs and Feet. Last, additional adjustments must be made for difficulty with simultaneous tasks (different usages of the hands may be incompatible). A standardized coding system for different activities has been adopted. Table 13.3 gives examples of this coding system.

An example MTM job analysis is given in Table 13.4 using the standard coding notations for lifting a tote box from a pallet to a workbench and then removing a part. The brackets indicate simultaneous motions and thus only the greatest time is entered in the center column, which in turn is used to predict the total time of the task. In this example, the tote box weighs 42.9 lbs (200 newtons). The individual part which weighs 2.1 lbs (10 newtons) is retrieved from among a collection of other parts jumbled together in the tote box.

TABLE 13.2
Therbligs used to Describe Worker Actions

Operation	*Abbreviation*	*Operation*	*Abbreviation*
Gilbreths		American Society of Mechanical Engineers	
	Physical Basic Elements		
Transport empty	TE	Reach	R
Transport loaded	TL	Move	M
		Change direction[a]	CD
Grasp	G	Grasp	G
Hold	H	Hold[a]	H
Release load	RL	Release load	RL
Pre-position	PP	Pre-position[a]	PP
Assemble	A		
Disassemble	DA	Disengage	D
	Semimental basic elements		
Position	P	Position[a]	P
Search	Sh	Search[a]	S
Select	St	Select[a]	SE
	Mental basic elements		
Plan	Pn	Plan[a]	PL
Inspect	I	Examine	E
	Objective basic element		
Use	U	Do	DO
	Delay basic elements		
Avoidable delay	AD	Avoidable delay[a]	AD
Unavoidable delay	UD	Unavoidable delay[a]	UD
		Balancing delay[a]	BD
Rest for overcoming fatigue	R	Rest for overcoming fatigue	F

[a]These are ineffective elements of movement; wherever they occur, an attempt should be made to eliminate them completely if at all possible.

Source: Courtesy of International Labour Office, Introduction to Work Study, Atar, Geneva, 1960.

TABLE 13.3
Coding Conventions for MTM Motion Analysis

Code Example	*Interpretation of Codes*
R8C	Reach, 8 inches, Case C
R12Am	Reach, 12 inches, Case A, hand in motion at end
M6A	Move, 6 inches, Case A, object weighs less than 2.5 pounds
M10C	Move, 10 inches, Case C, hand in motion at the beginning, object less than 2.5 pounds
M16B15	Move, 16 inches, Case B, object weighs 15 pounds
T30	Turn hand 30 degrees
T90L	Turn object weighing more than 10 pounds 90 degrees
AP1	Apply pressure, includes regrasp
G1A	Grasp, Case G1A
P1NSD	Position, Class 1 fit, nonsymmetrical part, difficult to handle
RL1	Release, Case 1
D2E	Disengage, Class 2 fit, easy to handle
EF	Eye focus
ET14/10	Eye travel between points 14 inches apart where line of travel is 10 inches from eyes
FM	Foot motion
SS16C1	Sidestep, 16 inches, Case 1
TBC1	Turn body, Case 1
W4P	Walk four paces

Source: R.M. Barnes, *Motion and Time Study*, New York: John Wiley and Sons, Inc., 1968, p. 503.

Benefits and Limitations of Synthetic Time Systems

In reviewing the example in Table 13.4, it should be noted that it provides much essential information.

- The workplace documentation includes a sketch of the layout with major hand motions and distances noted. Such a sketch is essential for planning changes in the workplace layout.
- The analysis reveals the relative balance of work between the two hands. If the imbalance were great, it could contribute to asymmetric symptoms in workers (e.g., dominant hand disorders and syndromes).
- The analysis describes the extent of static holding activities required of

TABLE 13.4
Example MTM Analysis of Lifting a Tote Box from a Pallet to a Workbench and Removing a Part from the Tote Box

Activity Description	*Left Hand Elements*	*Motion Time (TMU)*	*Right Hand Elements or Body Motions*
Sidestep to a pallet.		17.0	SS12G1
Stoop to a tote box, and		29.0	S
Reach to tote box during stoop.			
Grasp handles on tote box,	G1A	2.0	G1A
Arise, and			AS
Lift tote box during arise	$\frac{22}{2}\}$	31.9	$\{\frac{22}{2}$
Sidestep towards bench with tote box, and		23.7	SS12C
Move tote box during sidestep	M12B $\frac{22}{2}\}$		M12B $\frac{22}{2}$
Release tote box on bench.	RL1	2.0	RL1
Reach into tote box		12.9	R10C
Grasp part in box.		7.3	G4A
Move part to bench.		10.6	M18B
Release part on bench.		2.0	RL1

Total Time Required: 138.4 TMU or 4.98 seconds
Courtesy J. Foulke as presented in Chaffin and Anderson, 1984.

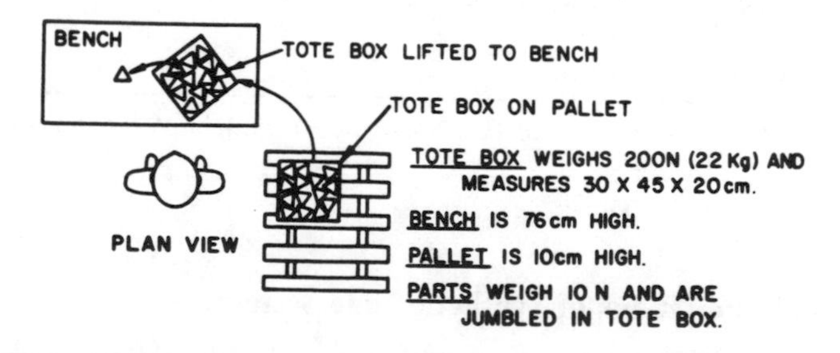
BENCH
TOTE BOX LIFTED TO BENCH
TOTE BOX ON PALLET
TOTE BOX WEIGHS 200N (22 Kg) AND MEASURES 30 X 45 X 20cm.
BENCH IS 76cm HIGH.
PALLET IS 10cm HIGH.
PARTS WEIGH 10 N AND ARE JUMBLED IN TOTE BOX.
PLAN VIEW

each hand (as required in using the hand to hold an object in lieu of a jig). Such postural requirements may contribute to localized muscle fatigue.

Despite these benefits, however, several limitations are apparent.

- Conventional systems rarely record the "variability" in terms of loads handled, distances moved, etc.; hence extreme deviations about the atypical "average" are ignored. The fact that the total time required to complete a particular task is relatively insensitive (or robust) to this variability does not lessen the importance of this aspect in the etiology of LBP.

- Conventional systems virtually ignore posture as a predictor variable. Except for the grossest classifications (such as standing versus seated work) these systems ignore one of the Gilbreths' major contributions. Their early practice of filming motions was designed to identify systematically awkward postures and inefficient motions. The recent emphasis on developing simpler and quicker "shortcut" techniques and procedures for job analysis may, indeed, be self defeating, especially in view of a concern for prevention of LBP.
- Perhaps the most important shortcoming of these methods is that they are designed *to account for time*. As such, no attention is given to infrequent or non-routine work. Typically, tasks that require less than 5 percent of the total workday such as occasional machine setups, maintenance or cleanup, or emergency procedures, are neither rigorously documented nor analyzed. Yet many follow-up accident/injury reports suggest that these undocumented, infrequent, yet required job tasks account for a large number of injuries.

Despite these limitations, however, this type of analysis forms a foundation for methods aimed at enriching the evaluation of manual jobs. For routine, repetitive work it is very common in industry to have analyses such as MTM available.

For nonrepetitive, unstructured work it is more common to conduct Work Sampling Studies to document job requirements. These, too, serve as a valuable starting place for improved job evaluation methods. Once the general work methods are documented, whether by a detailed synthetic time prediction system or a work sampling or time study method, special initiatives are necessary to: identify all tasks (including rare elements) that are required or expected of workers on the job; identify work postures which are potentially hazardous to the musculoskeletal system; and measure the actual loads that are handled, with special attention to the extremes of loads, distances, etc.

CONTEMPORARY JOB ANALYSIS SYSTEMS

These above mentioned shortcomings of conventional time and motion analysis systems, coupled with a perceived need for more specific job data, has led to several more contemporary approaches to improved documentation relative to musculoskeletal stress on jobs. What follows is a description of a few of these systems. For a more extensive discussion see Chaffin and Andersson (1984).

Physical Stress Checklists and Surveys

A checklist of physical tasks sometimes can be used to document the job's general physical requirements. In such a procedure a job analyst observes the worker and "checks" those activities performed from a list of 35 or more common manual tasks (e.g., bend trunk, pull with arm(s), lift with back, etc.). Such a list has been used in at least one study to improve job placement procedures for individuals with physical impairments (Smith, Armstrong and Lizza 1982).

Another informative survey approach has been developed which documents the average frequency of occurrence and weight handled during different standard manual tasks (USDOL 1972). This procedure was modified by Koyl and Marsters-Hanson (1973) to include the recording of the hours during the day that each activity is performed. By comparing of these data with a clinical assessment of a person's mobility and strength, Koyl and Hanson proposed that placement of individuals on specific jobs could be improved. The Department of Labor has used such data to classify jobs subjectively according to the general criteria given in Table 13.5.

It should be clear that such physical stress survey data are important when identifying jobs that could be potentially hazardous to a worker's musculoskeletal system. They are also applicable to jobs that are *not* highly repetitive, wherein traditional methods are weak. In this sense the surveys indicate when special in-depth studies may be warranted. Such surveys are especially useful when combined with injury data analysis to motivate more intensive evaluations.

Lifting Limits in Manual Materials Handling

A recently published federal guide entitled, *Work Practices Guide to Manual Lifting* (NIOSH 1981) offers specific recommendations for evaluating human *lifting* limitations in particular. The *Guide* is intentionally very simplistic in approach and application since it was intended for general industry use. Further the criteria include muscle strength and physical work capacities (aerobic) in addition to biomechanical aspects of disc compression.

The *Guide* specifically calls for the measurement of:

1. *Object weight* (L)—measured in pounds (or kilograms). If this varies from time to time, the average and maximum are necessary.
2. *Horizontal location of the hands* (H)—measured in inches (or centimeters) forward of the midpoint between the ankles. This is measured at the origin of the lifting task. A rule of thumb for this dimension is H = (W/2 + 6) where W is the horizontal width of the object (W/2 implies

TABLE 13.5
Strength Requirement Classification Criteria adopted by U.S. Department of Labor

Degree of Strength	*Amount of Lifting/Carrying*	*Posture; Other Activities*
Sedentary Work	Occasional: 10 pounds maximum	Primarily sitting; walking and standing at most occasionally
Light work	20 pounds maximum; 10 (or less) pounds frequently	Significant amount of walking or standing or Primarily sitting, but requiring pushing and pulling of arm and/or leg controls
Medium work	50 pounds maximum; 25 (or less) pounds frequently	Unspecified
Heavy work	100 pounds maximum; 50 (or less) pounds frequently	Unspecified
Very heavy work	Over 100 pounds allowed; 50 (or more) pounds frequently	Unspecified

Source: Smith, Armstrong and Lizza 1982.

that the hands grip the object at center and 6 inches is assumed for foot and body clearance).
3. *Vertical location of the hands* (V)—measured in inches (or centimeters) above the midpoint between the ankles.
4. *Vertical travel distance* (D)—measured in inches (or centimeters) from the origin to destination of the lift.
5. *Frequency of lifting* (F)—measured in lifts per minute assuming continuous lifting for a period of one or eight hours.
6. *Duration of period* (P)—assumed to be either occasional (less than one hour) or continuous (for eight hours).

The *Guide* specifically ignores a number of important variables. In particular, the *Guide* is only applicable to:

1. *Smooth lifting.* The *Guide* assumes no accelerations of the load or human body. Obviously, with acceleration or jerk the stress to the low back can be substantially higher than suggested by the guide.
2. *Two-handed, symmetric* lifting in the sagittal plane. The *Guide* assumes

that lifting is only two-dimensional with both hands directly in front of the body and does not allow twisting throughout the lift.

3. *Moderate width object.* The hands are assumed to be separated no more than shoulder width. The handling of extremely wide objects (such as a standard sheet of plywood) would be outside the scope of the *Guide.*
4. *Unrestricted, standing posture.* The *Guide* assumes that there are no obstructions to interfere with the movement of the object nor are there props or aids to facilitate the individual. Lifting style is not considered beyond simple stoop versus squat techniques.
5. *Good handles* on the object. The *Guide* specifically ignores grip strength, which has been extensively studied beyond just the lifting context.
6. *Favorable environmental conditions.* It is assumed that heat and cold stress do not contribute to the burden on the individual, nor is a slippery floor important.
7. *No other work.* It is assumed that when not engaged in lifting, the individual is essentially at rest (i.e., no significant carrying, pushing/pulling, holding, etc.).

The *Guide* presents *two limits* defined as the "Action Limit" and "Maximum Permissible Limits." The range between these limits is designed to reflect the variability between individuals in the U.S. working population. These limits are defined as:

1. *Maximum Permissible Limit* (MPL). This limit would reflect a lift that produces 1,430 pounds of compression on the L5-S1 disc, or creates a metabolic load of 5.0 kcal per minute for a healthy young male, or is only within the strength capabilities of 25 percent of men and virtually no women.
2. *Action Limit* (AL). This limit is algebraically equal to one-third of the MPL. This situation would create 770 pounds (3,593 N) of compression on the L5-S1 disc, or require 3.5 kcal per minute of a healthy young female, or would only be within the strength capabilities of 75 percent of women and virtually all men.

These two limits are used to define three categories of lifting stress within jobs. For each, a different organizational response is required as follows:

1. *Above MPL*—those tasks that require engineering controls (e.g., fundamental job/task modifications in terms of H,V,D,F,P, or L such as use of a hoist or modification of the pace of the work, etc.).
2. *Below the AL*—which presumably is of nominal stress and risk to most people.
3. *Between the AL and MPL*—which require either engineering controls or

administrative controls. Administrative controls would include workforce selection (functional capacity testing) and training (fitness and awareness).

Guideline Limits

With the large number of task variables and multiple criteria for what is reasonable (biomechanically, epidemiologically, physiologically, and psychophysically), it is not surprising that the guideline has many shortcomings. It is also apparent why there is little agreement among the 300 + articles in the scientific literature on "how much can be safely lifted."

In algebraic form the two limits are:

$$AL = 90\ (6/H)\ (1 - 01\ [v - 30])\ (.7 + 3/D)\ (1 - F/Fmax)$$

$$MPL = 3\ (MPL)$$

where

H ranges from 6″ to 32″ (15.2 − 81.3 cm)
V ranges from 0″ to 70″ (0 − 177.8 cm)
D ranges from 10″ to (80 − V)″ (25.4 − (203.2 − V)) cm
F ranges from .2 to Fmax

and

Fmax = 12 for continuous low lifting (P = 8 hr. and V less than 30″).
= 18 for occasional high lifting (P = 1 hr. and V less than 30″).
= 15 otherwise.

To record job lifting data, a Physical Stress Job Analysis Sheet such as shown in Figure 13.1 should be used. This form shows the analysis for a single task of loading a reel of stock into a machine as illustrated in Figure 13.2.

After the lifting evaluation is complete for *each* lifting task in a job, then resulting predicted *Action Limits* and *Maximum Permissible Limits* are computed and added to the coding forms in the columns provided. These Limits then can be compared line by line to determine the relative stressfulness of each lifting task.

Because the H distance is so critical in determining the amount of low back stress, it is prudent to measure H at both the origin and destination. The origin is usually associated with the greater inertia needed to accelerate the mass upward. If a great deal of control is also required in placing the load

PHYSICAL STRESS JOB ANALYSIS SHEET

DEPARTMENT Fabrication DATE 2-18-82

JOB TITLE Punch Press ANALYST'S NAME E. J. B.

TASK DESCRIPTION	OBJECT WEIGHT (NEWTONS)		HAND LOCATION				TASK FREQ	AL	MPL	REMARKS
			ORIGIN		DESTINATION					
	AVE	MAX	H cm	V cm	H cm	V cm				
Load Stock	200	200	53	38	53	160	0			

Figure 13.1
NIOSH Job Lifting Analysis Form filled in for lifting task depicted in Figure 13.2. Note: The zero entry for task frequency denotes the stock reel is loaded at a frequency of less than once every five minutes.

at the destination (i.e., in precise location or with a fragile load), it may be more appropriate to use the destination H value.

The NIOSH *Guide* is a recent effort to control one type of manual materials handling, namely the act of lifting in the sagittal plane. It is an attempt to be more comprehensive than previous efforts relative to job evaluation methods; criteria used for limits; and control strategies. It is too early to discern its effect on controlling musculoskeletal injuries in industry. Since all available copies of the *Guide* were distributed within months of its publication, reports of implementations should be forthcoming shortly. Copies of the *Guide* are now available from The American Industrial Hygiene Association in Akron, Ohio.

Static Strength Analysis

Because the NIOSH *Guide* only applies to symmetric lifting in the sagittal plane, a more generalized analysis scheme is often necessary to describe and evaluate manual work. One such scheme concentrates on static strength predictions.

A model developed at the University of Michigan (Garg and Chaffin 1975) compares the load moments produced at various body joints during any task to population capabilities. The job analysis requirements are comparable to the NIOSH procedure but can be extended to pushing and pulling in other than orthogonal directions. A complete description of the

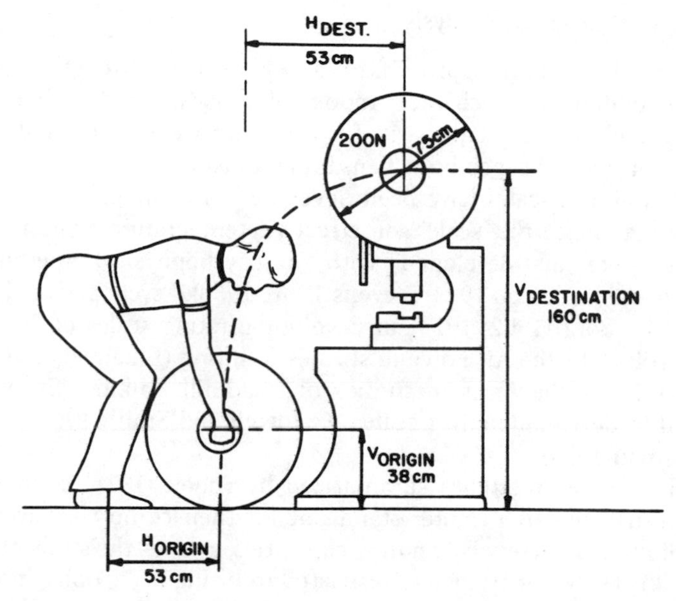

Figure 13.2
Example of lifting stock into punch press. It is assumed that the worker steps forward with the load to place it atop press, i.e., H. remains constant, while V changes (Chaffin and Andersson, 1984).

job analysis method and strength prediction model is given in Chaffin and Andersson (1984).

In general, the forces acting on the hands are viewed as vector quantities (in magnitude and direction) acting separately on each hand. The model combines the effects of these external loads with the effects of each body segment's weight for a given body posture, in order to predict the moment load on each major body articulation. The resultant strength requirement for each major muscle group (to counteract this moment load) is then predicted for various strata of the population.

The model also predicts the compressive force acting on the L5-S1 disc for each posture and load condition being analyzed. For each task of interest, a general body posture of the legs and torso can either be assumed (via a computerized postural optimization scheme) or described by the job analyst (e.g., standing, stooping, squatting, leaning forward, sitting, etc.), and the three-dimensional coordinates of each hand are recorded.

A detailed discussion of the strength prediction and L5-S1 compression force model and computer algorithm are available in Garg and Chaffin (1975) and Chaffin and Andersson (1984). The methods for assessing isometric strength in industry and predicting population norms are discussed in Stobbe (1982) and Herrin (1982).

Psychophysical Strength Analysis

The psychophysical approach to assessment of strength does not assume an underlying mechanical model of a body nor does it relate to spinal loading *per se*. Rather, it focuses directly on the individual's perception of manual capability when doing a task.

Psychophysical scales have been developed in many practical problem areas. For example, the scales of effective temperature, loudness, and brightness were all developed with a psychophysical methodology (Houghton and Yagloglou 1923; Stevens 1956, 1960). Psychophysics has also been used by Borg (1962, 1973) in developing rating scales of "perceived exertion" (PRE); by the Air Force in studies of lifting (Emanuel *et al.*, 1956; Switzer 1962); by the Army in studies of treadmill walking (Evans 1961, 1962); and in developing effort scales (Caldwell and Smith 1967, Caldwell and Grossman 1973).

A series of seven studies summarized by Snook (1978) conducted at Liberty Mutual Research Center stand out for their comprehensiveness in the area of manual materials handling capacities, as does the study of Ayoub *et al.* (1978). The latter study was restricted to lifting tasks only. Snook and his associates also provided the first analysis to relate psychophysical strength predictions to the increased risk of low back injury.

In order to evaluate the effectiveness of job design in the reduction of injuries (as well as the effectiveness of selection techniques and training procedures), 191 cases of low back injury from 32 states were investigated (Snook, Campanelli, and Hart, 1978). Questionnaire results revealed that about 25 percent of policy holder jobs involve manual materials handling tasks that (on a psychophysical rating scale) are acceptable to less than 75 percent of the workers. However, one-half of the low-back injuries were associated with these jobs indicating that a worker is three times more susceptible to low-back injury if performing a manual handling task that is acceptable to less than 75 percent of the working population. This suggests that, at best, two out of every three low-back injuries associated with heavy manual handling tasks can be prevented if the tasks are designed to fit at least 75 percent of the population. The third injury will apparently occur anyway, regardless of the job. The other low-back injuries not associated with heavy manual handling tasks will also occur. It can be concluded that up to one-third of industrial back injuries can be obviated by the careful and proper design of manual handling tasks.

In order to evaluate the relevant percentages of the population, Snook (1978) presented summary tables for industrial males and females, itemizing psychophysical limits on lifting, lowering, pushing, pulling and carrying for each of the following task variables:

1. Vertical region (knuckle height to floor level, or shoulder height to knuckle height, or arm reach to shoulder height).

2. Object width (horizontal dimension).
3. Distance moved (vertical dimension for lift, horizontal dimensions for pushing and carrying).
4. Frequency of exertion (in repetitions per minute).

One shortcoming of this analysis procedure is the referencing of the vertical dimensions of the task to a particular individual's anthropometry rather than absolute coordinates.

A similar study by Ayoub, *et al.* (1978) provided prediction models for maximum psychophysical lifting capacity as a function of worker and task variables. Six regression equations were generated for each lifting region, namely: floor to knuckle; floor to shoulder; floor to reach; knuckle to shoulder; knuckle to reach; and shoulder to reach. Experiments were conducted at frequencies of 2, 4, 6, and 8 lifts per minute and box sizes of 12, 18, and 24 inches (30.5, 45.7 and 61.0 cm respectively) in the sagittal plane. A series of job stress indices (variations of the ratio of weight lifted to individual lifting capacities aggregated across tasks) were developed in an effort to predict lost time resulting from injury . None of these indices was strongly correlated with the injury experience, due in part to a small sample size.

Load Pushing and Pulling

Pushing and pulling capabilities have, in general, been studied within a more limited scope. Furthermore, estimates of the number of injuries that occur during pushing or pulling of loads are not complete, although approximately 20 percent of overexertion injuries have been associated with such acts (NIOSH, 1981).

It would appear from previous work of Fox (1967) and Kroemer and Robinson (1972) that the effect of friction on static strength capability when pushing and pulling is of primary importance. They collectively showed that healthy young males could exert an average force of approximately 42.9 lbs (200 N) when the coefficient of friction was about 0.3. With a friction coefficient greater than 0.6, the mean push or pull strength capability increased to 64.3 lbs (300 N) for the same group, according to Kroemer and Robinson (1971). Bracing one foot with the use of the back to apply force (rather than using the hands) further increased the static push force capabilities (Kroemer 1969; Fox 1967).

Martin and Chaffin (1972), Ayoub and McDaniel (1974), Lee (1982) and Davis and Stubbs (1979) reported that the vertical height of the handle against which one pushes and pulls is of critical importance. In an experimental study of strength by Ayoub and McDaniel (1974) the elbows and rearward knee were kept straight for the exertions. This resulted in the recommendation that the optimal height for a handle to be pushed or pulled should be approximately 35.8 to 44 inches (91 to 114 cm) (i.e., about hip height) above the floor. Davis and Stubbs (1979) developed recommendations for pushing and pulling limits based on abdominal pressure measurements. Martin and Chaffin (1972) used a biomechanical model to predict the maximum push and pull forces for extreme postures. In this regard, Lee (1982) performed a set of dynamic push and pull experiments and found that the predicted compression forces were less when the hands were approximately 42.9 (109 cm) above the floor than either 59.8 ins (152 cm) or 26.0 ins (66 cm).

In short, the need to understand pushing and pulling activities is recognized since many overexertion injuries appear to be related to such activities. It is also clear that many biomechanical factors interact to alter push/pull capabilities. At present, however, only a limited amount of data and modeling of these common activities have been completed, and thus any limits must be interpreted carefully for job evaluation purposes.

Asymmetric Load Handling

Based on both biomechanical models and experimental strength studies, symmetric load handling (i.e., when the load is moved in the mid-sagittal plane with both hands) is recommended. Unfortunately, the studies of asymmetric load handling are few in number, because of the experimental and biomechanical modelling complexities associated with three-dimensional force analysis. It is clear, however, that asymmetric load handling increases the stress on the spine and most muscle groups and can result in poor postural stability. Unfortunately, asymmetric motions are often most efficient in terms of performance time and hence constitute a dilemma.

One recent study by Warwick *et al.* (1980) compared symmetric and asymmetric pushing and pulling. In general, the activities in asymmetric postures resulted in decreased strengths (of about 20 to 25 percent). The effects were very dependent on the direction of motion and postures as might be expected.

Job Posture Evaluation

An approach to recording potentially stressful postures was developed by Corlett, Madeley and Manenica (1979) and is referred to as *posture targeting*. This procedure requires the job analyst to observe a worker at random times during the workday and record the angular configuration of various body segments with the aid of the "body diagram" displayed in Figure 13.3. The angular data are recorded by simply placing a small "x" on each postural target when a body segment deviates from the erect

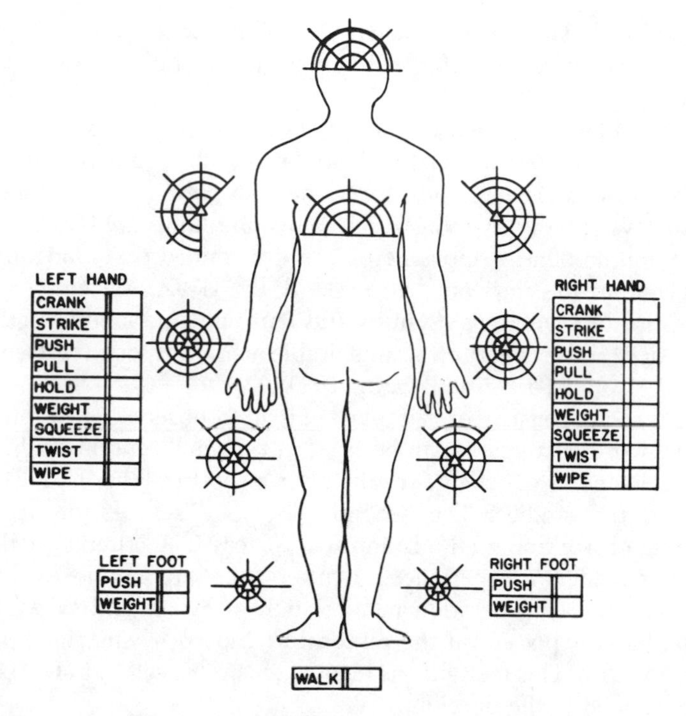

Figure 13.3
Body diagram, showing each target adjacent to its associated body part. Deviations from the standard position shown are marked on the neighboring diagram, otherwise no mark is made with each observation of the worker (Corlett, Madeley, and Manenica, 1979).

anatomical position shown. The concentric circles on each target represent 45, 90, and 135 degree angular deviations of a joint from that shown, with the arrow at the center of the target indicating the front of the body. The radial lines indicate the amount of deviation from the sagittal plane as viewed from overhead. Counting the number of "x" marks in a certain zone of a diagram, or simply observing how the marks cluster together, provides insight into possible stressful postures.

Although Corlett *et al.* did not combine these postural data with external load data for biomechanical analysis, such would be possible by noting the load magnitudes on the activity lists provided with the body diagram. In its simplest form, the procedure documents job postures in such a way that the analyst can easily identify the most frequent and potentially stressful ones for more detailed biomechanical analysis. Corlett and Manenica (1980) have also demonstrated that this procedure is useful in evaluating workplace layouts when combined with worker reports of localized musculoskeletal pain obtained at several intervals during a workday.

It should be mentioned that Priel (1974) proposed a system to allow postures to be numerically defined and recorded. From repeated observations of workers, the basic posture of the body within a three-dimensional coordinate system, the levels at which joints and limbs are located, and the direction and amount of movements are determined on a "posturegram." Another system was developed in Finland, the OWAS system. The Ovaco Working Posture Analysis System (OWAS) is a practical method for identification and evaluation of unsuitable working postures (see Karhu, Kansi and Kuorinka 1977 and Karhu 1981).

The method consists of two parts. First, an observation technique for evaluating work postures. It can be used in time and motion studies in the daily routine and gives reliable results after a short training period of those performing the studies. The second part is a set of criteria for the redesigning of working methods and workplaces. The criteria are based on evaluations made by experienced workers and ergonomic experts. The criteria take into consideration factors such as health and safety, but the main emphasis is placed on the discomfort caused by sustained awkward working postures. This method has been extensively used by a steel company that participated in the development.

A method similar to OWAS is a technique developed by Berns and Milner (1980) for analyzing moving work postures (TRAM). Both OWAS and TRAM record the postures at regular intervals on a recording sheet. By using various combinations of the basic posture, a wide range of postures can be described. Included in the recording is an approximation of a weight or force acting externally on the body as well as longer static positions.

A new Swedish system called ARBAN was developed by Holzmann (1982). ARBAN is a method for ergonomic analysis of work, including work situations involving greatly differing body postures and loads. The method consists of four different steps: recording the workplace on videotape or film; coding the posture and load situation in a number of closely spaced "frozen" situations; computerization; and evaluation of results.

A computer routine determines the total ergonomic stress for the whole body (as well as specified body parts) based on heuristic rules regarding the relative stress of specific acts. The results are presented as ergonomic stress/time curves, with the heavy load situation occurring as the peak of the curve.

The observation techniques can be used together with discomfort/comfort scales (see Corlett and Bishop 1976) and different biomechanical analyses techniques. Finally, Grieve (1979) has suggested a method to determine potential mechanical constraints during static exertions based on equilibrium (body balance) considerations reflected in a postural stability diagram.

Trunk Flexion Analysis

A measurement device for recording movements and postures in the sagittal plane has been developed by Andersson's group. The instrument consists of a pendulum potentiometer as a transducer, a five-level analog to digital converter, control circuits, and nine digital registers. Together the unit forms a portable battery powered system that weighs 2.1 lbs (1 kg) and can be worn on the back in a small harness. Although the analyzer has the potential for measuring movements of any body segment, it has so far only been adapted for measuring trunk movements in the sagittal plane (forward flexion).

The range of flexion is divided into five intervals. Flexion greater than 90 is recorded in the interval 73 to 90. The analyzer also records the total amount of time the worker spends in each interval of flexion as well as the number of times that the amount of flexion changes between one interval and another in one direction of flexion. The flexion analysis has been tested and found to be accurate and simple to use (Nordin, Ortengren and Andersson 1983; Hultman, Nordin and Ortengren 1983).

Changes in work techniques and work design can easily be measured and quantified. Continuous recording can be made without an observer. This is not recommended, however, since valuable information about the work cycle will be lost. The advantage of the flexion analyzer is that it is noninvasive, reliable, easily transported, and results are obtained immediately. The limitations of the instrument are obvious, it measures only flexion and does not measure external body loading.

EVALUATION OF WORKPLACE VIBRATION

The U.S. standard for vibration measurement is given by SAE J1013 (SAE). Accelerations are measured by a disc 8 inches (250 mm) in diameter made of molded rubber as shown in Figure 13.4. The disc has a maximum height of 12 mm and contains a cavity for the accelerometers. The accelerometers should be capable of 0.1 ms-2 to 10 ms-2 (acceleration) measurements. The disc is designed to sit comfortably between the ischial tuberosities. The acceleration is usually analyzed in 1/3 octave band levels or by means of a ride meter. Most ride meters compute and assign a weighting factor to the overall RMS acceleration based on the published SAE tolerance curves. This gives one number that is representative of the vibration

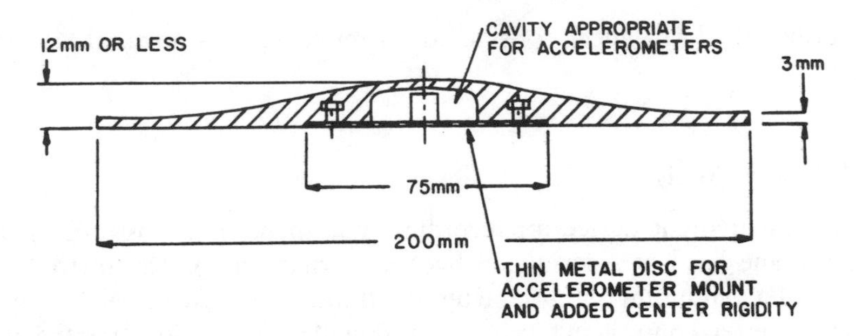

Figure 13.4.
Example MTM analysis of lifting a tote box from a pallet to a workbench and removing a part from the tote box (courtesy of J. Foulke as presented in Chaffin and Andersson, 1984).

exposure. The advantage of 1/3 octave band analysis is that it shows intensity as a function of frequency and is thus more useful for analytical and design purposes.

Typical vibration levels as a function of frequency are given in Chapter 6. It should be noted that extremely high levels are found in earth moving equipment and off-highway (military, agricultural, and recreational) vehicles.

Experience has shown attention to some special protocols will help to ensure good data. A low pass filter, with a cutoff frequency of less than 100 Hz should be used. Turning the accelerometer by 90 degrees from the vertical to the horizontal plane will provide a 1 g calibration line if the accelerometer is not a piezoelectric type. This should be done prior to and following the test to allow scaling of the data. Prior to taking the data, the operator (test subject) should sit with the disc in place for 30 minutes. This time period will allow for mechanical and temperature stabilization. In special cases in which the performance of the tasks may be affected by vibration, it may be necessary to establish vibration transmission to the head. In these cases, a bite bar or helmet-mounted accelerometer may be employed.

The test site should be carefully specified and the grade, speed and distance to be traveled specified. The road surface (or other surface) should be chosen to be representative of those used during the normal work week.

SUMMARY

It should be clear from the preceding that traditional work measurement systems are limited in their ability to provide the data necessary to evaluate and improve the biomechanical aspects of various jobs (Chaffin and Andersson 1984). Even the most detailed predetermined motion time systems such as MTM-1 do not include data on posture and load handling factors. Despite these limitations, however, it must be realized that these traditional motion-time analysis systems provide the fundamental data structure and procedures necessary to evaluate potential job-related biomechanical problems. The traditional systems emphasize a rigorous analysis. By including postural and load data, excellent biomechanical evaluations can be performed. Vibration measurements should be done using a disc-mounted accelerometer placed between the ischial tuberosities. The terrain and driving conditions should be carefully selected.

Unfortunately, it is not possible at this time to specify an optimal analysis method for future studies. In fact, from the limited attempts to develop such a system it would appear that several different job analysis schemes will be desirable in the future. Some considerations proposed by Chaffin and Andersson (1984) to assist in the choice of future analysis procedures are:

1. The degree of standardization of the tasks comprising a job.
2. The length of repetitive work cycles in a job.
3. The proportion of a workday involved in manual effort that is potentially overstressful to the musculoskeletal system.
4. The available time to observe and record manual activities (i.e., an MTM-1 analysis may require an analyst to spend 350 times the length of a job's cycle time to perform, according to Magnusson [1972]).
5. The degree of sophistication to be utilized in evaluating the biomechanical job data.
6. The expected end use of the analysis, e.g., to redesign a machine, tool, layout, or to select and place workers, or to modify work methods, or set productivity standards.

REFERENCES

Astrand, P.O.; and Rohdahl, R. 1970. *Textbook on work physiology.* New York: McGraw-Hill.

Ayoub, M.M.; Bethea, N.J.; Deivanayagem, S.; Asfour, S.; Bakken, G.; Liles, D.; Mital, A.; and Sherif, M. 1978. *Determination and modeling of lifting capacity.* DHEW(NIOSH) Final Report RO1 OH 00545–02.

Ayoub, M.M., and McDaniel. J.W. 1974. Effect of operator stance on pushing and pulling tasks. *AIIE Transactions*, 6 (3):185–95.

Ayoub, M.M.; Mital, A.; Bakken, G.M.; Asfour, S.S.; and Bethea, N.J. 1980. Development of strength and capacity norms for manual materials handling activities: The state of the art. *Human factors*, 22(3):271–83.

Barnes, R.M. 1968. *Motion and time study: Design and measurement of work*. 6th edition. New York: Wiley and Sons.

Bonjer, F.H. 1971. Temporal factors and physiological load. In *Measurement of man at work: An appraisal of physiological and psychological criteria in man-machine systems*. W.T. Singleton, J.G. Fox, and D. Whitfield (eds). London: Taylor and Francis.

Borg, G.A.V. 1962. *Physical performance and perceived exertion*. Copenhagen: Ejnar M. Munksgoard.

Borg, G.A.V. 1973. Perceived exertion: A note on "history" and methods. *Med. Sci. Sports*, 5:90–93.

Berns, T.A.R.; and Milner, N.P. TRAM—A technique for the recording and analysis of moving work posture. In *Methods to study work posture*, edited by N.P. Milner. ERGO-LAB, Stockholm, Sweden, Report 80:23, 22–26.

Brisley, C.L., and Eady, K. 1982. Predetermined motion time systems. *Handbook of industrial engineering*. G. Salvendy (ed.). New York: Wiley and Sons.

Caldwell, L.S., and Grossman, E.E. 1973. Effort scaling of isometric muscle contractions. *Journal Motor Behavior* 5 (1):9–16.

Caldwell, L.S.; and Smith. R.P. 1967. *Subjective estimation of effort, reserve, and ischemic pain*, U.S. Army Medical Research Lab Report No. 730. Ft. Knox, Ky.

Chaffin, D.B.; and Andersson, G.B.J. 1984. *Occupational biomechanics*. New York: Wiley and Sons.

Chaffin, D.B.; Herrin, G.D.; Keyserling, W.M.; and Garg, A. 1977. A method for evaluating the biomechanical stresses resulting from manual materials handling jobs. *AIHA* 38 (12):662–75.

Chaffin, D.B., and Park, K.S. 1973. A longitudinal study of low-back pain as associated with occupational weight lifting factors. *AIHAJ* 34 (12):513–25.

Chaffin, D.B.; Olson, M.; and Garg, A. 1981. Volitional postures during maximal push/pull exertions in the sagittal plane. *Proceedings of the Human Factors Society*.

Copley, F.B. 1923. *Frederick W. Taylor, Vol. I*. New York: Harper and Bros.

Corlett, E.N.; and Bishop, R.P. 1976. A technique for assessing postural discomfort. *Ergonomics* 19 (2):175–82.

Corlett, E.N.; Madeley, S.J.; and Manenica, I. 1979. Postural targetting: A technique for recording working postures. *Ergonomics* 22 (3):357–66.

Corlett, E.N., and Manenica, I. 1980. The effects and measurement of working postures. *Applied Ergonomics*. 11(1):7–16.

Davies, B.T. 1978. Training in manual handling and lifting. In *Safety in manual materials handling*. National Institute for Occupational Safety and Health, No. 78–185, pp. 175–78 (HE 20.7102:Sa1).

Davies, P.R.; and Stubbs, D.A. 1978. Safe levels of manual forces for young males. 3. Performance capacity limits. *Applied Ergonomics* 9 (1):33–37.

Emanuel, I.; Chafee, J.; and Wing, J. 1959. *A study of human weight lifting capabilities for loading ammunition into the F-86 ^{H} aircraft*. U.S. Air Force (WADC-TR 056–367), Wright Patterson Air Force Base, Ohio.

Evans, W.O. 1961. A titration schedule on a treadmill. U.S. Army Medical Rpt. No. 525, Fort Knox, Ky.

Evans, W.O. 1962. The effect of treadmill grade on performance decrement using a titration schedule. U.S. Army Medical Research Laboratory, Report No. 535. Ft. Knox, Ky.

Fox, W.F. 1967. Body weight and coefficient of friction determinants of pushing capability. *Human engineering special studies series*. No. 17, Lockheed Co., Marietta, Ga.

Garg, A.; Chaffin, D.B. 1975. A biomechanical computerized simulation of human strength. *AIIE Tr.* 7 (1):1–15.

Garg, A.; Chaffin, D.B.; and Herrin, G.D. 1978. Prediction of metabolic rates for manual materials handling. *AIHAJ* 39(8):661–74.

Garg, A., and Herrin, G.D. 1979. Stoop or squat: A biomechanical and metabolic evaluation. *AIIE Tr.* 11(4):293–302.

Gilbreth, F.B. 1911. *Motion study: A method for increasing the efficiency of the workman*. New York: D. Van Nostrand.

Gilbreth, F.B. 1912. The present state of the art of industrial management. *Trans. of ASME* 34:1224–26.

Grieve, D.W. 1979. The postural stability diagram (PSD): Personal constraints on the static exertion of force. *Ergonomics* 22 (10):1155–64.

Hanman, B. 1959. Clues in evaluating physical ability. *J. Occup. Med* (Comp. Ed.) 47(11):595–602.

Herrin, G.D. 1978. A taxonomy of manual materials handling hazards. In *Safety in manual materials handling*, C. Drury (ed.). National Institute for Occupational Safety and Health, Publ. No. 78–185, pp. 6–15 (HE20.7102:Sa1).

______. 1982. Standardized strength testing methods for population descriptions. In *Anthropometry and biomechanics—Theory and application*, R. Easterby, K.H.E. Kroemer, and D.B. Chaffin (eds.). NATO Conference Series Vol. 16. New York: Plenum Press.

Holzmann, P. 1982. ARBAN—A new method for analysis of ergonomic effort. *Applied Ergonomics* 13 (2):82–86.

Houghton, F.C.; and Yagloglou, C.P. 1923. Determination of the comfort zone. *J. Am. Soc. Heating Ventil. Eng.* 29:515–36.

Hultman, G.; Nordin, M.; and Ortengren, R. 1983. The influence of a educational program on trunk flexion in janitors. *Applied Ergonomics*. In press.

Karger, D.W.; and Bayha, F.H. 1966. *Engineered work measurement*. New York: Industrial Press.

Karger, D.W.; and Hancock, W.M. 1982. *Advanced work measurement*. New York: Industrial Press, Inc.

Karhu, O.; Harkonen, R.; Sorvali, P.; and Vepsalainen, P. 1981. Observing working postures in industry: Examples of OWAS application. *Applied Ergonomics* 12 (1):13–17.

Karhu, O.; Kansi, P.; and Kuorinka, I. 1977. Correcting working postures in industry: Practical method for analysis. *Applied Ergonomics* 18:199–201.

Konz, S. 1979. *Work design*. Columbus, Oh.: Grid Pub. Co.

Koyl, F.F.; and Marsters-Hanson, P. 1973. *Age, physical ability and work potential*. Washington, D.C.: Manpower Administration, U.S. Dept. of Labor.

Kroemer, K.H.E. 1969. *Push forces exerted in 65 common work positions*. AMRL–T–68–143. Aerospace Medical Research Laboratory, Wright Patterson Air Force

Base, Ohio.

Kroemer, K.H.E.; and Robinson, D.E. 1971. *Horizontal static forces exerted by men standing in common working postures on surfaces of various tractions.* AMRL–TR–70–114, Aerospace Medical Research Laboratory, Wright Patterson Air Force Base, Ohio.

Kumar, S. 1980. Physiological responses to weight lifting in different planes. *Ergonomics* 23(10):987–93.

Laubach, L.L. Human muscular strength. 1978. In *Anthropometric source book*, NASA No. 1024, U.S. National Aeronautics and Space Administration, Washington, D.C. pp. VII–I–VII–55 (NAS 1.61:1024/v.1).

Lee, K. 1982. *Biomechanical modelling of cart pushing and pulling.* Unpublished doctoral dissertation, Univ. of Michigan, Ann Arbor.

Magnusson, K. 1972. The development of MTM-2, MTM-V, and MTM-3. *Journal of Methods Time Measurement* 17:11–23.

Martin, J.B.; and Chaffin, D.B. 1972. Biomechanical computerized simulation of human strength in sagittal-plane activities. *AIIE Tr.* 4(1):19–28.

Mundel, M.E. 1978. *Motion and time study.* 5th edition. New Jersey: Prentice-Hall.

NASA, 1978. *Anthropometric source book—Volume II: A handbook of anthropometric data*, NASA Reference publication 1024. Washington, D.C.: National Aeronautics and Space Administration (NAS 1.61:1024/v.2).

National Institute for Occupational Safety and Health. 1981. *Work practices guide for manual lifting.* Technical report 81–122, Cincinnati, Ohio: Division of Biomedical and Behavioral Science, NIOSH.

Niebel, B.W. 1972. *Motion and time study.* 5th ed. Homewood, Il.: R.D. Irwin, Co.

Nordin, M.; Ortengren, R.; and Andersson, G.B.J. 1983. Measurement of trunk movement during work. *Spine.* In press.

Priel, V.Z. 1974. A numerical definition of posture. *Human factors* 16 (6):576–84.

Smith, P.; Armstrong, T.J.; and Lizza, G.D. 1982. IE's can play crucial role in enabling handicapped employees to work safely, productively. *Industrial Engineering* 14 (4):98–105.

Snook, S.H. 1978. The design of manual handling tasks. *Ergonomics* 21(12):963–85.

Snook, S.H.; Campanelli, R.A.; and Hart, J.W. 1978. A study of three preventative approaches to low back injury. *J. Occup. Med.* 20 (1):478–81.

Stevens, J.C.; and Cain, W.S. 1970. Effort in muscular contractions related to force level and duration. *Perception psychophysics* 8 (4):240–44.

Stevens, S.S. 1956. The direct estimation of sensory magnitudes loudness. *Am. J. Psychology* 69 (1):1–25.

———. 1960. The psychophysics of sensory function. *American Scientist* 48 (1):226–53.

Stobbe, T.J. 1982. *The development of a practical strength testing program for industry.* Unpublished doctoral dissertation. Univ. of Michigan.

Switzer, S.A. 1962. *Weight lifting capabilities of a selected sample of human males.* Aerospace Med. Res. Lab. Report No. AD–284054, Wright Patterson Air Force Base, Ohio. MRL–TDR–62–57.

Taylor, F.W. 1929. *The principles of scientific management*. New York: Harper and Bros.

Tichauer, E.R. 1978. *The biomechanical basis of ergonomics: Anatomy applied to the design of work situations*. New York:Wiley-Intersciences.

U.S. Department of Labor. 1972. *Handbook for analyzing jobs*. Supt. of documents, U.S. Govt. Printing Office, Washington, D.C., No. 2900–0131.

Warwick, D.; Novack, G.; and Schultz, A. 1980. Maximum voluntary strengths of male adults in some lifting, pushing, and pulling activities. *Ergonomics* 23(1):49–54.

14 Workplace Design

Don B. Chaffin, Malcolm H. Pope, and Gunnar B. J. Andersson

INTRODUCTION

Approximately one-third of the workforce in the United States is required to exert near their maximum strength in the daily performance of their jobs (NIOSH, 1981). This illustrates that gross manual activities, even if only infrequently performed in a job, are still required of a large number of people, placing them at high risk of incurring LBP. It also has been shown that LBP is often caused by slipping or tripping (Manning 1983), emphasizing the need for greater care in the design and specification of floors and shoes. Lastly, growing evidence has accumulated to indicate that sitting in prolonged postures without appropriate low back support, or when subjected to vibration, increases the risk of LBP (Kelsey and Hardy 1975).

This chapter emphasizes the role of workplace design in the prevention of low back pain (LBP) in industry. As discussed in the preceding chapter, the magnitude of the load one handles, the postures assumed in such material handling, and the frequency of strenuous exertions combine in a complex fashion to increase the risk of LBP. In addition, the period of exposure must also be included in the injury model.

One method of preventing LBP is to design the workplace in a manner that minimizes conditions believed to increase the mechanical stresses on low back tissues beyond the physiological limits. Although the precise relationship between a specific set of work conditions and resulting risk of LBP are not available at this time, the gross biomechanical models and epidemiological data discussed earlier provide the basis for some general workplace design guidelines.

To design a workplace so that stresses on the low back are minimized the following factors must be controlled: posture; external load (material and tools); movements; and vibration. Each of these will be discussed in the chapter, but the reader is also referred to Chapter 2, in which the occupational biomechanics of the spine is discussed.

POSTURE

Practical Considerations in Seat Design

Because there are many different opinions and user requirements, available chairs vary widely. Compare, for example, the driver's seat in an automobile with an office chair, or a work chair in a factory. Regardless of use, it is important to be able to adjust any chair to meet basic anthropometric dimensions of the worker. A number of recommendations have been published which are different for various populations. This is not surprising, since anthropometric dimensions vary greatly between populations of different countries. Rather than providing a list of all these data, an approach that explains why different dimensions are critical will be used, with the main emphasis on the specification of office and industrial chairs. (See also Kroemer 1971; Kroemer and Robinette 1969; Chaffin and Andersson 1974.)

Some important chair dimensions are shown in Figure 14.1. A proper seat height is desired and should be adjustable for the individual user. The seat surface should be 1.2–2" (3–5 cm) below the knee fold when the lower limb is vertical. Foot supports can be used with higher than normal chairs. The width of the seat should be sufficient to accommodate the user population. The edges of the seat should not be detectable during ordinary sitting work.

The depth of the seat is also important. It must be possible to use the backrest, so the seat pan must not be too deep. In particular, pressure should be avoided on the back of the thigh near the knees. A free area between the back of the lower limb and the seat pan is also useful to facilitate rising and leg movements. About 4" (10 cm) is suggested as the minimum clearance. The front part of the seat should be contoured. In some cases a forward slope can be advantageous in reducing the load on the spine. This is particularly important for semi-sitting postures or a raised chair. More often, a backward slope of about 5° is suggested for normal sitting posture. This slope facilitates the use of the backrest and prevents sliding on the seat surface when one moves around in the chair. Sliding can also be prevented by the choice of seat contour and seat cover materials with high surface friction. A proper choice of seat cover materials is important to climatic comfort (i.e. they should breathe in hot environments) but also in the industrial setting where specific materials may be necessary (e.g., to avoid static electricity).

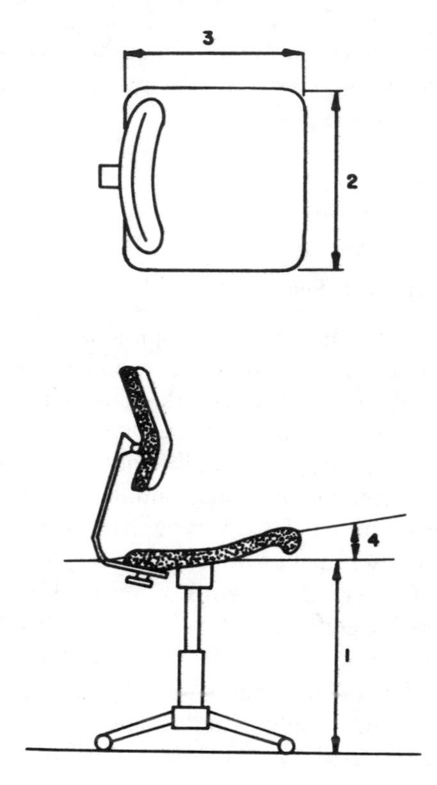

Figure 14.1
a. Seat-dimensions; (1) Seat height, (2) seat width (breadth), (3) seat length (depth), (4) seat slope.

For the backrest the following dimensions are important: 1) top height, 2) bottom height, 3) center height, 4) height, 5) backrest width, 6) horizontal radius, 7) vertical radius, and 8) seat angles (Fig. 14.1). Other factors are 9) pivoting and recline possibility, 10) softness, 11) adjustability and 12) climatic comfort. The backrest should be adjustable both in horizontal and vertical directions. It should support the lumbar spine without restricting movements of the spine and arms. The width should allow all users to be supported without arm interference. The shape should be convex to the normal lumbar lordosis, and concave from side to side to conform to human anatomy and support the occupant in the chair. A spring loaded pivoting action can allow the backrest to follow natural body movements while maintaining body support, but can be impractical in jobs that demand great precision. When reclining is permitted, a control system must be used so that a firm support is given when the backrest is in the reclined position.

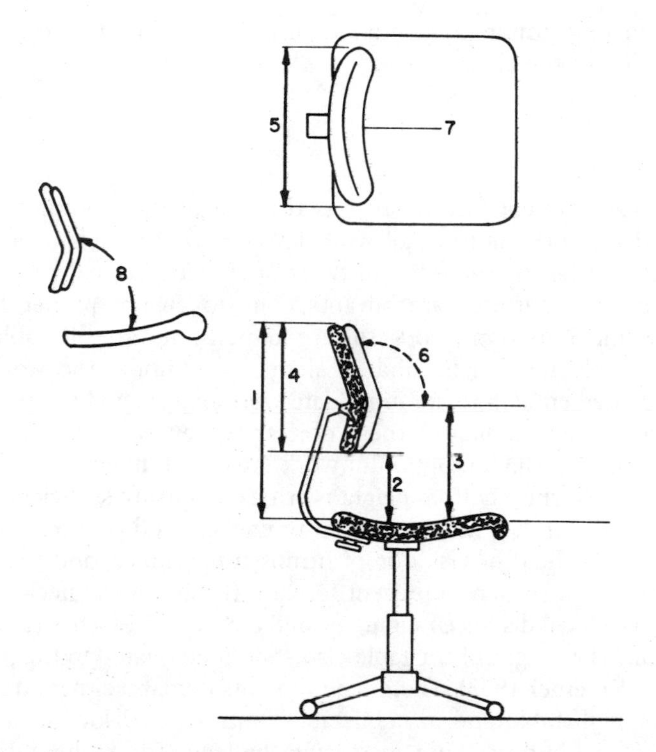

Figure 14.1
b. Backrest dimensions; (1) Backrest top height, (2) Backrest bottom height, (3) Backrest center height, (4) backrest height, (5) Backrest width (breadth), (6) Backrest horizontal radius, (7) Backrest vertical radius, (8) Backrest seat angle.

Another chair feature is the armrest. Design considerations for the armrest include: 1) length, 2) width, 3) height, 4) width between armrests and 5) distance from armrest front to seat. When the armrests are too high the occupant must raise the shoulders and abduct the arms. This is also the case when the armrest-to-armrest width is too large. Too low an armrest, on the other hand can result in the occupant sliding forward or being forced to lean to one side to use it. In addition, the length and placement of an armrest should allow the chair to slide forward enough that the worker can sit erect using the backrest for support. In general, placement of the armrests is important to reduce pressure on the seat surface, load on the spine, and facilitate rising from the chair.

Factors such as the number of feet on the chair, the base diameter and the use of wheels or casters are necessary safety considerations. Five casters or feet are suggested with a minimum radius of 12″ (30 cm) to prevent tipping and a maximum horizontal radius of 14" (35 cm) to prevent tripping over the base of the chair.

The Table

Because different factors such as the size of the workpiece, motions demanded by work, and overall work layout vary, the height of the work surface cannot be the same for all types of activity. For each type of work adjustable work surfaces are advantageous to ensure proper fit to the worker. Some important work surface dimensions are: 1) table bottom height, 2) table top height, and 3) slope. In addition, the work surface should be large enough to accommodate work objects and the friction high enough to prevent sliding of these objects. When controls are used they must be placed within an optimum work area in all planes.

The work surface bottom height is critical to ensure sufficient leg room, while the table top height is important to ensure good posture and optimal conditions. The field of vision is of utmost importance, and the table top height should be chosen to prevent forward flexion of the neck and of the trunk. An eye focal distance ranging from 8 inches to 15 inches (20 to 40 cm) is common. The height of the table also should be related to the position of the elbow (Kroemer 1963). These requirements demand either an adjustable table or a high table and an adjustable footrest. Too low a table causes kyphosis of the lumbar spine, increasing the load. Too high a table, on the other hand, causes abduction of the arms and elevation of the shoulder as well as kyphosis of the neck causing fatigue in the shoulder and neck muscles. Chaffin (1973) found a 15° angle at the neck to be acceptable. It is important to remember that the work surface height is not always the table height. In using a typewriter, word processor or computer, for example, it is the keyboard height that is important. Tilting the work surface toward the worker is a good method of preventing unnecessary forward flexion of neck and back.

MATERIAL HANDLING

Although several material handling principles are reviewed in earlier chapters, some other aspects of manual materials handling that are related to the design of the workplace now will be discussed.

General Manual Material Handling Considerations

When an object is large and heavy it is often desirable to use a hoist or crane to assist in its movement. If a hoist is used which is manually moved from one location to another, sufficient time must be provided to the operator to avoid sudden jerking motions that will result in an uncontrolled swinging of the object being moved. It is not uncommon for hoist operators to suffer back injuries by attempting to stop an uncontrolled swing motion of an object supported by the hoist. It is also possible for an object to slip from the hoist or overload the hoist when swinging. For this latter reason, specific structural and safety specifications must be met if a hoist or crane is to be used, and the reader is referred to *American National Standard B30.2.0* and *OSHA Regulation 1910.180*.

Carts and manually powered trucks are also used to move heavy objects in the workplace. The risk of low back pain being induced by the use of such devices arises from two types of hazard. First is the overexertion hazard associated with pushing or pulling on too heavy a load. The compressive forces on the L5-S1 disc become quite high, particularly when pulling a load (Lee, 1982). To avoid this situation, it is important to assure: that the pushing or pulling hand force requirements are below 50 pounds (225N) and that the hands are at about hip to waist level when making a maximum exertion, thus minimizing the spinal load moments.

These requirements can be best satisfied in the design of a workplace by specifying a cart that 1) has vertical handles that can be grasped at varying heights, 2) uses large rubber tires with good bearings that will not "hang up" on irregular surfaces, 3) has two wheels which easily pivot, and 4) is designed to handle the intended load. Furthermore, it is just as important to assure that the floor surface be smooth, kept clean, and have a grade inclination of no greater than 4 degrees. These concepts are illustrated in Figure 14.2.

The second hazard to the back when pushing or pulling carts comes from the increased risk of slipping during such activities. It is not uncommon for the required coefficient of friction (ratio of foot shear force divided by normal force) to exceed 1.0 during pushing and pulling activities. It is for this reason that the floor surface in areas wherein pushing and pulling activities take place must be kept clean and dry and that workers should be instructed to use or be provided high traction shoes.

Container Design Considerations

The weight and dimensions of a load to be lifted are primary risk factors. In this context, *container dimensions* are important.

In general, if a container can be designed to be compact a worker can minimize the spinal load moment by keeping the object's center-of-mass close to the body. The effect of the center of gravity of an object being lifted on spinal stress has been approximated in the NIOSH (1981) *Workpractices Guide to Manual Lifting*, (Figure 14.3). The figure illustrates the necessity of

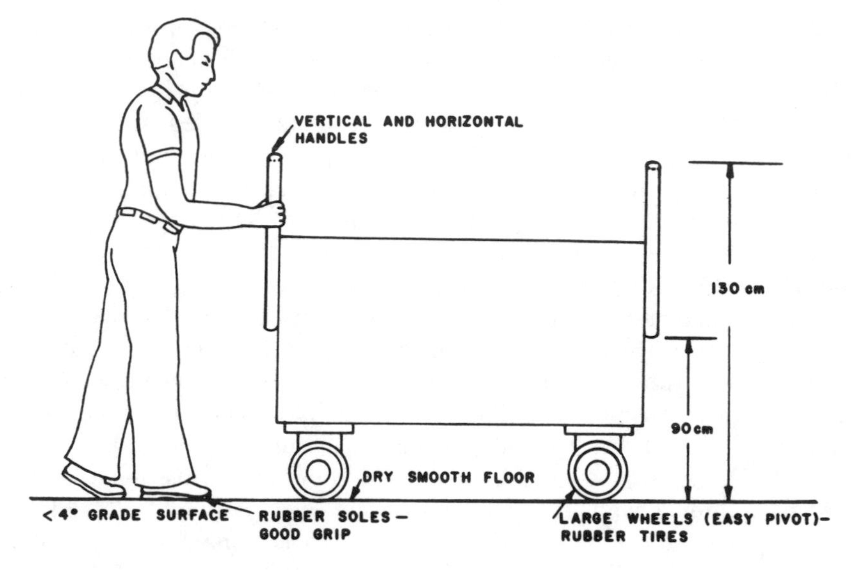

Figure 14.2
Important specifications in a cart.

having compact containers; the load moment arm "H" should be kept as small as possible.

The dimensions of a container are also important if a person must ascend or descend stairs while carrying the object. If the container is too large, it will impair vision, resulting in an increased risk of a trip or fall as shown in Fig. 14.4. Clearly, in stair descent the hazard is even greater. In general, stair climbing should never be done with large or heavy containers in the workplace. As a general safety rule, one hand should always be able to quickly grasp the stair handrail to prevent a fall injury. This is especially important during descent. Carrying any containers on a stair seriously compromises this rule. Therefore, to move goods from one level to another, hoists, platforms, lifts, and elevators should be used.

If the container must be lifted from the floor, and it is too large to pass between the knees, it will require the person to lift the object in front of the knees. This will cause a larger spinal load moment than would be the case if the object could be lifted between the knees. (This concept was discussed earlier in Chapter 2 on biomechanics of the spine.) Any object larger than about 12″ (30 cm) will not easily be lifted between the knees, thus increasing the H distance presented earlier in Figure 14.3.

Container design must also include a means to firmly grasp it. This is important not only to lift the container properly but also to avoid sudden spinal inertial loads that occur when one attempts to regain control of an object slipping from the hands. In general, a forceful grasp of an object is

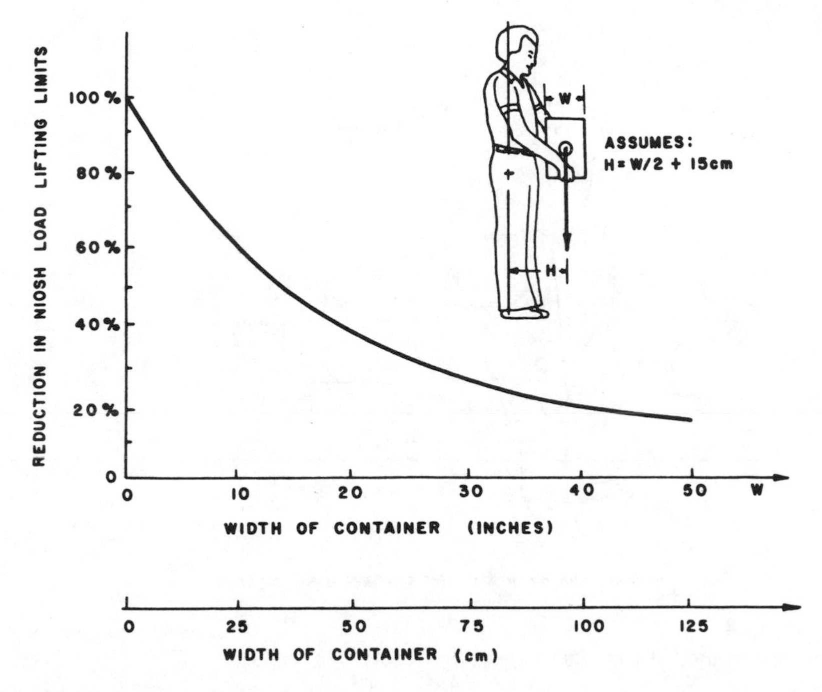

Figure 14.3
Minimizing load moment H: The importance of compact containers.

provided by either a "hook grip" whereby the fingers wrap around the object but without thumb opposition, or a "power grip" whereby the thumb assists in retention of the object by overlapping the fingers. The hook grip is used when grasping the bottom edge of a container and lifting it. This is often adequate for moderate loads lifted for short periods, but it entails both finger flexor muscles as well as wrist flexor muscles and other upper extremity muscles to maintain the coupling between the hand and container.

If a proper handle is provided, a power grip can be used. This requires less upper muscle action to maintain the coupling because the handle is fully secured between the fingers and the thumb. To provide such a handle requires consideration of hand anthropometry. Figure 14.5 illustrates the principle dimensions of concern. Finger depth clearance requires a minimum space of 1.2 inches (3.0 cm) for the bare hand of a large person. An additional 1.0 inch (2.5 cm) is suggested for cold weather work gloves (Damon *et al.* 1966). Hand breadth minimum clearance of 5.0 inches (12.7 cm) is recommended for a larger person's bare hand. A heavy work glove could increase this by 1.0 inch (2.5 cm) also. The minimum radius of the handle to provide a low pressure bearing surface for the fingers is about 0.75 inch (1.9 cm).

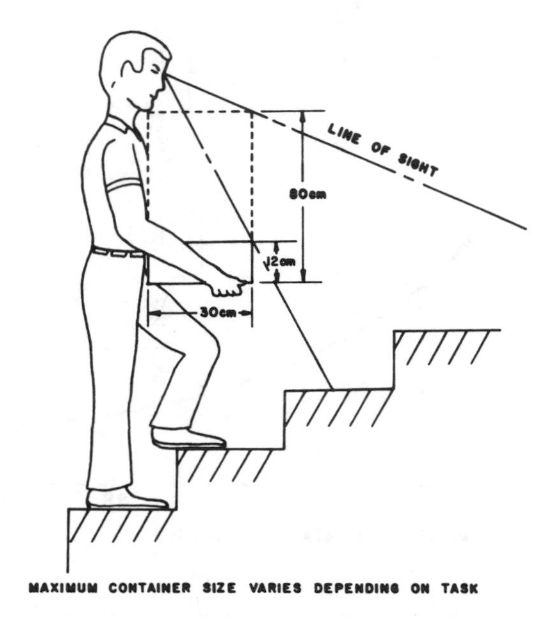

Figure 14.4
The importance of container size while negotiating stairs.

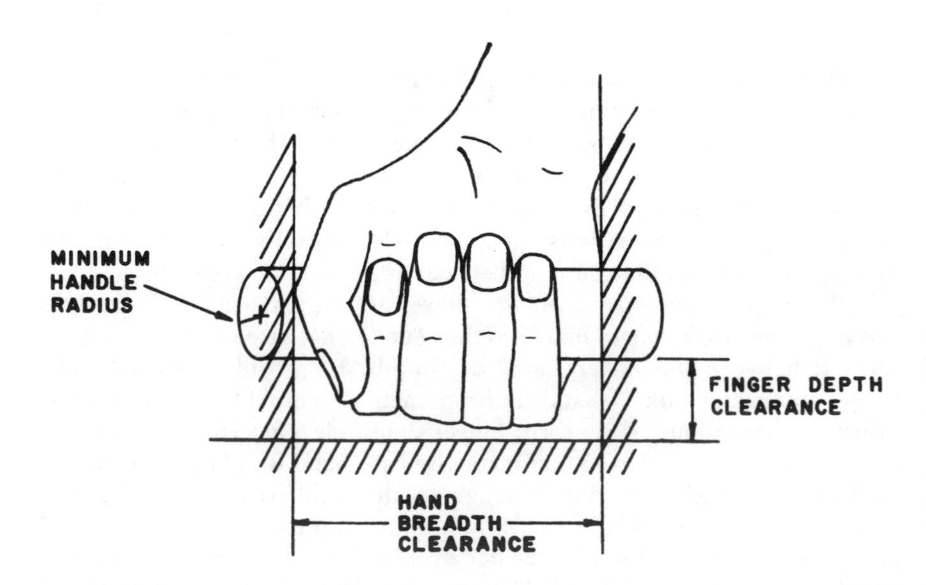

Figure 14.5
Dimensions in handle design.

WORKSPACE DESIGN CONSIDERATIONS

One of the primary concerns when designing a workspace to prevent low-back disorders is to allow the person to stand or sit erect while performing manual activities. This requirement means that workers should not have to reach beyond a comfortable arm reach in front and to the side of the body, especially if the manual activity required is lifting of heavy materials.

As was discussed in the preceding section on container design, the horizontal distance between the worker and the center of mass of an object being lifted (H) should be minimized. This means that obstructions over which a worker must lean to reach an object or place an object must be eliminated. If a heavy object is to be lifted, it should be located so that the worker can lift it without bending down or leaning forward. In practice, this may require:

1. Roller conveyors which allow objects to be pulled in toward the body before lifting (Figure 14.6a) or powered conveyors to bring containers to the worker.
2. Gravity-fed slides (or shelves) which present items to the worker such as used in airline baggage claim areas (Figure 14.6b).
3. Machine designs which have the workpiece or part being manufactured located close to the operator (Figure 14.7).
4. Tilting of large stock bins to present parts closer to the operator (Figure 14.8).
5. Enough room for a worker to walk around large bins or pallets of parts to avoid having to reach and lean forward to remove parts from the opposite side.

It should also be remembered that when lifting loads above the shoulder, not only does fatigue in the shoulder and upper back muscles become important, but a person is more unstable because of the higher center-of-mass of the body and load combination. Concern was shown for this situation in the NIOSH *Workpractices Guide for Manual Lifting* by inclusion of a vertical height factor in computing the load lifting limits. The effect of this vertical correction is shown in Figure 14.9; it clearly indicates the disadvantages of lifting loads above shoulder height (the suggested limits will be as much as 50 percent lower than if lifting loads are at knuckle height).

It is also clear in Figure 14.9 that loads should not be lifted from below standing knuckle height: the lifting limits are 30 percent less for a lift from the floor than for a lift at knuckle height. In this regard, thought should be given to the use of stock support tables that adjust vertically to allow the operator to maintain an erect trunk posture regardless of the remaining

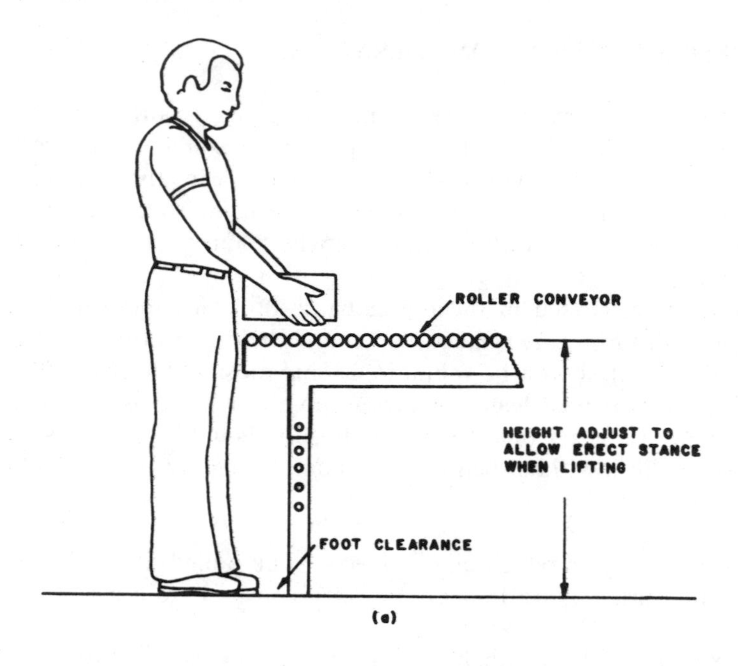

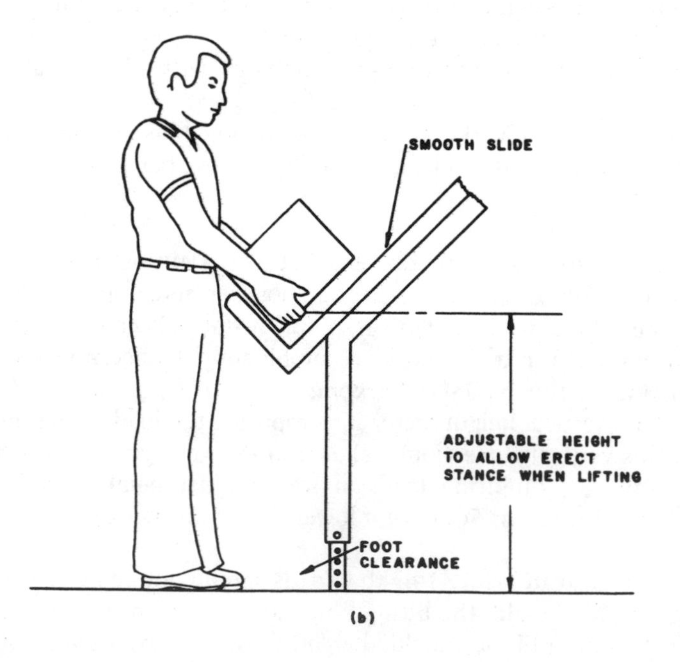

Figure 14.6
Improvements of workplace design a) roller conveyors, b) gravity fed slide.

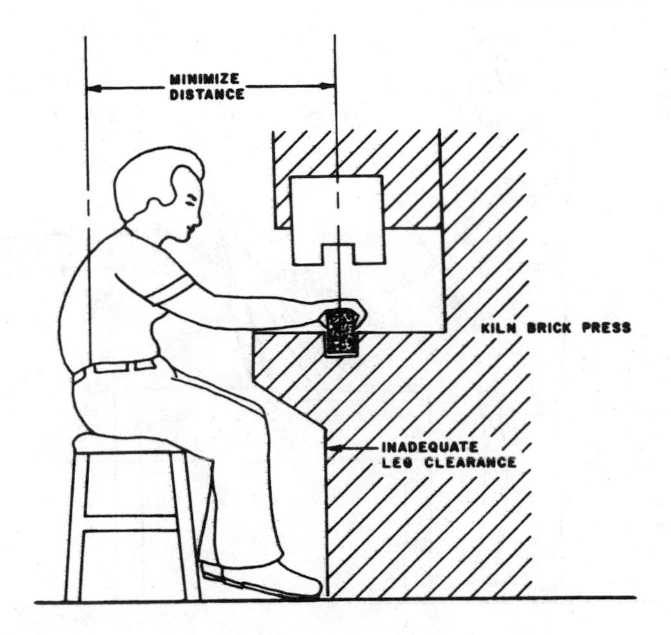

Figure 14.7
Considerations in making task close to worker.

stock in a bin or on a pallet (Figure 14.10); and the use of work benches and tables with adjustable height when a person is required to perform prolonged work on such surfaces. These would normally be adjusted to about elbow height.

WHOLE BODY VIBRATION

As discussed in Chapter 7, workers who drive vehicles are more likely to have LBP. It is probable that vibration is a major factor in the etiology, but postural stress, muscular effort, and shock and impact forces are also important. It will be remembered that many vehicles subject the driver to vibrations in the range of resonance of the spine and that muscle fatigue also occurs in a vibrational environment. Recently, work has been carried out to reduce these risks, including the effects of postural supports, cushions, and dampening devices.

One method of reducing whole body vibration is to reduce the vibration input. This can be done by the choice of vehicle and by training operators to choose the terrain over which they drive (if possible) with the objective of reducing vibrations and by changing their driving speed and style. Much of this is within the control of the operator.

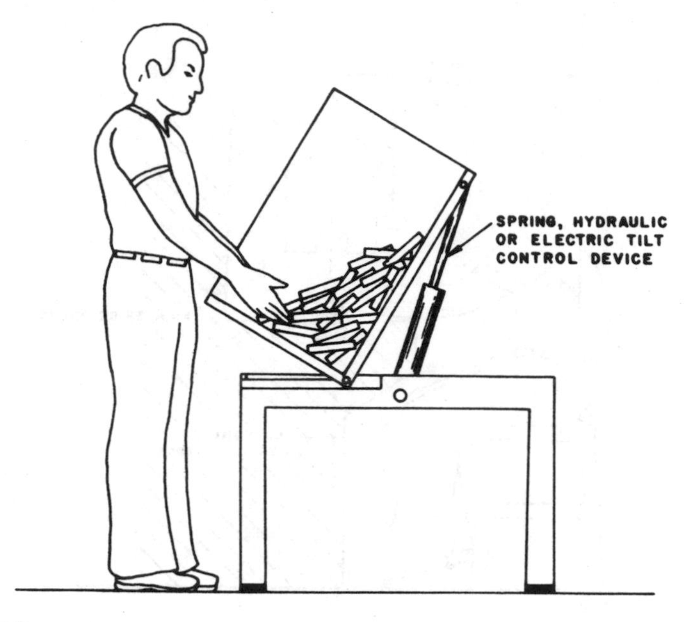

Figure 14.8
Tilting of bins gets work close to worker.

Andersson *et al.* (1975) showed that spinal stress could be minimized in the seated subject. The inclination of backrest was 110° and the seat pan angle was 6°; there was additional curved support for the lumbar spine level with the third lumbar vertebra also.

There are other advantages to inclining the backrest at 110°. If the trunk is upright or leaning forward when accelerated vertically, the head and shoulders will bend further forward. This flexor torque adds to the net spinal stress. The psoas muscle activity is reduced, and on vertical acceleration the load is distributed over the backrest as well as the seat.

It appears that a position of extension would be favorable in vibration as it is in static situations, although this could result in the imposition of bending vibrations ("backslap") directly to the spine.

Wilder *et al.* (1983) have shown that the various postural supports can markedly affect the spine's vibrational response. In general, these postural supports shift loads from the lumbar area to other spinal areas. The lumbar, arm and foot supports were generally helpful, but did affect adjacent structures. The commercially available vibration damping seats that have been tested reduced deleterious vibrations.

To reduce vibrations to less harmful frequencies, soft cushions should be replaced with firm ones. The seat should be suspended to give it a natural frequency of less than 1.5 Hz (Fig. 14.11). Suspension seats made to this

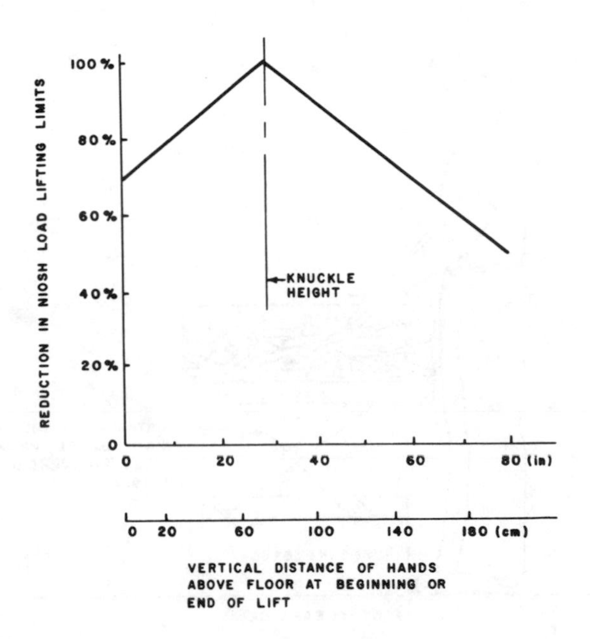

Figure 14.9
The effect of vertical height in NIOSH load lifting limits.

specification are readily available commercially, and most of them reduce vibration levels. The marked isolation of a suspension seat as compared to a conventional seat is shown in Figure 14.12. The intensity of vibration from unsafe to safe levels is defined in the guide published by the International Standards Organization (ISO 2631) (see Fig. 14.13). See also Chapter 6. It is logical to attempt to reduce vibration levels to below the levels in ISO 2631.

The following guidelines are based on ISO-2631-1978 and ISO 2631/DAM 1–1980. The level of vibration in the ISO guidelines is expressed as acceleration (meters/second^{-2}) RMS. The limits are given in terms of 1) reduced comfort boundary, 2) fatigue-decreased proficiency boundary and, 3) exposure limit. Different values exist for vertical and horizontal vibrations.

The reduced comfort boundaries are obtained by dividing the fatigue-decreased proficiency limits by 3.15 (a decrease of 10 dB). Exposure limits are obtained by multiplying the proficiency limits by 2 (an increase of 6 dB). The 1980 supplement of the standard suggests that to estimate reduced comfort and proficiency a filtered vibration signal over the entire frequency range should be used. Since prolonged exposure to a vibration environment leads to muscle fatigue, lifting directly following a long period of driving

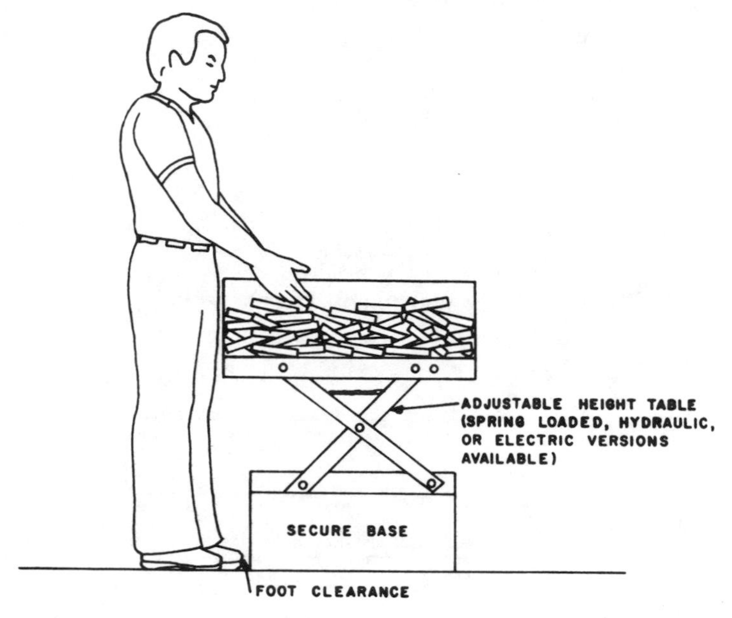

Figure 14.10
The use of an adjustable height table.

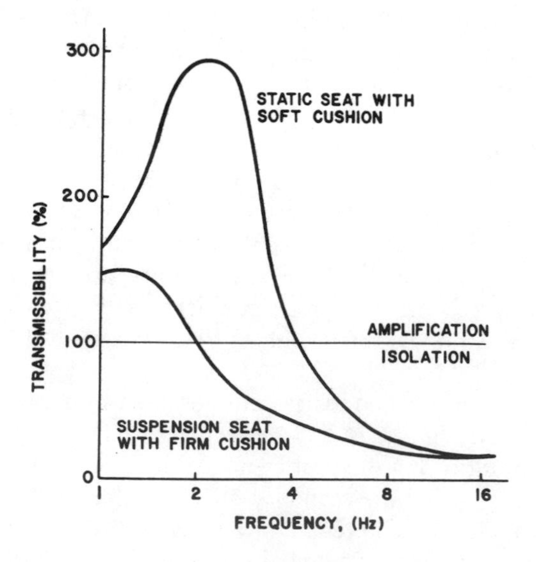

Figure 14.11
Natural frequency of typical suspension seat.

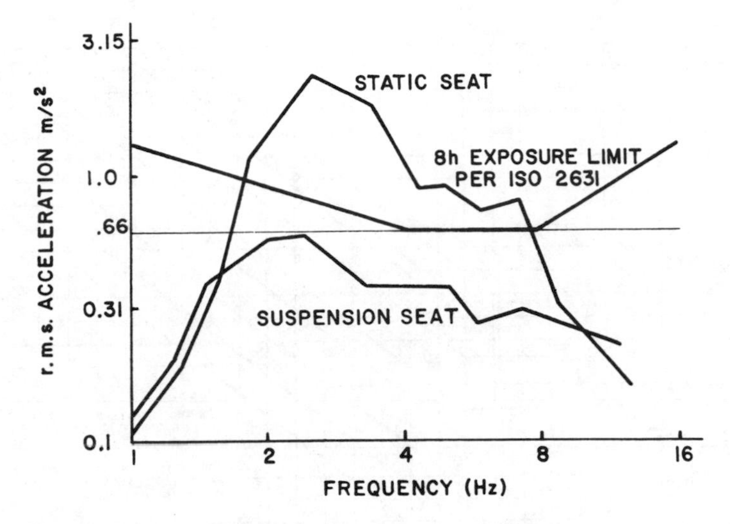

Figure 14.12
Suspension seat behavior as compared to ISO standard.

should be avoided. Drivers should be encouraged to stretch and exercise during their trip to reduce postural fatigue.

These data suggest that the following features are desirable in a vehicle seat:

1. Three-axis damping of seat pan to levels below ISL 8-hour exposure;
2. Firm rather than soft seat cushions;
3. Lateral support to reduce lateral flexor activity;
4. Damped but rigid lumbar support;
5. Conforming thoracic support of graded stiffness caudally;
6. No neck rest;
7. Damped arm support;
8. Tilted seat pan;
9. Seat back in a position of extension.

SUMMARY

Sitting in the workplace is currently the most common posture in the industrial world. Although sitting offers many excellent advantages, the seat design needs to be thought out carefully to avoid causing musculoskeletal problems to the workers. Seat vibration increases the complexity of the problem. The seat itself is only one part of the sitting workplace. This chapter offers some guidance in this area, but needs to be supplemented by

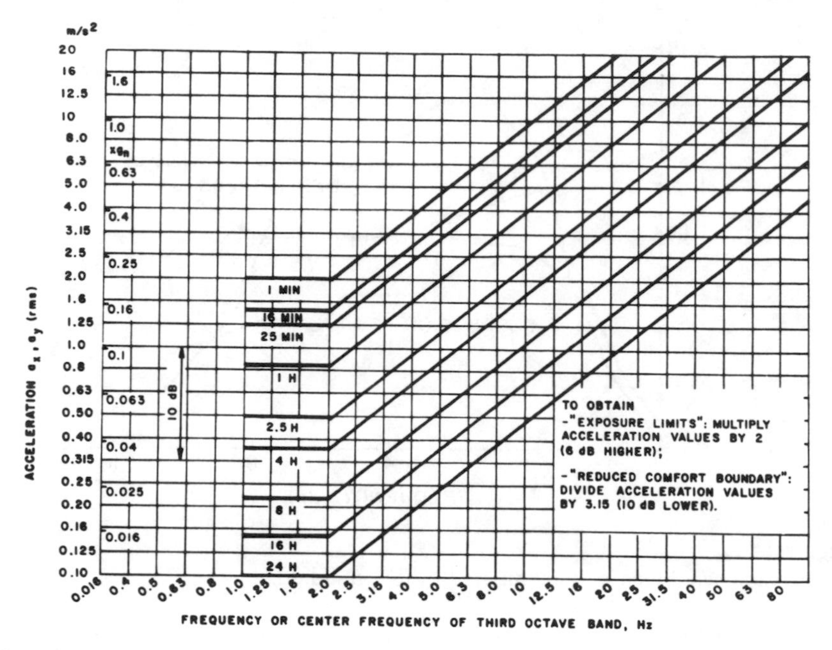

Figure 14.13
The ISO standard for horizontal vibration.

the information on workplace design and function available in many ergonomics texts to be fully effective. With attention to detail in design, layout and work method seated work can be made much safer than is currently the case in many offices and factories.

It should also be clear that manual exertions in a standing posture can stress the low back if care is not taken in the design and layout of the workplace and objects that are handled (containers and carts). Once again, allowing the worker to remain in a stable, upright posture with any load held close to the body is essential. A high proportion of the U.S. workforce has to exert forces close to their maximum strength daily. Container dimensions should be minimized to reduce spinal load moments and to improve visibility. Hoists, cranes and carts should be used when possible. However, operators should not control the swinging of loads supported by the hoist. Pushing and pulling forces of the carts should be below 225N, and the carts should be designed to minimize spine load moments. Workplace design should allow the worker to stand or sit erect and not force the worker to reach beyond a comfortable distance.

Cyclic loadings should not exceed the ISO requirements. Vibration isolation seats should be used when possible and firm cushions should be used. Support of the lumbar region and arms should be employed. In general, there is no reason that, with the present technology, the workplace should not be adjustable to "fit" an individual. The expense is often justified if the worker no longer is subjected to a hazardous working environment.

REFERENCES

Frymoyer, J.W.; Pope, M.H.; Costanza, M.C.; Rosen, J.C.; Goggin, J.; and Wilder, D.G. 1980. Epidemiologic studies of low back pain. *Spine* 5 (5):419–23.

Hall, M.A.W. 1972. Back pain and car-seat comfort. *Applied Ergonomics* 3 (2): 82–89.

Herrin, G.D.; Chaffin D.B.; and Mach, R.S. 1974. *Criteria for research on the hazards of manual materials handling*. Workshop proceedings on contract CDC–99–74–118, U.S. Dept. of Health and Human Services (NIOSH), Cincinnati, Oh.

Kelsey, J.L.; and Hardy, E.J. 1975. Driving of motor vehicles as a risk factor for acute herniated lumbar intervertebral disc. *Am. J. Epid.* 102:63–73.

Kroemer, K.H.E. 1963. Uber die Hohe von Schreibtischen. *Arbeitswissenschaft* 2 (4):132–40.

______. 1971. Seating in plant and office. *Am. Ind. Hyg. Ass. J.*, 32 (10):633–52.

Kroemer, K.H.E.; and Robinette, J.C. 1969. Ergonomics in the design of office furniture. *Ind. Med. Surg.* 38 (4):115–25.

Laurig, W. 1969. Der Stehsitz als physiologisch gunstige Alternative zum reinen Steharbeitsplatz. *Arbeitsmed Sozialmed Arbeitshyg* 4:219.

Lee, K. 1982. *Biomechanical modeling of cart pushing and pulling*, Doctoral dissertation, Univ. of Michigan, Ann Arbor, Mi.

National Institute for Occupational Safety and Health. 1981. *A work practices guide for manual lifting*, Tech. Report No. 81-122, U. S. Dept. of Health and Human Services (NIOSH), Cincinnati, Oh.

Okushima, H. 1970. Study on hydrodynamic pressure of lumbar intervertebral disc. *Arch. Jap. Chir.* 39:45–57 (In Japanese).

Pope, M.H.; Wilder, D.G.; and Frymoyer, J.W. 1980. Vibration as an aetiologic factor in low back pain. *Proc. of Inst. Mech. Engs. Conf. on Low Back Pain*. Inst. Mech. Engs. Paper #C121/80.

Rosemeyer, B 1971. Elektromyographische Untersuchungen der Rucken-und Schultermuskulatur im Stehen und Sitzen unter Berucksichtigung der Haltung des Autofahrers. *Arch. Orthop. Unfallchir.* 71:59–70.

PART VI
Legal Aspects

15 Impairment Rating

Richard A. Brand, Jr.

INTRODUCTION

The process of evaluation or assessment is used in determining the severity of the disability but does not imply the identification of a diagnosis or a cause of disability. This evaluative process is central to determining the level of compensation.

In particular, assessment of disability is required by statutory programs (e.g., Supplemental Security Income, Workers' Compensation) which require an assessment of disability to establish eligibility and, in the case of Workers' Compensation, to determine degree of compensability. Assessment is required in litigated cases involving liability for injury.

Assessment is also critical to the chronic LBP patient because inadequate assessment techniques in many cases lead to the conclusion that "nothing is wrong," thus unfairly preventing individuals from being helped by statutory programs to which they might be entitled.

It must be recognized that any disability determinations are somewhat arbitrary and may be contaminated by philosophical, ethical, and even political notions. First, the existence of Workers' Compensation implies that society believes the cost of injury to a worker should be included in the cost of production. Second, the existence of a statutory program such as Social Security Indemnity implies that society should bear some of the cost of supporting disabled persons. Third, the existence of evaluation schemes that are based solely on objective criteria implies that society should not, or does not, trust the subjective information from a patient when disability determination is made.

Any assessment technique must be designed with a particular purpose in mind. It is clear that a given technique does not suit all purposes. In fact, there is evidence (Lehmann, Brand, and O'Gorman 1982) that different techniques might even be important for different chronic low back problems. The design, choice and application of a given technique is determined not only by the purpose, but also by the philosophical biases of the assessor. For example, the assessor with a strong individualistic "work ethic" may choose or apply a scheme that will result in a lesser disability rating than the person who has more egalitarian leanings.

This chapter will review some current notions about disability assessment, review the role of the physicians in light of current statutory programs, review current low back rating practices, and finally will present one approach with its assumptions and philosophical biases. Chronic or permanent, rather than acute or temporary disability, will be the focus of this chapter.

CURRENT CONCERNS OF DISABILITY ASSESSMENT

A clear perception of current disability assessment practices is dependent upon an understanding of several important concepts that have developed over the past few decades. We have previously discussed the concepts of impairment and disability. To recapitulate, impairment is quantifiable anatomic or functional loss which may be temporary or permanent and is solely a result of a medical condition.

Disability, on the other hand, is the loss of capacity to engage in gainful employment caused by impairment and takes into account one's education, training, experience, and such relatively unquantifiable factors such as motivation, psychosocial situation, and perhaps other factors.

Most statutory programs clearly distinguish "impairment" from "disability" (or similar terms). The legal distinctions imply that impairment can, in fact, be determined objectively (when, in fact, it sometimes cannot). Nonetheless, the legal distinctions are important to introduce a sense of fairness and uniformity to the evaluation process.

Since impairment is purely the result of a medical condition, the rating of impairment is solely the responsibility of a medical expert. However, disability is an administrative decision rather than a medical decision.

A third concept is that impairment rating is based on the "whole person" concept. A person's ability to engage in gainful employment is dependent upon the sum of his anatomic parts and their functional capabilities. The partial or even complete loss of a single function therefore creates only partial impairment. Thus, a hypothetical person with 100 percent spine impairment has only partial impairment based on the "whole person." The *AMA Guides to the Evaluation of Permanent Impairment* (1971) provides tables to estimate whole man impairment owing to impairment of a given part or several parts.

Also central to this process is the concept of a "healing period." The healing period ends when the medical problem has resolved or when there is no longer reasonable progress towards resolution. We know that many low back conditions may continue to improve slowly over many months or even years (Weber 1978a and b). However, the legal healing period should be ended when there is no reasonable progress toward resolution rather than when there is no progress at all.

Role of the Physician

Physicians have four general responsibilities in the evaluation process. In some cases these responsibilities are established by law, while in some situations the responsibilities are established by tradition.

The first responsibility under Workers' Compensation is to establish a causal relationship between the injury event and an existing impairment. (For cases under consideration for disability programs, such as SSI or private plans, there is ordinarily no need to establish a causal relationship between work and impairment, whereas in statutory or litigation cases this is central.) This causal relationship is easily shown when the impairment is the result of a single obvious event (e.g., a fall from a height). In contrast, the task may be quite difficult when the impairment is as a result of a repetitive event (e.g., repetitive twisting). For some purposes, compensability applies to a work-aggravated condition as well as a work-caused condition, thereby increasing the difficulty in establishing a causal relationship. This relationship between a work event and impairment is established by a patient's history and physical findings and other studies if appropriate. A relationship ordinarily exists if the findings (impairment) could have been caused by the injury event with reasonable certainty. It is not the physician's role to question the patient's perception of the injury event but rather to record that perception and to assess the reasonableness of that perception with regard to the impairment. The employer or insurance carrier may challenge the patient's story for a variety of reasons.

The second responsibility of the physician is to determine the end of a healing period. It is wise to give any injured person an estimate of a healing period at the outset. In the case of a work-injured patient the physician should also inform the patient whether there will be only temporary impairment or some permanent impairment and to give the person an estimate of the length of each. The patient should understand that the healing period will end when there is no longer reasonable progress toward resolution of the problem. Ideally, the healing period should end when a maximum anatomic or physiologic recovery occurs. However, we know that injured tissues continue to remodel to some degree over a period of one to two years, and we know that in some low back conditions (e.g., herniated nucleus pulposus (Weber 1978a and b), the patient may continue to improve over two to five years. There is a very low chance that a patient will return to work if he is "disabled" for as long as two years, and the fact is that the greatest amount of recovery occurs within six months of an injury. For these reasons, it is better to make an early permanent impairment rating and get the patient back to some gainful employment, even with temporary restrictions. This aggressive approach may reduce the devastating psychosocial effects of a long time out of work, preserve greater self-esteem of the injured person, and save society money. The doctor who unnecessarily prolongs the healing period (and thus total temporary disability) is often doing the patient a disservice. This must be balanced against the possibility that too early a return to work may lead to reinjury and can increase impairment. In these circumstances, frustration or anger with the doctor may compound and confuse the picture. Therefore, the doctor must seek to terminate the healing period at some optimal time that is neither too early nor too late. Since they are advocates of the patient, most doctors tend to err on the side of a prolonged healing period.

The third responsibility of the physician is to establish whether permanent impairment exists at the end of the healing period and, if so, to rate that impairment. A percentage rating, based on the whole man concept, is the most common way of rating impairment, although there is no uniform scheme for all purposes. In general, impairment is related to the injury and not to response to treatment or complications of treatment. However, there is considerable variation in rating impairment since legal guidelines have not been established. The impairment rating process is often confounded by the presence of some pre-existing condition. Although difficult, it may be important to attempt to distinguish that portion of impairment caused by the pre-existing condition from that portion caused by the injury event. This distinction is required in legal cases whereas in Workers' Compensation cases, both work-caused and work-aggravated conditions are ordinarily compensable.

In most Workers' Compensation programs, either the injured worker, employer, or carrier has the legal right to challenge an impairment rating and to insist on a second or third determination. The physician should not feel insulted when his rating is challenged. In fact, the doctor is wise to encourage the patient to seek a second evaluation, particularly when the doctor is unsure of the rating.

The final responsibility of the physician is to estimate work capacity or restrictions. Often this is done with only a general knowledge of the person's usual work. The physician will then outline any work restrictions and should state whether those restrictions are temporary or permanent. Such restrictions obviously are based not only upon the nature and extent of the impairment, but also on the nature of the person's work, even though the impairment rating itself is not based on the person's work. In the case of temporary restrictions, the physician generally estimates the length of time the restrictions will be necessary.

CURRENT LOW BACK IMPAIRMENT RATING PRACTICES

Impairment rating methods exist principally because of statutory programs for injured and disabled persons. Initial rating schedules are simple and often based on accident insurance statistics (McBride 1936). However, the distinctions between impairment and disability as well as the whole man concept evolved early. The changes in rating schemes that have occurred over the past fifty years have been evolutionary modifications rather than conceptual changes.

Physicians currently use a wide variety of methods to rate impairment (McBride 1936; Kessler 1970; American Academy of Orthopaedic Surgeons 1966; American Medical Association 1971; U.S. Bureau of Disability Insurance 1970). No method is universally satisfactory to all people or for all purposes, even in the evaluation of persons with similar disabilities (e.g., low back pain). Each rating is, at best, an estimate of impairment. Figures 15.1, 15.2 and Tables 15.1 and 15.2 give the most commonly used methods and the AMA guides to the Evaluation of Permanent Impairment. It should be noted that a second separate impairment rating may be required for associated neurologic dysfunctions in the lower extremities. In these

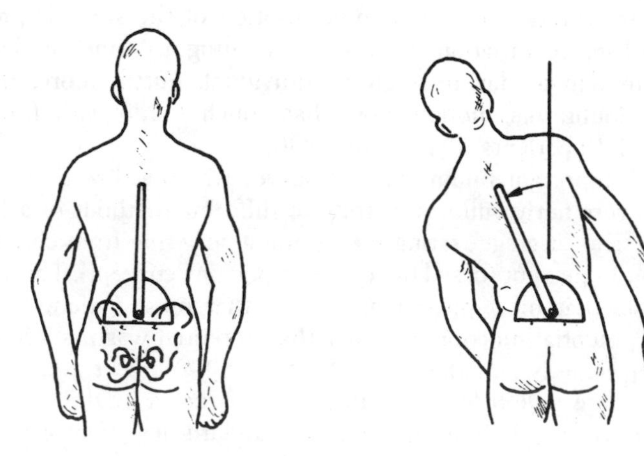

Figure 15.1
Measurement of impairment in lateral flexion.

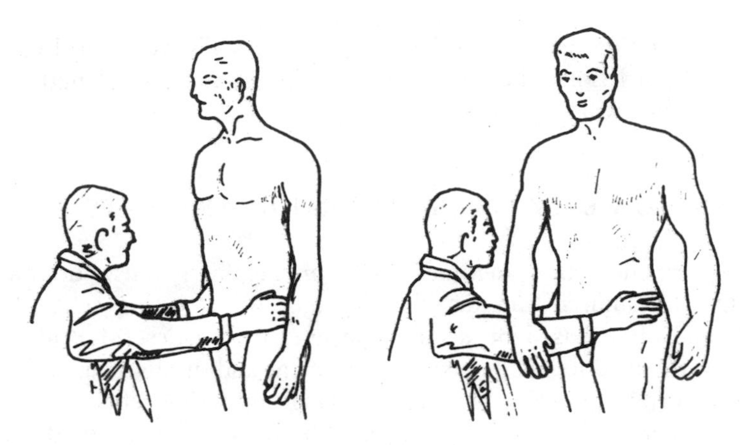

Figure 15.2
Measurement of impairment in trunk rotation.

instances, the "whole person" impairment is the sum of the low back and the neurologic impairments.

In a survey conducted to study rating practices, it was found that 96 percent of responding orthopaedists used multiple criteria including subjective factors (Brand and Lehmann 1983). Although the rating of impairment solely on objective criteria is conceptually attractive and a desirable goal, at the present time it would be unfair to patients to ignore subjective criteria for the following reasons: 1) little is known about the specific causes of low back pain and any selected criteria may not be related to the cause or effect of the LBP; (2) currently used so-called "objective" criteria are subject to considerable interpretation. For example, motion of the spine is prone to certain interobserver variation, variation according to time of day, and variation from day to day in a given individual. Furthermore, there is considerable doubt that motion alone has much to do with functional limitations in LBP patients (Pope *et al.* 1980).

A second important finding of that survey was that low back impairment ratings were fairly uniform in spite of differing methods of achieving those ratings. The "average" rating for a patient suffering from chronic low back pain was 13 percent (based on a "whole person" concept). This average rating suggests that most physicians believe that most patients still had considerable potential function, whether that potential was realized or not. (This rating average is remarkably similar to the 14 percent average rating reported by White [1969] for "disability" awards given by the Workmen's Compensation Board of Ontario.) The low standard deviation of 8 percent suggests reasonable uniformity of ratings (at least within one state). The range of ratings was from 0 to 50 percent. Even the most severely disabled did not exceed 50 percent of "whole person" function.

TABLE 15.1
Impairment in Dorsolumbar Region as Measured by Lateral Flexion

Restricted Motion
Average range of LATERAL FLEXION (lateral bending) = 40 degrees
Value to total range of dorsolumbar motion = 25%

	Degrees of Dorsolumbar Motion		*Impairment of Whole Man*
	Lost	*Retained*	
Right lateral flexion from neutral position (0°) to:			
0°	20	0	4%
10°	10	10	2
20°	0	20	0
Left lateral flexion from neutral position (0°) to:			
0°	20	0	4%
10°	10	10	2
20°	0	20	0
	Ankylosis		
Region anyklosed at:			
*0° (neutral position)			30%
10°			45
20° (full right lat. flex.)			60
Region ankylosed at:			
*0° (neutral position)			30%
10°			45
20° (full left lat. flex.)			60

*position of function
Source: AMA guide to the Evaluation of Permanent Impairment.

AN APPROACH TO RATING LOW BACK IMPAIRMENT

We have developed one approach to rating low back impairment that is based on the following facts. The likelihood of return to work decreases with increasing time after the injury. Doctors and patients often misunderstand the disability process and this often leads to increased disability (Lehmann and Brand 1982). The outcome of a given case is difficult to predict. It often seems convenient to defer decisions hoping that the problem will resolve or

Table 15.2:
Impairment in Dorsolumbar Region as Measured by Trunk Rotation

Restricted Motion
Average range of ROTATION = 60 degrees
Value to total range of motion = 35%

	Degrees of Dorsolumbar Motion		
	Lost	*Retained*	*Impairment of Whole Man*
Right rotation from neutral position (0°) to:			
0°	30	0	5%
10°	20	10	4
20°	10	20	2
30°	0	30	0
Left rotation from neutral position (0°) to:			
0°	30	0	5%
10°	20	10	4
20°	10	20	2
30°	0	30	0
	Ankylosis		
Region ankylosed at:			
*0° (neutral position)			30%
10°			40
20°			50
30° (full right rotation)			60
Region ankylosed at:			
*0° (neutral position)			30%
10°			40
20°			50
30° (full left rotation)			60

*position of function
Source: AMA guide to the Evaluation of Permanent Impairment.

that the clinical picture will somehow change so as to make a decision easier. Furthermore, it is difficult for the doctor to tell a patient, and even more difficult for a patient to accept, that he has only 10 percent impairment after he has been out of work for two years. In the mind of the patient, he is 100 percent disabled.

The approach to be described has four critical elements: 1) early medical decisions; 2) avoidance of delay in administrative decisions; 3) frank, open discussions with the patient about his medical conditions and the disability process; and 4) insuring that the patient understands his role and the doctor's role in the disability process.

Most industrially injured LBP patients have recovered and returned to work by six to twelve weeks. In almost all cases in which recovery had not occurred in six to twelve weeks, a definitive course of treatment should be outlined and an initial impairment rating estimate should be given to the patient. In establishing a course of treatment, a recommendation may be made to follow a nonoperative rehabilitation course of treatment which accepts some permanent partial impairment. The alternative may be a surgical intervention that may have the same result. The "healing period" concept is explained to the patient along with the fact that temporary benefits will terminate at the end of the healing period. The patient must understand that when he is no longer getting better, the healing period will end and this may provide some motivation for continued improvement. It is helpful to explain simply the whole man concept, particularly in relationship to the initial impairment rating estimate. This lets the patient know that his impairment rating will be a small portion of the "whole person" function. It also suggests that he will not get a large settlement. The patient should be referred to another specialist at the six to twelve week time if the doctor is unable or unwilling to make this decision, if the doctor is not familiar with the impairment rating and the disability process, or if the patient (or employer) wishes a second opinion.

This relatively rapid sequence has the effect of requiring the patient to take a more active role in the rehabilitation process, since prolonged temporary benefits may be a disincentive to recovery. Another important outcome of this effort is that the patient can begin early planning for future work or vocational rehabilitation.

SUMMARY

Evaluation is the process of determining the severity of the disability and is required by statutory programs and by litigation. Evaluations are imprecise and may be affected by the attitudes of the examiner. Statutory programs distinguish between impairment and disability. Impairment ratings are based on the impairment of the whole person. Even if different methods are used, the ratings are quite uniform. The average whole person rating for a chronic LBP patient is 13 percent. The physician's role in the evaluation process is to evaluate the relationship between injury and impairment, the end of the healing period, the existence of permanent impairment, and restrictions on working capacity. Evaluation should be rapid, avoid administrative delays, and include open discussions with the patient.

REFERENCES

American Academy of Orthopaedic Surgeons. 1966. *Manual for orthopaedic surgeons in evaluating permanent impairment*. Chicago.

The American Medical Association Committee on Rating of Mental and Physical Impairment. 1971. *Guides to the evaluation of permanent impairment*. American Medical Association, Chicago.

Brand, R.A., and Lehmann, T.R. 1983. Low back impairment rating practices of orthopaedic surgeons. *Spine* 8 (1):75–78.

Kessler, Henry H. 1970. *Disability-determination and evaluation*. Philadelphia: Lea and Febiger.

Lehmann, Thomas R., and Brand, Richard A. 1982. Disability in the patient with low back pain. *Orthop. Clin. North Am.* 13 (3):559–68.

Lehmann, Thomas R.; Brand, R.A.; and O'Gorman, T.W. 1982. A low back rating scale. Unpublished manuscript.

McBride, Earl D. 1936. *Disability evaluation and principles of treatment of compensable injuries*. 6th ed., Philadelphia: J.B. Lippincott.

Pope, M.H.; Rosen, J.C.; Wilder, D.G.; and Frymoyer, J.W. 1980. The relation between biomechanical and psychological factors in patients with low-back pain. *Spine* 5 (2):173–78.

U. S. Bureau of Disability Insurance. 1970. *Disability evaluation under social security. A handbook for physicians*. U.S. Government Printing Office, Washington.

Weber, Henrik 1978a. Lumbar disc herniation. A prospective study of prognostic factors including a controlled trial. Part I. *J. Oslo City Hosp.* 28 (3-4):33–61.

Weber, Henrik 1978b. Lumbar disc herniation. A prospective study of prognostic factors including a controlled trial. Part II. *J. Oslo City Hosp.* 28 (7-8):89–113.

White, A.W.M. 1969. Low back pain in men receiving Workmen's Compensation: A follow-up study. *Canad. Med. Assoc. J.* 101 (2):61–67.

16 Workman's Compensation

John D. Kemp and Richard M. Keane

HISTORY

Workers' Compensation represents social legislation that creates a compromise between the right of the injured worker to sue his employer and the right of the employer to use common law defenses against the employee. Each State and Federal law is written with the intent that it shall be construed liberally in favor of the injured worker.

Before the present system was introduced, workers who suffered injuries on the job had to sue their employers to recover their lost wages and medical expenses. Employers used the common law defenses of Contributory Negligence, Assumption of Risk, and the Fellow Servant Rule (the negligence of an employee is not passed to the employer) (Larson 1979).

The consequences of the old system were unsatisfactory in the burgeoning economy of the United States in the late nineteenth century. Employees were required to prove the negligence of their employers, often an expensive and time consuming legal battle. Often the worker could not afford these expenses, and if the employee won the case, the amount recovered was usually small. From that award, attorney's fees were deducted and medical bills had to be paid. It was also common that an injured worker who won the legal battle lost his job. Those who lost their legal battles, often became burdens on their families and society. Employers were also unhappy and considered the system disruptive and expensive. Although it provided them with strong defenses against many of the claims, it left them guessing about their true liability.

The movement toward industrialization brought ever greater pressure on society to change the system. In 1884, Germany passed the Sickness and Accident Law, the modern foundation for Workers' Compensation laws. A number of other European countries followed suit before the United States passed its first attempt at a Workers' Compensation law. Unlike other nations, the United States adopted laws state by state rather than on a national basis.

A number of Workers' Compensation laws were tested in the U.S. courts for constitutionality. The issues litigated concerned the loss of the rights of the parties to a liability suit. Both the plaintiffs and the defendants compromised their legal rights in exchange for a quick, reliable, and, for the most part, fair solution for the injured worker.

In the early part of the century many occupations were excluded from coverage in United States (Millus and Gentile 1976). The types of injury and disease that were covered were narrowly defined. Injuries had to arise out of, and occur in the course of employment (Horowitz, 1944). Injuries without a specific accident were not covered. A man who injured his back as a result of continuous trauma would not be compensated. Many diseases were arbitrarily excluded.

The report of the National Commission on State Workmen's Compensation Laws (Mitchell 1972) recommended that states should have both medical and claim benefits unlimited as to "duration and sum."

WORKERS' COMPENSATION BENEFITS

Claim

A common problem for injured workers is their relative lack of knowledge about Workers' Compensation benefits. Since 1960 unions have been making a concerted effort to make members aware of their rights. Attorneys began representing large numbers of injured workers and many court decisions have liberalized the application of the law. Although many states require employers to help injured workers understand their benefits, most workers look to their physician, their friends, or to an attorney to help them determine "what they should get" (Tebb, 1974). Currently, the increase in the number of lawyers, state-assisted legal aid programs, and increased worker sophistication may make legal help more available.

Temporary Total (TT) Benefits

A worker who sustains a compensable injury or disease is entitled to compensation for wages lost as a result of the injury or disease. A waiting period of three to six days is usually required, but the benefit is then payable from the first day, in most states. The TT benefit is equal to two-thirds of the injured worker's salary up to a maximum set by law. Some states tie the maximum to the state's average weekly wage while others set the amount by statute. In 1982, maximum temporary benefits ranged from $112 per week in Mississippi to $940 per week in Alaska (U.S. Chamber of Commerce, 1982). Since 1975, many states have been changing their maximum benefit annually to keep pace with inflation.

Temporary total benefits are payable for as long as a person is totally disabled and for as long as his condition has not been declared permanent. The nature of back injuries makes it difficult to determine when a condition has become permanent. As a consequence both employers and insurance companies are anxious to have doctors report any change in status as soon as that information is available.

Permanent Partial (PP) Benefits

When an injured worker, in the opinion of the attending physician, will not improve further, but a partial disability remains, he may be entitled to a Permanent Partial benefit. These benefits have been a source of controversy since their inception.

Permanent partial benefits are usually classified as scheduled or nonscheduled. Scheduled benefits require a physician to rate the permanency of an injury in terms of percentage loss of function of that body part. The doctor's estimate is applied to the state schedules and an award is calculated. Despite a wide variety of attempts to simplify the administration of this benefit, tremendous variations exist in benefits paid for very similar injuries (Larson, 1982). Rating the loss of function "to the body as a whole" is a difficult if not impossible task, yet one routinely demanded of doctors treating back-injured workers (Mitchell, 1973). This problem is discussed in detail from a medical perspective in the preceding chapter.

Nonscheduled permanent partial benefits attempt to compensate an injured worker for an earnings impairment as opposed to a physical impairment. These benefits may be based on actual lost wages, on an estimate of lost earning capacity, or on the degree of disability (Larson, 1982).

Florida has made an attempt to defuse the PP problem by using the concept of wage loss. A number of states, including Louisiana and Pennsylvania, have a loss of earning capacity provision. This provision is more closely tied to loss of function than actual loss of earning capacity. Florida also requires that rehabilitation services be provided for disabled workers who are unable to do their usual and customary work. Following rehabilitation services, benefits are paid as necessary to make up for the difference between the average wages earned prior to the injury and the average earned after rehabilitation. Although this benefit is touted as being infinitely more fair to the injured worker and the employer, catastrophic injuries may pose special problems. For example, a paraplegic who is rehabilitated and goes to work as a computer programmer may earn the same amount of money he did prior to the injury. The insured workers would be entitled to no further wage loss benefits, despite the fact that he may have to hire someone to paint the house or mow the lawn, both tasks well within the disabled person's capabilities prior to the injury.

Temporary Partial (TP) Benefits

The concept of wage loss is not new. Most states require that compensation be paid to injured employees who return to work at wages less than pre-injury wages. This benefit provides an incentive for LBP patients to attempt part-time or slower-paced work. Unfortunately, this incentive is frequently overlooked. Doctors who are anxious to have their disabled, but clinically improving, patients return to some type of activity should consider this benefit. In most states, injured workers who return to a job with lower wages are entitled to two-thirds of the difference between preinjury and postinjury wages up to a maximum normally equal to their temporary total rate. This nontaxable benefit is usually paid when proof of earnings is submitted.

There are a number of ways to employ this benefit in helping low back injured patients return to work. A worker may be released to return to work the normal number of hours but with frequent or prolonged rest breaks; another option is to have a shorter work week to allow the worker to recover. Yet another way is to return the injured worker to a lighter or modified job. The worker may earn less than he did before the injury, but modified work may strengthen the muscles and confidence while reducing his financial burden. Doctors who consider helping their patients back to light or modified work are encouraged to seek information about such jobs directly from the employer or from the insurance company.

A detailed job description should provide the physician with sufficient information to make a judgment about the appropriateness of the work. Recommendations for changes in the job should be made when needed to allow injured workers to return to the safest possible job. The job should be one that allows the patient to return to work as quickly as possible.

Permanent Total Disability (PTD) Benefits

Workers whose injuries permanently preclude their return to employment are entitled to PTD benefits. In some states, the injury must rule out a return to customary employment. In other states, it must preclude a return to any job for which the injured worker is qualified. Finally, in other states, a worker may be entitled to PTD benefits if rehabilitation services are unsuccessful in returning a disabled worker to work.

Many disability cases could result in successful outcomes if the doctor, employee, employer and insurance carrier worked together. Too many back injury cases are ignored until the injured worker's chances of returning to work are nonexistent. A worker who believes he is permanently disabled usually will not return to work! The point is stressed in Chapter 7.

PTD cases are serious social and economic problems in the United States. Injured workers receive a periodic payment that is equal to a percentage of the preinjury wages. In some states this payment will include a cost of living increase. Many who have been declared permanently totally disabled are also eligible for Social Security Disability Income. With the two benefits together, it has been assumed that such a worker is financially secure. Although this may be true, many of these "secure" people suffer from the emotional trauma that follows when a person is labeled permanently and totally "useless" in our working community. It is a challenge both for the Workers' Compensation system and injured workers. Employers, doctors, lawyers, insurance companies and legislators have failed to come to grips with this problem.

VOCATIONAL REHABILITATION

Vocation rehabilitation merits serious consideration by the injured worker, the employer, treating physician and insurance carrier. This benefit is not limited to long term training programs. It may involve counseling and assistance in developing job seeking skills which will help the worker to help himself to find work. It may involve a counselor working with an injured worker to define interesting job opportunities within his capabilities. On-the-job training may become the first step in a new career. These vocational rehabilitation services may be tailored to the very special set of circumstances of the injured worker.

Most people identify with the work ethic (Nemiah, 1963). Injured workers who receive compensation benefits are stigmatized, at least in their own minds, because they are not part of the work force or therefore, the dominant segment of the community. Sometimes the failure to consider work as therapy results in the need for psychological and psychiatric treatment (Ruben, 1978).

Even in states where vocational rehabilitation benefits are not widely publicized, most insurance companies are anxious to provide rehabilitation services for an injured worker who is unable to return to his usual employment.

MEDICAL BENEFITS

All Workers' Compensation laws provide for medical care for compensable injuries. Many states, however, qualify which medical expenses must be paid with phrases such as "reasonable and necessary," "to cure and relieve," "tend to lessen the period of disability," and "usual and customary." Medical professionals dealing with Workers' Compensation cases for the first time may find the law arbitrary.

Generally, bills for standard medical and surgical treatments will be paid at a rate that approximates the average within that community. New experimental techniques will be challenged as will fees substantially above the norm. In some states, for example, the employer or insurer may challenge a repeat surgery on the back as excessive. In most states, a hearing may be held with the Workers' Compensation administrator to decide the appropriateness of the additional surgery. Although relatively few such cases occur, it is important to understand the potential that exists in some jurisdictions for medical services to be challenged by the insurer.

Since the first Workers' Compensation laws, the list of approved medical services and providers has grown substantially. Medical doctors, podiatrists, chiropractors, physical therapists, psychotherapists, nurses and many more specialties are reimbursed directly for services provided to injured workers. Arkansas, Connecticut and Delaware are among a short list of states that recognize treatment by prayer or spiritual means!

A small number of states allow the employer or insurance carrier to choose the doctor who will provide services. In these states, an employee who chooses a doctor other than the one designated may be required to pay for the services (except in emergency situations). Georgia is an example of a state that provides the employer or insurance carrier with the choice of medical professionals.

New York is an example of the opposite end of the spectrum. It is a misdemeanor for an employer or insurer to interfere with an injured worker's right to free choice of medical doctors. Most state laws fall somewhere in between these two extremes.

In every state the insurance carriers maintain the right to obtain an independent medical examination. These examinations have become a routine part of Workers' Compensation claims for low back pain. Attending physicians who are frustrated by chronic low back conditions often welcome an additional opinion.

Unfortunately, many independent medical examinations may appear to be biased rather than unbiased (Horowitz, 1944). It was reportedly common in Florida, for example, for a Workers' Compensation supervisor to find his "hired gun" to offset the opinion of a very liberal physician concerning the extent of disability. Doctors were categorized and used for claims litigation by both sides. Ultimately, a judge would weigh both opinions and make a decision as to the percentage of disability that would be used to calculate a permanent partial benefit award. Florida's law was substantially revised in 1979 to end the battle zone mentality of Workers' Compensation claims administration. However, a number of states continue to operate in a litigious manner.

Generally, acute care for injured workers is covered by Workers' Compensation Insurance without question if the claim is compensable under the law. That is, unless there is a question about whether the injury occurred within the jurisdiction of the Workers' Compensation law, acute care coverage is virtually guaranteed. Services such as ambulance transportation, emergency room care, first aid, doctors' office visits, and medication will be paid without question for compensable injuries.

Coverage for chronic care is less straightforward. The patient who has undergone various conservative medical treatments and finally surgical intervention may find that the employer or insurer is beginning to scrutinize medical reports and bills. Doctors may be asked to be more specific about a prognosis or about the necessity of certain medications or tests. The chronicity of the patient's condition becomes the focal point of the frustrations of both the insurers and the doctors (Enelow, 1967).

Medical rehabilitation is generally payable if it is provided or prescribed by a medical doctor. Rehabilitation hospitals may be regarded as an appropriate step after acute care. Out-patient physical and occupational therapy for a back injured worker may also be strongly encouraged, at the appropriate time, following back surgery. Memberships at YMCAs have been paid to allow a back-injured person an opportunity to use swimming as a physical therapy.

In rare cases, a low back injury case is so severe that it involves a variety of overwhelming symptoms. In these relatively rare cases, attendant care may be necessary for a few hours each day.

Medical supplies and equipment such as braces and Transcutaneous Nerve Stimulators (TNS) prescribed by a medical doctor to provide for a compensable injury will usually be paid for under Workers' Compensation policies. Similarly, medications necessary for the treatment of injuries are covered under Workers' Compensation law. While this provision may seem rather obvious, there is a considerable gray area. Take for example, a low back injury that requires long-term use of a certain medication. The medication may cause side effects which require additional treatment. It is relatively common for patients with long-term chronic back problems to experience reactive depression. Antidepressive drugs are covered by Workers' Compensation if the depression is linked to the back injury.

Psychiatric and psychological care associated with LBP are covered under almost every Workers' Compensation law. The law requires that the employer or insurer must accept an injured worker "as you find him." If pre-existing emotional problems are exacerbated by a compensable Workers' Compensation injury, then treatment will be covered (Wales, 1977).

A final medical expense is reimbursement for travel care. Each state has its own set of rules regarding what travel to receive medical care will be paid and at what rate. Generally, reasonable travel and other expenses for covered medical service will be reimbursed to the injured worker. A low back injured worker referred to an outpatient clinic some distance from his home may receive an advance for mileage, meals and room expenses, as may be appropriate.

INJURY TYPES

A wide variety of causes for LBP have been discussed in previous chapters of the book. A single trauma, cumulative trauma, diseases and psychological strains may all relate to LBP. At one time, a compensable back injury was one related to a specific "accident." Over the years, the scope of compensable injuries has grown considerably. Today, workers with LBP primarily caused by aging or disease are receiving Workers' Compensation benefits because their condition was "made worse" by their work (Burton, 1981).

THE CLAIMS PROCESS

The injured worker must make a report to his employer to be compensated. At one time the injuries that were not reported within a specific time period following the accident were not compensable. If it can be shown that the employee's failure to notify the employer prejudiced the employer's case or if the employee intended to mislead the employer, the claim may be barred (Wales, 1977). The courts are more lenient today, but it is always advisable to report all injuries as soon as possible.

The second step in this process is the employer's report to the insurance company. Insurance company is used here to refer to both companies that provide insurance coverages and companies that only provide claim services. The difference lies in the answer to the questions, "Who actually pays the claim?" If insurance is provided, the insurance company pays the claim. If only service is provided, the employer is self-insured and will pay all claims.

The 1973 Report of the National Commission on State Workers' Compensation Laws provided the impetus for many states to impose penalties for errors in claim processing caused by late reporting (Wales,

1977). Employers who fail to file reports quickly may be fined in some states if the insurance carrier had no notice of the injury.

Rehabilitation services should be provided as needed to accomplish the highest priority outcome (Mattingly, 1983). First, workers should be helped to return to the same job with the same employer. Often, a misunderstanding of the requirements of the job results in a failure to consider this option. A poorly written or loosely interpreted job description may cause the attending physician to rule out the "heavy lifting," "excessive bending or stooping," etc.

The second priority is to return the employee to the same employer in a similar job or perhaps a modification of the same job. On-the-job training with the same employer should be considered at this stage of rehabilitation. An employee who knows the physical plant and the people will generally experience a less stressful return to work. An employer who takes back an injured worker avoids the costly period of indoctrination that inevitably takes place with all new employees.

The third priority is to search for a similar job with a new employer. An injured worker who can use transferable skills to return to work will experience less turmoil than the worker requiring formal training. This alternative may require a slight job modification, but such expenses can be paid as a vocational rehabilitation expense, particularly if the benefit-to-cost ratio is considered.

The fourth priority is to utilize on-the-job training with a new employer. If an injured worker has specialized skills and experiences, it is easier to understand why a new company may be anxious to employ him/her. It is ironic that a worker who may be considered marginal or worse by one employer may be very highly rated by a new employer.

The fifth priority involves extensive evaluation and consideration of available jobs requiring formal training. Such training may be accomplished at a technical school, a college, or special skill-training programs set up by a government agency. In addition to the other difficulties, an injured worker will frequently complete a formal training program only to find that jobs are not readily available for that particular skill. Rehabilitation plans that involve formal training should be geared to provide the injured worker with skills that make him a good candidate for jobs that will have openings.

DOCUMENTING CONTINUING DISABILITY

A claim supervisor or adjuster must periodically document the file to substantiate continued payments. In several states, a worker who is no longer disabled but refuses to return to work may have a right to a hearing before compensation benefits may be discontinued. A hearing may take several months to schedule.

At each review all involved should optimize return to work with the hierarchy of structures given above.

LITIGATION

Workers' Compensation was developed as a nonadversarial social program. An injured worker was to report an injury to his employer and the employer would report it to an insurance company for prompt payment. In fact, the system does work well for most injured workers and appears to be equitable for most injuries.

Litigation of Workers' Compensation claims has increased sharply over the past 15 years. A study conducted by the California Workers' Compensation Institute was in response to a 20% increase from 1973 to 1974 (Tebb, 1974). This study showed that the most critical element in reducing litigation was information. Most injured workers went to an attorney because they felt they did not properly understand their state's Workers' Compensation system. The increase in complexity of Workers' Compensation laws has left many people, including doctors, confused about what benefits an injured worker should receive. In an effort to protect themselves, injured workers turn to lawyers. What some injured workers do not realize, however, is that the attorney's fees represent a significant percentage of the settlement.

Workers' Compensation issues are usually initially litigated in an administrative forum with appellate procedures into the state's judicial system. A hearing officer or commissioner may decide cases in an informal manner. The appellate procedure may on occasion travel to the state's highest Court. Considerable "new law" is made annually from these judicial decisions, which complicates this issue.

CLOSING FILES

There is considerable pressure to close Workers' Compensation cases as soon as possible. Large employers are liable for additional Workers' Compensation premiums based on estimates of claim and medical benefits to be paid on a case. In addition, employers and insurance companies have come to realize that the longer an injured worker is paid compensation benefits, the more difficult it will be for that person ever to return to work.

The usual closing of a file involves the injured worker recovering and returning to his job. In some cases, he may be entitled to permanent partial benefits for loss of function. If the injured worker is found to be able to return to work but refuses, the file may be closed because no further disability exists (Millus, 1976). Cases may also be closed because an injured worker dies of injuries or disease unrelated to the Workers' Compensation case.

Other files are closed as a result of settlements. Cases may be settled by a compromise in questions of compensability or to provide the injured worker with capital to begin self-employment. Most files settled in this manner terminate the insurance company's liability for claims or medical benefits.

A number of states bar settlements of some or all types of benefits. California, for example, does not recognize settlement of the vocational rehabilitation benefits. In some cases where settlements were arranged, injured workers subsequently were awarded rehabilitation benefits that included a weekly check equal to temporary total benefits. This way an inequity arose which allowed the injured worker to collect benefits twice.

SUMMARY

A problem of worker's compensation is that the injured worker tends to be unaware of the complexities of the system. A worker who sustains a compensable low back injury can receive temporary total benefits. This benefit is equal to two-thirds of the injured worker's salary up to a maximum defined by law. If a partial disability remains and there is no likelihood of further improvement, the patient may be entitled to a permanent partial benefit. Some jurisdictions use the concept of lost wages or lost earning capacity. Temporary partial benefits are designed to provide an incentive for chronic LBP patients to return to lighter work. Permanent total disability benefits are available for workers with low back injuries that permanently preclude a return to work. Medical rehabilitation and vocational rehabilitation benefits are usually paid at the average rate in the community. Acute care is usually covered without dispute. Continued disability must be documented. There is considerable pressure by the carriers to close worker's compensation cases as soon as possible.

REFERENCES

Burton, C.V. 1981. Conservative management of low back pain. *Low Back Pain* 168–83.

Enelow, A.J. 1967. The compensable injury. *Cal. Med.* 106 (3):179–82.

Horowitz, S.B. 1944. *Injury and death under Workmen's Compensation laws*. Boston: Wright & Potter Printing Co.

Larson, A. 1979. *Workmen's Compensation for occupational injuries and death.* New York: Matthew Bender.

Larson, A. 1982. *The law of Workmen's Compensation.* New York: Matthew Bender.

Mattingly, L. 1983. Role of the private rehabilitation supplier. Presented at the Institute of Continuing Legal Education Company. Atlanta, Ga., April 8, 15, and 29.

Millus, A.J.; and Gentile, W.J. 1976. *Workers' Compensation law and insurance.* New York: The Roberts Publishing Corp.

Mitchell, F.L. 1973. Some medical issues in Workmen's Compensation. *Supplemental studies for the National Commission on State Workmen's Compensation Laws.* Vol. II, Govt. Printing Office, Washington, D.C. pp. 354–62.

Nemiah, J.C. 1963. Psychological complications in industrial injuries. *Arch. Environ. Health* 7(10):481–86.

Rubin, S.E.; and Roessler, R.T. 1978. Guidelines for successful vocational rehabilitation of the psychiatrically disabled. *Rehabilitation Literature* 39 (3):70–74.

Tebb, A. 1974. Litigation in Workers' Compensation. Presented at the State Workmen's Compensation Advisory Committee. Oct. 22, San Diego, Ca.

U.S. Chamber of Commerce. 1982. Analysis of Workers' Compensation laws. Washington. D.C.

Wales, J.U.; and Ideson, H.A. 1977. LLB Claims law Worker's Compensation. Basking Ridge, NJ, American Educational Institute.

17 Hiring Practices

John D. Kemp

INTRODUCTION

The handicapped employment discrimination regulations are intended to remove the discriminatory aspects of hiring disabled persons and, at the same time, provide employers with workers who are qualified to perform essential job requirements. These rights are founded in the Rehabilitation Act of 1973 (16a), and recently amended state laws.

Prior to Congressional passage of Sections 501, 503, and 504 of the Rehabilitation Act of 1973, and Section 502 of the Vietnam Veterans Readjustment Act of 1974 (16b), handicapped persons who experienced employment discrimination were left with no recourse. Statistical evidence indicates the disparity in employment opportunities for disabled and non-disabled workers, even during the past decade. For example, labor force participation by deaf people declined from 65.5 percent in 1972 to 61.3 percent in 1977, and their personal income, as a proportion of the national average, declined from 74.6 percent in 1971 to 64.2 percent in 1976.[1]

COVERED EMPLOYERS AND REGULATIONS

The 503 and 504 requirements for Affirmative Action and Equal Opportunity for qualified handicapped individuals apply to federal contractors and every recipient of federal financial assistance. A federal contractor is any company that holds a contract or subcontract to provide goods or services in excess of $2,500 to the federal government or its contractor. A recipient that operates a health, welfare or any social service program funded with federal assistance also must comply with these provisions. Thus, almost all employers meet one of these definitions and are affected by this legislation. In this chapter, we will use the terminology "employer," rather than the more precise and legal terminology of "recipient."

The regulations do not protect all handicapped applicants and employees from discrimination on the basis of handicap. In order to be protected, a handicapped person must be designated as a "qualified handicapped person" which, as defined by the law, means: A handicapped person who, with reasonable accommodation, can perform the essential functions of the job in question.[2]

In light of this definition, the key questions concerning whether a handicapped person is qualified to perform a particular job are:

1. What are the *essential* functions of the job (i.e., what basic qualifications are necessary to perform the job)?
   As stated in Section 503 and 504, as *such sections relate to employment*, "such term does not include any individual who is an alcoholic or drug abuser, whose current use of alcohol or drugs prevents such individual from performing the duties of the job in question, and whose employment, by reason of such current alcohol or drug abuse, would constitute a direct threat to property or the safety of others."
2. Are there *reasonable accommodations* available that would enable the handicapped person to perform *all* the *essential* functions?

To date, the meaning of the term "essential" has not been clarified. Because of the unlimited variety of circumstances under which an individual may be expected to perform a particular job, the determination must be made on a case-by-case basis. When a question arises concerning the "essential" nature of a job function, it has been determined that the *burden* is *on the employer* to demonstrate that the job function is "essential."[3]

Protected Employment-Related Activities

A contractor or recipient of federal financial assistance may not deny equal employment opportunities to qualified handicapped individuals. This requirement applies to *all* full-time and part-time employment situations and affects practices such as recruitment, hiring, promotion, job assignment, and sick leave.[4]

The *final* 504 regulations stated, "Section 84.11 simply bars discrimination in providing fringe benefits and does not address the issue of actuarial differences. The government believes that currently available data and experience do not demonstrate a basis for promulgating a regulation specifically allowing for differences in benefits or contributions."[5] An employee may not be asked to waive his rights to insurance or to bring an action against the employer for any injury he incurs regardless of a disability.

In addition to requiring that an employer make employment-related decisions free from discrimination, "an employer may not participate in contractual or other relationships that have the effect of subjecting qualified handicapped applicants or employees to discrimination on the basis of handicap."[6] This provision is based on the premise that a recipient or contractor may not do indirectly that which he cannot do directly.

For example, an employer has a contract arrangement with a private organization to provide a training program. The employer requires completion of the training program in order for the worker to be eligible for promotion. A LBP-impaired employee cannot take the program because the organization's training site requires a long drive which aggravates his condition. The employer cannot thereafter deny that employee a promotion based solely on his failure to complete the training program.

A contractor or employer cannot excuse noncompliance with the regulations because of the existence of a state law or policy that limits the eligibility of qualified handicapped persons to practice an occupation. Neither is an employer exempted because of the existence of a collective bargaining agreement that contains provisions that conflict with the requirements of the regulations.

Reasonable Accommodation

An employer must provide and pay for the accommodation in order to ensure equal employment opportunity for a qualified handicapped person. A covered employer, who is excused from providing the accommodation, may not discriminate against a handicapped applicant or employee who is able to make his own arrangements to provide the necessary accommodation to perform the essential functions of the job.

The regulations provide that a covered employer shall make reasonable accommodation to the known, but only the known, physical or mental limitations of an otherwise qualified handicapped job applicant or employee, unless he can demonstrate that the accommodation would impose an undue hardship on the operation of its program.[7] The employer is under no obligation to hire or retain the handicapped person, and is therefore not required to provide the reasonable accommodation. For example, an applicant with a low back condition applies for a particular job that requires a) several years' experience working with institutionalized persons, b) significant travel, and c) lifting persons or packages weighing in excess of 50 pounds. It would be reasonable to arrange an accommodation in order to permit other persons to do the lifting; however, the applicant does not have to have essential qualifications of the job. Even with an accommodation to meet one of the requirements, the applicant would still not be "otherwise qualified"; thus, the employer is not obligated to hire the person or provide the accommodation.

A covered employer must notify applicants and employees of its obligation under the regulations.[8] This includes the obligation to make a reasonable accommodation, unless such accommodation would impose an undue hardship. The burden then shifts to the handicapped person to inform the employer that he is handicapped and requires an accommodation. If the handicapped person does not explain the nature of his or her handicap and does not request an accommodation, there is no violation of the regulations if the accommodation is not provided.[9] For example if an employee has a low back condition and has difficulty carrying weights of 25 pounds (11 kg), the employer is under no obligation to provide accommodations if he has not been informed of this situation.

Once the qualified handicapped person informs an employer that he requires a reasonable accommodation, the employer may wish to solicit proposed accommodations from the handicapped person. However, he is not required to provide the reasonable accommodation proposed by the handicapped person if an alternative accommodation still meets the requirements of the regulation. Examples include: A low-back injured worker indicates that he needs a special type of chair costing $400. The employer locates another chair costing $100 that meets the needs of the handicapped worker.

The reasonable accommodation provision applies to all employment decisions, not simply hiring and promotion decisions. For example, an employer may be required to make a reasonable accommodation with respect to policies such as job assignments, transfers, sickness and other absence leaves.

The government has received several inquiries concerning the applicability of the reasonable accommodation provision. One was from a teacher who was losing his eyesight. He was employed by the school district in a high school that was not close to his home. A job for which he was qualified opened up at a high school that was within walking distance of his home. He was told that the school must make reasonable accommodations with respect to job transfers and reassignments.[10]

In a separate letter, the Office of Civil Rights which enforces Section 504 explained that the regulation does not require that a employer give preference during a reduction in the work force to a handicapped individual employee unless it can be shown that the employer had previously engaged in systematic discrimination against handicapped applicant and/or employees.[11]

Even when an employer has knowledge of a handicapping condition, it is important to ascertain from the handicapped person whether an accommodation is required. The Education Department's Inflationary Impact Statement, which accompanied the Department's "Notice of Intent" to publish a proposed Section 504 Regulation, explained that "for most combinations of types of handicapped conditions and job categories, *reasonable accommodation will require either no or only minor outlays.*"[12] A survey, conducted by the U.S. Civil Service Commission and the DuPont Company, as well as the experience of the Office of Federal Contract Compliance Programs in the Department of Labor in implementing Section 503, concluded "To gratuitously provide an accommodation may be both unnecessary and costly and, in some cases, discriminatory." (The circumstances under which the gratuitous provision of accomodations are considered discriminatory are discussed in the next sections of this chapter.)[13]

The "undue hardship" provision attempts to place a reasonable cap on the costs that an employer may incur in order to ensure equal opportunity for handicapped applicants and employees. The section-by-section analysis accompanying the 504 regulations indicates that standards for "undue hardship" are similar to the "business necessity" standard of the regulations implementing Section 503. The use of the term "business necessity" would not be appropriate to the majority of the programs and activities covered by the Section 504 regulations. The Education Department felt that the factors listed to determine "undue hardship" would make the "business necessity" standard applicable to and understandable by recipients of its funds.[14] Another plausible explanation is that the Secretary was concerned that the regulations strike a balance between the rights of the handicapped applicants and employees.

It must be noted that this is the first time that the issue of costs associated with providing the accommodation is raised. During the discussion of "reasonable accommodation," it is never raised as a factor in determining the reasonableness of an accommodation. Certainly, the variety of accommodations listed in the regulations indicates that cost is not a determinant of "reasonableness." Accommodations that are discussed include job restructuring (which usually requires no additional costs), the provision of readers and interpreters (which may require moderate costs), and making facilities readily accessible and usable (which, if structural modifications are required, may involve significant costs).

If an "undue hardship" can be shown, the employer is excused from making the reasonable accommodation.[15] In determining whether an accommodation would impose an undue hardship on the employer, factors to be considered include:

1. the size of the employer's program with respect to number of employees, number and type of facilities, and size of budget;

2. the type of the employer's operation, including the composition and structure of the employer's workforce; and
3. the nature and cost of the accommodation needed.[16]

Although cost is not a factor in determining "reasonableness," it is clearly one of the factors to be considered in determining "undue hardship." Even when cost is considered, it is not, in and of itself, the sole determinant of "undue hardship." The *size* and the *type* of the recipient's program or activity must also be considered.

If an employer can demonstrate that the provision of an accommodation would impose undue hardship, he is excused from providing the necessary accommodation without violating Section 504. However, the employer would be violating the regulations if "undue hardship" were used as an excuse to deny the handicapped individual the *opportunity* to be employed if the individual is able to provide his own accommodation. These arrangements might include the provision of the accommodation by public or private nonprofit agency or, even as a last resort, the provision of the accommodation by the handicapped individual. For example, assume that a small research organization receiving a grant from HEW advertises an opening for a position as a computer programmer and a LBP-impaired person applies for the job. The employer is able to demonstrate that it would impose an undue hardship on the operation of its program if it were to acquire special seating to enable the person to operate the computer. Assume that the LBP-handicapped individual expressed a willingness to pay for the necessary modification or purchase the special seat. Although the employer need *not* provide the accommodation, or in this case the seat, it would be a violation of Section 84.12 (d) if he were to deny a job to a qualified handicapped person who is willing to provide his own accommodation.

DEFENSES AND LIMITATIONS

The Bona Fide Occupational Qualification (BFOQ) Defense

The general rule invoking the "Bona Fide Occupational Qualification (BFOQ) Defense" is that an employer must look at the qualifications of *each* handicapped person to determine whether, with reasonable accommodation, that *individual* is a "qualified handicapped person." However, there is one limited situation. Congress expressly recognized that, under Title VII of the Civil Rights Act, there are certain limited circumstances where an employer would be permitted to base a selection decision on characteristics common to the entire class, rather than on the basis of individual qualifications. The concept is commonly referred to as the "BFOQ" provision.[17] Based on a 1975 memorandum, the General Counsel's Office of HEW concludes that there is a BFOQ concept implicit in Section 504.

> The language "otherwise qualified" implies that differentiating on the basis of handicap is *not per se violative of Section 504 if the handicap goes to the essence of the job*. In such a case, a Section 504 BFOQ, which requires the absence of a particular handicap, would have to be narrowly construed in order to be consistent with the statute. *Thus, it would be applicable only to those situations in which the employer could demonstrate that the BFOQ would not preclude any handicapped individual from employment who would "otherwise qualify"* (i.e., who, with reasonable accommodation would perform the essential functions of the job) *if individually tested.*[18]

Consistent with this conclusion, Sections 503 or 504 would not prohibit, for example, a bus company from adopting a rule barring all blind persons from applying for a job as a bus driver.[19] The Section BFOQ essentially presumes that there is no reasonable accommodation to overcome the limitations of this class of persons.

In sum, the BFOQ exception is only applicable to those situations in which the employer can demonstrate that the exception would not preclude *any* handicapped individual from employment who could otherwise qualify. Because the use of the BFOQ defense results in the exclusion of a class of handicapped persons from a particular job, it is extremely important to look at the reasonable accommodations available before determining that a BFOQ exists for that job. Even in situations in which accommodations are not currently available, rapid technological advances may alter the exclusions in the near future. Thus, BFOQs are likely to become less common.

EMPLOYMENT TESTS AND CRITERIA

The regulations contain a specific standard applicable to decisions concerning hiring and promotion. The provisions state that: "An employer may not make use of any employment test or other *selection* criteria that *screens out or tends to screen handicapped persons* or any class of handicapped persons unless: 1) the test score or other selection criterion, as used by the employer, is shown to be job-related for the position in question, or 2) *if there are alternative job-related tests or criteria* that do not screen out the individuals, the more discriminatory tests may not be used.[20] Based on this section, there is a clear method for determining whether a particular selection test or criterion is acceptable:

1. It must be shown that the test or criterion "screens out or tends to screen out handicapped persons." In *proposed* regulations, the employer is obligated to martial statistical evidence that show that employment criteria and qualifications are necessary. This requirement was changed in the *final* regulation because the small number of handicapped persons taking tests would make statistical showings of "disproportionate, adverse affect" difficult and burdensome.[21]
2. Once it is shown that an employment test substantially limited the opportunities of handicapped persons, the employer must show the test or criteria is *job-related.*
3. Once the employer has shown the test or employment criteria to be job-related, the burden shifts to the government to identify alternative job-related tests that do not screen out as many handicapped persons.

The government relies heavily upon the principle established under Title VII of the Civil Rights Act of 1964 in *Griggs v. Duke Power Company.*[22] The plaintiffs in *Griggs* challenged the defendant's employment criteria that an applicant have a high school diploma or attain a predetermined score on an intelligence test. The plaintiffs contended, among other things, that such criteria were not geared to measure the level of potential performance. The court concluded that since the company's testing and educational requirements had been shown to disqualify blacks at a substantially higher rate than whites, the burden shifted to the company to show that each requirement was demonstrably related to successful job performance.[23]

In addition to specifying the procedure for determining the legality of the selection criteria or tests used by the employer, we must take into account that some tests and criteria depend upon sensory, manual, or speaking skills that may not themselves be necessary to the job in question, but that may make it impossible for the handicapped person to pass the test. To overcome this problem, the statutes specify that the employer must select and administer tests so as best to ensure that the test will measure the handicapped person's ability to perform on the job, rather than the person's ability to see, hear, speak, or perform manual tasks, except, of course, when such skills are the factors that the test purports to measure.[24]

PREEMPLOYMENT INQUIRIES

Historically, employment application forms and employment interviews requested information concerning an applicant's physical or mental condition. This information was often used to exclude applicants with handicapping conditions, even before their ability to perform the job was determined.[25]

The regulations limit the circumstances under which employers may make preemployment inquiries, including medical examination. These regulations minimize the likelihood that employers will practice discrimination on the basis of handicap. The preemployment inquiry prohibition is used most often in limiting discrimination against persons with handicaps that are readily apparent, for example epilepsy, diabetes, emotional illness, heart disease, and cancer.

Preemployment inquiries are prohibited concerning the existence of a handicap and the nature and severity of the handicap. The term "preemployment" as used in this instance means the period prior to the point at which a *conditional offer of employment had been made*. For example, then HEW Secretary Califano, in a letter to the National Association of Chain Drug Stores concerning the applicability of Section 504 to the hiring of drug addicts, stated: "in order to ensure that addicts are not hired for jobs which require access to controlled substances, an employer *may routinely inquire, after offer of the job is made*, whether the individual is addicted."[26]

In other words, a "preemployment" inquiry is *only* prohibited prior to an offer of employment, not prior to the date of commencing employment. This offer of employment may be conditional on passing a medical examination or on determining that a "reasonable accommodation" will not cause an "undue hardship" to the operation of the employer's program.

Thus, the following questions traditionally appearing in application forms given to the applicant at the *initial stages of the interview process are unacceptable*:

1. "Are you handicapped? If so, to what extent?"
2. "Were you disabled in a war?"
3. "Have you ever had any of the following conditions . . . rupture, hernia, arthritis, low back pain?"[27]

Since the purpose of the provision is to minimize discrimination on the basis of handicap, rather than to require that covered employers hire unqualified persons, the regulations clearly state that an employer may make preemployment inquiry into an applicant's ability to perform job-related functions. In addition, it may make inquiries concerning a person's handicap if the person voluntarily includes reference to his handicap in a resumé.

There are three exceptions to the general rule prohibiting preemployment inquiries. Preemployment inquiries can be made:

1. in connection with remedial action obligations under Section 84.6(a) of the 504 regulations;
2. when an employer is taking *affirmative action* under Section 503 of the Rehabilitation Act of 1973; and

3. when an employer is taking *voluntary action* as outlined in Section 84.6(a) of the 504 regulations to overcome the effects of conditions limiting opportunities for handicapped persons.[28]

If an employer makes inquiries for any of the reasons described, it must also satisfy the following conditions. First, the employer-recipient must include a paragraph *preceding* any questions concerning the existence of a handicapping condition stating that the information is intended for use solely in connection with one of the three purposes listed above. In addition, the paragraph must state that a) the information is being requested on a *voluntary basis*; b) that it will be kept *confidential*; and c) refusal to provide it *will not subject the applicant to any adverse treatment.*

In addition, there are three types of inquiries that are acceptable after a conditional offer of employment has been made:

1. *Medical examinations.*

Section 84.14(c) of the 504 regulations expressly permits an employer to conduct a medical examination after a conditional offer of employment has been made. Nothing in this section shall prohibit an employer from conditioning an offer of employment on the results of a medical examination conducted prior to the employee's entrance on duty, *provided that*:

(i) all entering employees are subjected to such an examination regardless of handicap; and

(ii) the results of such an examination are used only to determine whether the applicant still satisfies the nondiscriminatory (job-related) employment criteria established by the employer.

2. *Inquiries related to the need for a reasonable accommodation.*

In addition to permitting a medical examination *after a conditional offer of employment*, an employer may also make inquiries concerning whether the person is handicapped to determine:

(i) whether the person required a reasonable accommodation; and

(ii) the nature and extent of the accommodation, if one is required.

Although the government has never expressly stated that such inquiries are permissible after a conditional offer of employment, it is logical to ascertain prior to employment the nature of reasonable accommodation required.[29]

3. *Inquiries related to job requiring direct access to controlled substances.*

The BFOQ allows all employers to refuse to hire drug addicts for jobs requiring access to controlled substances:

> In order to ensure that addicts are not hired for such jobs, an employer may routinely inquire, after a conditional offer of a job is made, whether the individual is addicted. The job offer may be

withdrawn if the answer is affirmative. If addiction is denied and subsequently discovered, the offer may be withdrawn, or the person discharged.[30]

4. *Procedures governing use of information concerning handicapping conditions.*

The regulations require that information provided or collected during preemployment inquiries concerning the history or medical conditions of the handicapped applicant remain confidential. This information must be maintained on separate forms and "shall be accorded confidentiality as medical records, except that:

(i) supervisors and managers may be informed regarding restrictions on the work or duties of handicapped persons and accommodations required;
(ii) first aid and safety personnel may be informed, where appropriate, if the condition might require emergency treatment; and
(iii) government officials investigating compliance with the Act shall be provided relevant information upon request."

SUMMARY

The handicapped employment regulations regulate the discriminatory aspects of hiring handicapped workers. Covered employers must be aware of their affirmative action and nondiscrimination responsibilities to disabled applicants and employees. This section discusses employers' major duties and techniques for achieving compliance. A qualified handicapped person is defined as one who can do the job provided reasonable accommodations are made. Reasonable accommodations must, in fact, be available. An employer may not discriminate against an applicant who makes his own adjustment or an applicant who makes his own arrangements to make the accommodation. The *bona fide* occupational qualification can be used to show that a particular class of handicapped persons is unable to perform the task. Employment tests cannot be used unless the test is job-related, there are no other job-related tests that are less discriminatory. Preemployment inquiries concerning the nature and severity of a handicap are generally prohibited under the law.

NOTES

1. Senate Committee on Labor & Human Resources, Report on Senate S.446, 1979.
2. Interview with John Wodatch.

3. The four points adopted from testimony by John D. Kemp of Kemp & Young, Inc., given before the S.446 Report Hearings, June, 1979.

4. 45 CFR 84.11 (b) (1977).

5. 45 CFR 84.11 (b) (1977).

6. 45 CFR 84.11 (b) (1977).

7. 45 CFR 84.12 (a) (1977).

8. 45 CFR 84.8 (a) (1977).

9. 45 CFR 84.12 (a) (1977)

10. Letter from David S. Tatel to Congressman Hawkins (June 22, 1977).

11. Letter from John Wodatch to Dolores D'Antonio (July 1, 1977).

12. "Discrimination Against Handicapped Persons: The Costs, Benefits and Inflationary Impact on Implementing Section 504 of the Rehabilitation Act of 1973 Covering Recipients of HEW Financial Assistance,@ 41 FR 20312, 20322 (May 17, 1976).

13. a) Rehabilitation Act of 1973, Statutes-at-Large, v. 87, Public Law 93-112, 355-394, 1973.

b) Vietnam Era Veterans' Readjustment Assistance Act of 1974, Statute-at-Large, v. 88, Public Law 93–508, 1578-1602, 1974.

14. Section-by-section analysis, 42 FR 22688, col. 3 (May 4, 1977).

15. 45 CFR 84.12(a)(1977).

16. 45 CFR 84.12(c)(1977).

17. U.S.C. Section 2000 e–2(e)(1970)(Section 703 of Title VII).

18. OCR Memorandum from Jeff Rosen to John Wodatch, "Employment Discrimination under Section 504–the Bona Fide Occupational Qualifications (BFOQ) Exception" (April 14, 1975).

19. Section-by-section analysis, 42 FR 22686, col. 3 (May 4, 1977).

20. 45 CFR 84.13(a)(1977) (emphasis added).

21. The discussion is based on the section-by-section analysis 42 FR 22688, col. 3 (May 4, 1977). (Emphasis added).

22. 401 U.S. 424 (1971). United States Reports Ju6.8:424.

23. 401 U.S. 424 (1971).

24. Section-by-section analysis, 42 FR 22689, col. 1 (May 4, 1977).

25. In *Cassias* v. *Industrial Commission of the State of Colorado*, No. 75-291 (Colo. Ct. App. Sept. 23, 1976), plaintiff appealed from the denial of unemployment compensation benefits based on the determination of the Industrial Commission that claimant was properly discharged for falsification of his employment application. Plaintiff failed to disclose his condition of epilepsy. He had been, with the aid of medication, free from seizures for several years. The employer's physician stated that the employer had a policy to hire epileptics. The court held that the applicant would be entitled to employment benefits unless the falsification of an employment application is found to be material to the applicant's job performance.

26. HEW letter from Secretary Califano to Robert Bolger (Feb. 28, 1978).

27. OCR letter from Michael Middleton to Vincent DeShazo (June 19, 1978).

28. 45 CFR 84.14(b)(1977).

29. Interview with John Wodatch.

30. 45 CFR 84.14(d)(1977).

PART VII
Future Directions

18 Low Back Pain—Future Trends

Kenneth Mitchell

INTRODUCTION

The citizens of the twentieth century share the experience of low back pain with the ancients as well as with future generations. Knowledge has improved regarding diagnosis and treatment. Although these improvements will continue, many future advances will occur in the socioeconomic, legal, cultural and political arenas. These arenas will dictate many aspects of low back pain (LBP) treatment. In fact, future responses to chronic disabilities will be dictated by the politics of incapacity, a process which is already underway. Social and political trends that are reviewed in this chapter are: 1) Legal and compensation trends; 2) Rehabilitation strategies; 3) Labor-management relations; 4) Corporate rehabilitation benefit plans; and 5) Health care delivery systems.

These five areas are not a definitive list, but do comprise the most potent forces that will influence the prevention and management of LBP disabilities and will be areas of major initiative for the development of a social and economic system that promotes restoration, equity, and independence for those people impaired by chronic pain.

PHILOSOPHICAL ASSUMPTIONS

Treatment of the future will promote restoration rather than compensation, and forge effective links between health professionals and employers, as well as encourage patient independence. It seems likely that certain events will occur:

1. Prevention programs for chronic disabling conditions will be required in primary health care delivery systems.
2. Medical management programs that are part of compensation systems will integrate chronic pain treatment programs.
3. Treatment philosophy will focus on both the psychological and physical aspects.
4. Nonspecific, unlimited or ill-defined treatment of painful and disabling conditions will be discouraged by third party payers.
5. The care of individuals with chronic pain will shift to industrial centers.
6. A reliable and valid disability scale, and possibly even a pain quantification index, will be used for compensation and disability determinations.
7. Management will assume an increasing role in the primary health care planning for their employees.

LEGAL-COMPENSATION SYSTEMS

The statutory definitions surrounding disability determination have been identified by some as one saboteur of the treatment process, and the compensation award as a disincentive to effective rehabilitation. The next decade will bring forth a strenuous effort by benefit managers, state and federal legislators, lawyers, and doctors to resolve the complex issues of the statutory definitions and quantitative nature of chronic disabling pain. Second, benefit packages will be revised to provide an equitable incentive to return to work.

We understand this truism: one man's pain is another man's delight. Future practitioners will have a greater capacity to quantify and define the impairment, functional limitations, and resulting disability created by chronic pain. Such quantification will be the outgrowth of current research in the area of rehabilitation indicators and computer simulations of disability. Future compensation systems will require a more systematic and objective data base for disability determination based upon an indexing system. This chronic pain, work disruption index will be a computer analogue of worksite characteristics, measurements of functional performances, objective measurements of pathophysiology, and measures of the affective response. One crucial component of this index will be the formulation of a reliable and valid pain descriptor or identifier. These pain "measures" will permit disability determinations to move away from the current arbitrary system that haunts the medical and legal practitioner and the employer. Thus, the benefit system will be less a barrier to restoration efforts.

The past decade's emphasis on greater economic payoffs for impairments will come to a close. The movement toward a benefit system that promotes rehabilitation has already begun. Benefit systems and insurance coverage in general will provide the impetus to reduce lag time between injury or symptom identification and the development of a comprehensive rehabilitation plan. Economic survival and solvency are the primary motivators for this trend. The active participation of the employer will dramatically accelerate this process.

Benefit systems will begin to recognize particularly effective treatment protocols for chronic pain. The payer of the service will apply more and more leverage to the practitioner to expose the patient to methods of pain management that are cost effective. Obviously, this will have a major impact on the logistics of medical care for the independent medical provider. The advent of the preferred provider organizations (P.P.O.) indicates the leverage that a health care payer can have in the development of practitioners' habits.

REHABILITATION STRATEGIES

Future treatment and rehabilitation will conceptually and programmatically go beyond the single modality model of treatment. For example, no longer will drug, exercise or bracing regimens suffice. Singular programs of job retraining, counseling, and selective placement will be obsolete. Total management will be necessary, replacing the traditional concepts. This conceptual shift recognizes the need for comprehensive care at critical junctures in the postinjury process. Such recognition couples the physiological and psychosocial characteristics of the individual.

Disability management to reduce impairment and functional limitations will go beyond traditional office and hospital based rehabilitation programs. At the heart of the concept of disability management will be the creation of a systematic, coordinated effort linking industry to the appropriate medical services. This linkage involves several key elements or assumptions. They are:

A. Rehabilitation will emphasize those modalities that promote 1) observable and measurable behavioral changes, 2) an informed and cooperating management, labor, and medical service, 3) an environment that promotes the use of substantiated multimodal treatment approaches, 4) the relationship between the diseases and the workplace, and 5) a benefit structure that promotes timely and effective treatment of chronic diseases in the workplace.
B. The recognition of the specific characteristics and effects of chronic low back pain in the workplace such as 1) the potential for disruption of work, 2) the potential to affect the worker's skills as they relate to endurance, flexibility and strength, 3) the potential for a slow, progres-

sive loss of productivity, 4) the uncertainty on the individual's part about improvement, 5) resistance of low back pain to complete cure, 6) the effect of low back pain on a broad range of social and emotional elements within the scope of the individual's life.

The use of a disability management program will be predicated on the linkage between industry and medical service providers. In the past, this relationship has often been noninteractive at best and often has been adversarial at its worst.

LABOR-MANAGEMENT RELATIONS

The work environment plays a unique role in the treatment of the individual with chronic low back pain. The workplace is often perceived as the cause of pain, though workplace modification can create an opportunity to permit the resumption of a productive lifestyle.

The worksite is controlled directly by interactions between the labor force and management. The labor force requires certain conditions and has particular expectations regarding the work environment. Management attempts to create an environment that optimizes productivity and return on the owner's investment. This complex and sometimes conflicting relationship may form the basis for success or failure in the restoration of an individual with chronic low back pain.

The next decade will bring forth an informed labor-management relationship. Management and labor will become more central to the health planning process for their employees. Ill-defined treatment protocols, unnecessary surgery, lengthy hospitalizations, and treatment programs that do not take the person's job into consideration will be discouraged by both parties. It is likely that organized labor will provide certain health care facilities. Low back injury and other occupationally related diseases are two very likely areas of action. Comprehensive approaches to pain problems will be negotiated. Labor-management negotiations will have as their central focus the opportunity to receive the most effective health care possible at the best possible cost.

CORPORATE REHABILITATION BENEFIT PLANS

The corporate plans for health care will diverge from their traditional efforts to offer standard coverage for sickness and preventative health care. The last several years have brought creative benefit plans that offer to the worker a virtual buffet of alternatives. The alternatives range from different degrees of copayment, and different deductible rates to different types of coverages for different health needs. Incentives are created for the worker in an effort to reinforce efficient use of the health care system, reinforcement of good health habits, and to reduce health care costs.

The corporate health care planner will create a benefit system that focuses on prevention, parameters for screening programs, on-site treatment within the plant, and treatment programs that will be reimbursed. Such a system will delineate the responsibilities of the worker and the company, and will expect the type of accountability that encourages equal participation and results from all parties.

HEALTH CARE SYSTEMS

Today's health care system will be changed by the economic factors which have forced reappraisal of the interaction between service providers and consumers. Hospitals will be de-emphasized and out-patient programs promoted. Care for the individual with low back pain will more often occur within the industrial setting. Ideally, the industrial rehabilitation centers will provide a comprehensive treatment program which is transitional to the work setting. In this setting, an individual who has chronic low back pain will no longer be considered as sick.

SUMMARY

Changes in the political and economic arenas will modify the treatment and rehabilitation of the low back injured worker during the next decade. Industry, labor, health care providers, and third party payers will work together to provide a comprehensive rehabilitation program that goes beyond single modes of treatment. Such rehabilitation facilities will be more closely linked to the workplace and stress early intervention and early return to work. Benefit plans will be reorganized to promote health behavior rather than disability or illness behavior. Potent economic forces will dictate these changes.

19 Summary

John W. Frymoyer, Malcolm H. Pope, and Gunnar B.J. Andersson

This book has been an attempt to bring together current knowledge regarding occupational low back pain, ranging from anatomy, structural function, load bearing, differential diagnosis, epidemiology, and prevention strategies, to the role of the law and society in the future of preventing this ubiquitous problem. Because there has been no previous systematic effort to bring together the diverse literatures and opinions regarding industrial low back pain (LBP), we have taken care to present the diverse, and sometimes conflicting scientific basis for our present understanding of LBP. Despite the diversities, it is hoped the reader has seen the evolution in understanding that has occurred in the past 50 years. Much has been learned and much remains to be learned; yet the scientific and practical bases exist today to reduce the human cost and suffering of LBP as it occurs in the workplace. Unfortunately, a negative attitude exists in certain quarters. A recent advertisement described industrial LBP as the nemesis of medicine and the albatross of industry. This view unfortunately still prevails in many quarters. Fortunately, others are recognizing the challenge to reduce the problem. Professor Alf Nachemson, to whom this book is dedicated, has a lecture entitled "Springtime of Backs." In this lecture he emphasizes the rapid accumulation in the scientific and practical knowledge regarding low back disease, and looks forward with enthusiasm to increased societal awareness, and to future scientific advances that will occur. It is on this note that the reader might ask, what have I derived from the knowledge available within this book? If nothing else, it would be the knowledge that LBP is a problem that will not be solved by engineers, chemists, safety engineers or doctors, but will include all members of society, most particularly the worker himself.

The magnitude of the problem is great. The epidemiologists have defined the problem, regardless of country, as involving 60 to 80 percent of the adult population. The cost is great, and although the greatest cost results from a relatively small number of "high cost cases," there exists no reliable means today to predict exactly who will fall into this small and costly group. Part of this problem rests with our present inability to make a precise diagnosis. Indeed, you have learned that no more than 50 percent of low back sufferers can be given a precise structural diagnosis. Perhaps as advances in scientific knowledge of the causation of low back disorders improves, treatment will become more precise and the success rate more predictable. It should not be forgotten that already such advances have occurred. For example, the CT scan when applied to properly selected patients, has given a whole new dimension of knowledge regarding the spinal canal, facets, and their relationships to the spinal nerves. A better understanding of the precise physiology of muscles and ligaments and the factors that improve or limit their functions is needed. The technology may be at hand through the combination of nuclear magnetic resonance imaging and improved knowledge relating to muscular electrical function and biochemistry. Even without these new advances, there are many practical lessons which can be learned. It is important for everyone to know that most low back disease tends to heal with time and that outcome usually is compatible with relatively normal function. The worker with LBP needs to be reassured regarding this point, while at the same time a timely, carefully orchestrated program of management is essential. If surgery is seen as the ultimate cure, there must be clear understanding of both reasonable expectations for its success as well as its limitations. Regardless of the mode of treatment chosen, the overwhelming message is that the educated individual ultimately must take responsibility for his own back care.

Obviously, a greater challenge is to prevent the worker from incurring a low back injury. Today's knowledge strongly suggests mechanical overloads are one basic and increasingly understood cause of low back disorder. Overexertion may not only involve simple lifting, but also pushing, pulling, and twisting. Loads may not only be applied in a single event. In some the basic structural problem may result from repetitive, chronic loads to the spine. Today there is sufficient knowledge to define in mechanical terms much of the load bearing capacity of the human spine. The computer age, combined with increasing scientific knowledge of the structural strength of the components of the spine will make it probable that job requirements can be described in more rigorous terms. Currently, one does not have to await more sophisticated computer models, although the information that they impart will be more precise. Three strategies for prevention, worker selection, worker training, and job redesign have been reviewed. Worker selection based on traditional methods such as radiography has little predictive value; but the evidence mounts that measures of physical function, particularly strength, should result in better worker selection. The balance between the capacity of the worker and the workplace demand is critical. Worker training is attractive because theoretically it can be applied to large numbers at low cost. Attractive as this strategy may be, formulae for proper lifting, applied indiscriminately, are not the answer. Well-conceived, reinforced educational programs that are sufficiently individualized to take into account differences in physical capacity, the workplace, and the type of loads being managed appear to be essential for worker training programs to be successful. Job redesign in many ways is most complex, but the message given here seems quite clear: The cost of redesign may in fact be more cost effective than other strategies, particularly when measured against the high cost of low back-incurred disability. In the broadest overview, there is no simple preventive formula that can be applied to meet the needs of all workers and all industries. Industries must apply those strategies that work best in their environment, and it is most likely that components of all three preventive strategies (selection, training and design) should be applied in many industrial settings. All of these preventive strategies require thought, and most importantly a commitment not only from workers but industry and labor as well. At the same time, there are other more passive preventive strategies that can be applied. Vehicular vibration and seating problems are common to many low back problems. Seats can be redesigned, and vibration can be dampened quite independent of worker compliance. Already trucking and busing industries are learning this lesson.

Lest the messages of this book seem too idealistic, there is one practical matter that cannot be ignored. LBP is costly. If the figures are correct, and there is little reason to doubt them, a 10 percent reduction in LBP in the United States alone would result in a saving of $2 to 3 billion. As these costs increase, the impetus for change likewise increases. Dependent on one's views, the industrial world in which we live is no longer one of worker, industry and labor, but a highly complex legal society. The complexities of the law as they apply to the worker with a disability should have been derived from the legal section of the book, as well as the complexities of the law that protect the disabled individual seeking work. The future as presented by Mitchell may appear to some as being overly idealistic, yet scrutiny of the changes that have occurred in our society suggest that many of the strategies he is proposing are already in progress and will become dominant forces over the next decade.

Cynical readers may believe that back pain is the greatest boondoggle of this century, if you will a mechanism for a paid vacation. If you are among the cynical, we respectfully suggest that your opinion may change if you become one of those who has an acute or chronic low back condition. If you have had LBP, your cynicism probably has been modified, and you may be particularly sensitive to the growing needs of society to solve this problem. If we continue to look at industrial LBP as the albatross of industry and the nemesis of medicine, progress will be impaired. If we look at LBP in the context of the accumulated knowledge, and the advancing basis for understanding, then Nachemson is correct; there is a springtime for the back.

Glossary

Accelerometer–Instrument that measures acceleration.
Amnestic–Pain killing medication.
Anaphylactic–Increasing the susceptibility to an infection.
Annular–Ring shaped; related to annulus of intervertebral disc.
Annulus–Ring shaped structure, outer part of intervertebral disc.
Anteroposterior–Pertaining to front and rear.
Anthropometric–Measurement of proportions of human body.
AP–Anteroposterior.
Apophyseal–Relating to projection from a bone; outgrowth without an independent center.
Arthropathy–Any disease affecting a joint.
Articular–Pertaining to a joint.
Autoimmune–Directed against the body's own tissues.
Axon–The axis of the nerve.

Biofeedback–The process of providing an individual with visual or auditory evidence of the status of an automatic body function.
Biomechanics–The application of principles of mechanics to the human body.

Caudally–Inferiorly, toward the lower spine.
Chemonucleolysis–The administration of the drug chymopapain to liquify.
Chronicity–State of being chronic (of long duration).
Claudication–Limping.
Cohort–Individuals who are members of the same group.
Chymopapain–Enzyme that dissolves soft tissue, especially intervertebral disc nucleus.
Compression Fracture–A fracture (failure) of a vertebral bone caused by a downwardly directed load.

Diaphragm–A partition wall; the wall of muscle separating thorax and abdomen.
Diaphyseal–Relating to shaft of a bone.
Discitis–Inflammatory disease of the disc.
Discogenic–(Pain) caused by derangement of an intervertebral disc.
Discograph–An x-ray photograph of an intervertebral disc containing an injected radiopaque contrast medium.
Dorsal–Related to back, posterior.

Electromyograph–Recording of electrical output of the contraction of a muscle.
Endocrine–Denotes gland that internally secretes body chemicals.
Endogenous–Originating or produced within the organism.
Endplate–The part of a vertebra where it joins the intervertebral disc.
Epidemiology–Sum of what is known regarding epidemics (of disease).
Erector spinae–Back muscles that extend the spine.
Ergonomics–Science that seeks to adapt working conditions to suit the worker.
Etiology–Study of causes (of disease).
Extensor–Any muscle that performs extension.

Facets–A pair of joints, at each vertebral level, important for spine stabilization and bending capacity.
Fascia–Sheet of fibrous tissue, envelops body beneath skin and encloses muscles.
Fibrosus–Formation of fibrous tissue.
Flexor–Any muscle that flexes a joint.

Ganglion–An aggregation of nerve cells.

Herniated disc–Rupture of a disc allowing the relatively fluid nucleus to push outward through the annulus.
HNP–Herniated nucleus pulposus.
Hypochondriasis–Morbid anxiety about the health.

Idiopathic–Denoting a primary disease, one originating without any apparent extrinsic cause.
Innervation–Distribution of nerves, or degree of nerve stimulation.
Intervertebral–Situated between 2 adjacent vertebrae of the spine.
Ischemic–Relating to local anemia caused by mechanical obstruction to the blood supply.
Isometric–Of equal or constant dimensions.

Kinematics–Science of motion, including movements of body.

Lamina–A thin flat plate. Part of the dorsal region of a vertebra.
Laminectomy–Excision of posterior arch (lamina) of a vertebra, commonly used approach to back surgery.
Latissimus dorsi–Broad flat muscles of the back.
LBP–Low back pain.
Lordosis–Curvature of spinal column as seen from the side (sway back).

Lumbar—Pertaining to the lower part of the spine. The last 5 vertebrae.
Lumbosacral—Pertaining to sacrum and lower back.

Mechanoreceptor—A receptor that is stimulated by mechanical pressure.
Morbidity—Proportion of disease to health in a community.

Necrosis—Death of one or more cells.
Neoplasms—New growth of cells or tissues, especially in cancerous conditions.
Nociceptor—A peripheral nerve organ or mechanism for the transmission of painful stimuli.
Nucleus pulposus—The central, more viscous portion of the intervertebral disc.

Orthosis—Orthopaedic device for assisting the function of part of the body, without replacing it.
Osteoarthritis—A "wear and tear" arthritis affecting any joint.

Pseudospondylolisthesis—"Deceptive resemblance to dislocation of vertebra." Usually means a forward slippage of one vertebra relative to another, secondary to degeneration.

Radiopaque—Impenetrable by x-rays.
Radiculopathy—Disease of the nerve roots ramus. One of primary divisions of blood or nerve vessel.
Retrospondylolisthesis—Condition in which the sacrum lies anterior to fifth lumbar vertebra (vertebra out of proper alignment).

Sacral—Related to sacrum (tailbone).
Sacroiliac—Relating to the joint between the pelvis and the vertebral column.
Sagittal—Anatomic plane passing vertically through body from front to back.
Scarification—Making a number of superficial incisions in the skin or other tissue.
Scoliosis—Abnormal curvature of vertebral column, as seen from the frontal plane.
Segmental Instability—An abnormal (excessive) motion between two adjacent vertebrae, associated with painful symptoms.
Somatization—Conversion of anxiety into physical symptoms.
Spasm—Involuntary muscular contraction.
Spinal stenosis—Reduction in the size of the spinal (neural) canal.
Spinous—Related to spine.

Spondylolisthesis–Forward displacement of one vertebra relative to the vertebra below.

Spondylotic–Dissolution of body of a vertebra. Usually refers to loss of disc height secondary to spinal degeneration.

Stenosis–Narrowing of any canal, especially the neural canal of the spine.

Stereophotograph–Two images viewed simultaneously to give perception of being in 3-D.

Index

Contributors

Gunnar B.J. Andersson, M.D., Ph.D.
Associate Professor
Department of Orthopaedics
Sahlgren Hospital
Goteborg S-413345
Sweden

Richard A. Brand, Jr., M.D.
Professor of Orthopaedic Surgery
University of Iowa Hospitals and Clinics
Iowa City, Iowa

Don B. Chaffin, Ph.D.
Professor and Director
Center for Occupational Health & Safety Engineering
University of Michigan
Ann Arbor, Michigan

John W. Frymoyer, M.D.
Professor and Chairman
Department of Orthopaedics and Rehabilitation
University of Vermont College of Medicine
Burlington, Vermont

Gary D. Herrin, Ph.D.
Director of Center for Ergonomics
University of Michigan
Ann Arbor, Michigan

James Howe, M.D.
Assistant Professor
Coordinator Residency Training Program
Department of Orthopaedics and Rehabilitation
University of Vermont College of Medicine
Burlington, Vermont

Roger C. Jensen, M.D.
Chief, Accident and Injury
Epidemiology Branch
National Institute for Occupational Safety & Health
Morgantown, West Virginia

Richard M. Keane
Assistant Director
Worker's Compensation
The Travelers Insurance Companies
Hartford, Connecticut 06115

John D. Kemp, J.D.
Director of Human Resources
National Easter Seal Society
Chicago, Illinois

Thomas R. Lehmann, M.D.
Associate Professor
Department of Orthopaedic Surgery
University of Iowa Hospitals and Clinics
Iowa City, Iowa

Raymond L. Milhous, M.D.
Professor and Chairman
Division of Rehabilitation
Department of Orthopaedics and Rehabilitation
University of Vermont College of Medicine
Burlington, Vermont

Kenneth Mitchell, Ph.D.
Director
Industrial Commission of Ohio
Rehabilitation Division
Clinical Assistant Professor
Ohio State University School of Medicine
Department Physical Medicine and Rehabilitation
Columbus, Ohio

Malcolm H. Pope, Ph.D.
Professor and Director of Research
Department of Orthopaedics and Rehabilitation
Professor, College of Engineering
University of Vermont College of Medicine
Burlington, Vermont

Stover H. Snook, Ph.D.
Project Director, Ergonomics
Liberty Mutual Research Center
Hopkinton, Massachusetts
Lecturer on Ergonomics
Harvard School of Public Health
Boston, Massachusetts

Arthur H. White, M.D.
Orthopaedic Surgery Department
St. Mary's Hospital
San Francisco, California

About the Editors

GUNNAR B. J. ANDERSSON is Associate Professor of Orthopaedic Surgery at the University of Göteborg, and head of the Section of Occupational Orthopaedics of the Department of Orthopaedic Surgery, Sahlgren Hospital, Göteborg, Sweden. During 1982–1983 he was distinguished visiting professor of Orthopaedic Surgery at Rush-Presbyterian-St. Luke's Medical Center, Chicago, Illinois.

Dr. Andersson's reputation in the study of the lumbar spine and industrial low back pain is worldwide. He has lectured extensively on low back pain and has been published widely. He is a member of the editorial board of *Spine*, and an associate editor of *The Journal of Orthopaedic Research*.

Dr. Andersson received his medical degree at the University of Göteborg, Sweden, and also holds a Ph.D. degree from the same university.

JOHN W. FRYMOYER is Professor of Orthopaedics and Chairman of the Department of Orthopaedics and Rehabilitation at the University of Vermont. He is Director of the Vermont Rehabilitation Engineering Center.

Dr. Frymoyer has performed extensive research efforts dealing with a variety of lumbar spine problems and was a recipient of the Volvo Award for the Study of Low Back Pain. Dr. Frymoyer is recognized worldwide for his efforts in low back pain research. He has been published extensively and serves as an associate editor of *The Journal of Bone and Joint Surgery*.

Dr. Frymoyer received his B.A. from Amherst College and holds an M.S. and M.D. degree from the University of Rochester, N.Y.

MALCOLM H. POPE is Professor of Orthopaedics and Professor of Mechanical Engineering at the University of Vermont. He is Director of Research at the University's Department of Orthopaedics and Rehabilitation and Co-Director of the Vermont Rehabilitation Engineering Center.

Dr. Pope is internationally known in the area of low back pain and spine biomechanics research and has been widely published. In addition, he has hosted two symposia on Industrial Low Back Pain and was a recipient of the Volvo Award for the Study of Low Back Pain. He is a member of the Editorial Board of *Spine*.

Dr. Pope holds an M.S. from the University of Bridgeport, Connecticut, and a Ph.D. from the University of Vermont, Burlington, Vermont.